T0297573

Science and the Global Environment

Science and the Global Environment

Case Studies for Integrating Science and the Global Environment

Alan McIntosh

Jennifer Pontius

ELSEVIER

AMSTERDAM • BOSTON • HEIDELBERG • LONDON
NEW YORK • OXFORD • PARIS • SAN DIEGO
SAN FRANCISCO • SINGAPORE • SYDNEY • TOKYO

Elsevier
Radarweg 29, PO Box 211, 1000 AE Amsterdam, Netherlands
The Boulevard, Langford Lane, Kidlington, Oxford OX5 1GB, United Kingdom
50 Hampshire Street, 5th Floor, Cambridge, MA 02139, United States

Library of Congress Cataloging-in-Publication Data
A catalog record for this book is available from the Library of Congress

British Library Cataloguing-in-Publication Data
A catalogue record for this book is available from the British Library

ISBN: 978-0-12-801712-8

For information on all Elsevier publications
visit our website at https://www.elsevier.com/

Publisher: Candice Janco
Acquisition Editor: Laura Kelleher
Editorial Project Manager: Emily Thomson
Production Project Manager: Mohanapriyan Rajendran
Designer: Mark Rogers

Typeset by TNQ Books and Journals

Contents

Online support materials for this title are available at http://booksite.elsevier.com/9780128017128/

Foreword

Many academic environmental science and natural resource programs feature introductory classes designed for students interested in science and environmental issues or who are planning to major in the field. Often already having completed AP science classes and, in many cases, having had hands-on experience with laboratory and/or field investigations, today's entering students are well prepared to tackle often complex environmental issues. This textbook is designed for students ready to take that "next" step and apply what they've learned in their science and math classes to real-world environmental problems.

This is not a typical environmental sciences textbook. We make no attempt to cover every environmental concept, process, or methodology. We've chosen instead to focus on some of today's most pressing environmental issues, linking them to their basic scientific underpinnings and providing opportunities for students to apply current methodologies and skills common to the profession using real data and cutting edge tools.

To ensure that students are prepared for this exploratory work, we begin with an introductory "Tools and Skills" section that reviews the techniques and approaches for collecting and evaluating environmental data that today's environmental scientist should be familiar with. All entries in this section of the text appear in later case studies to provide additional practice.

The heart of the book is a series of case studies focused in three areas: water resources, air quality and atmospheric sciences, and landscapes. Each case study includes an overview of the environmental issue, the specific problem being addressed, current research, and a solutions section. Embedded in each case study is a set of exercises designed to let students delve deeper into the issue through guided and independent inquiry. Students are asked to apply tools like Google Earth and online data visualization portals, analyze real laboratory and field data sets, critically think about the issues raised by the case study, evaluate alternative explanations for environmental outcomes, and explore the "systems connections" linking human activities to air, water, and land impacts.

We conclude each case study by asking students to consider some of the broader societal issues raised by the case study. The complexity of many of the global issues we focus on (e.g., climate change, e-waste, ocean acidification, and deforestation) helps students appreciate the importance of working with individuals from many different scientific disciplines and stakeholder perspectives. These issues also often represent perfect opportunities for students to explore societal aspects like environmental justice.

The book concludes on an upbeat note with a section looking toward a greener future and considering some advances that promise to reduce our impact on global resources. It's easy for students to become overwhelmed with the magnitude of some of today's environmental problems. Stressing some of the positive steps now being taken to reduce environmental impacts is an important way to end such a book.

While we hope that all students using the textbook will have sufficient background knowledge to tackle the case studies, we recognize that some individuals may need additional review. For each case study, we provide an online Resource Page that links to websites for a basic review of the scientific concepts as well as more in-depth material.

We believe that this textbook makes an important contribution by helping students apply the knowledge they've gained in their coursework to address real-world problems. In the process, we feel they'll be gaining invaluable experience that they will be able to apply as they begin their careers as professional environmental scientists.

Acknowledgment

Our thanks to the many colleagues who made helpful suggestions about the project and provided initial feedback on the concept. Dr. McIntosh is indebted to his wife Barbara for her comments about and contributions to the book, and her patience throughout the process. Dr. Pontius would like to thank the US Forest Service, Northern Research Station for their support in engaging and educating the next generation of environmental professionals, and, in particular, her husband John and children Grace and Danny for putting up with many late nights and leftovers.

CHAPTER 1

Tools and Skills

Science and the Global Environment. http://dx.doi.org/10.1016/B978-0-12-801712-8.00001-9

Chapter 1.1

Introduction

FIGURE 1.1.1
This is not your grandmother's toolbox. While some of the skills introduced in this section are tried and true, the field of environmental science has us constantly identifying new ways to view (think satellite technologies) and characterize (think risk assessment) ecosystems and the services they provide [think cost–benefit analysis (CBA)]. Here, the 40-year-old Landsat satellite program allows scientists to study, in ways they never imagined, the Earth's surface. *Source: NASA Goddard Flight Center.*

Back when I was in college (and no, dinosaurs did not rule the Earth), I found that the best way to prepare for a test was to take all my notes from the class and condense them into as few words as possible on as small a piece of paper as possible. Not only did this exercise itself help to solidify what were the most important concepts and force me to summarize the take-home message, it also provided a perfect study guide for last minute cramming before the test. I highly recommend this approach. But I digress. Here I am, an unnamed number of years later, doing the same thing, only this time for you.

In this chapter, we sort through the data toolbox used by environmental scientists, managers, and decision-makers to summarize a few of the key approaches (Fig. 1.1.1). These are the most common techniques used in the fields of environmental science and natural resource monitoring and management. At a minimum, you should understand what these techniques are and how they can be used to help understand, monitor, and solve environmental problems. And that is what this chapter will do, provide you just enough information to understand and practice a few of the most common techniques.

This is not a substitute for a full course in basic or applied statistics. If you hope to become an environmental professional, you must be armed with a more complete understanding of how data can be collected, analyzed, and interpreted. Nor is this a substitute for a full course in geospatial technologies (GST). Understanding the complexities of ecosystems involves consideration of the larger landscape that is best discovered using GST tools. While geospatial coursework is not offered at all universities and is required in only a few environmental degree programs, I can't recommend this skill set enough.

This Tools and Skills chapter provides a set of "crib notes" meant to serve as an introduction to these skills. While this introduction is not meant to make you an expert, it should be sufficient to allow you to practice these tools in the case studies that follow. For those of you who want more information on these techniques, we provide links to some very informative websites on your Resource Pages. With that said, I'm also amazed at what a simple Google search can turn up on many of these topics. For the rest of you, we'll provide, within these few pages, just enough information to make you dangerous.

Chapter 1.2

Critiquing Statistics You Encounter

FIGURE 1.2.1
Every day, we are bombarded by statistics that are intended to sway opinions and influence decision-making. *Photo by Myrealnameispete (CC by-SA-2.0) via Flickr.*

HOW TO SURVIVE IN SPIN CITY

Understanding nature requires that one describe it. While poets and philosophers use words to do that, researchers and land managers rely on numbers. From parts per million of pollutants in a water sample to acres of forest cleared, we must be able to quantify the characteristics of an ecosystem before we can uncover the patterns and investigate the drivers of those patterns.

But numbers can also be dangerous. Too few, and you run the risk of not accurately capturing the characteristics of the entire ecosystem. Too many, and you run the risk of identifying patterns that are spurious and random. Here we will cover the basics of using numbers to tell the story. But before we get into how to do that with our own data, we'll start with some tips on how to tell if others are telling the right story (Fig. 1.2.1).

In a Nutshell

We live in a world filled with uncertainty, and in order to navigate the chaos, we need some way to make sense of the patterns. Most people rely on past experience or gut instinct to do this, leaving our decision-making often driven by fear. This makes us notoriously bad at assessing risk.

But fear not, statistics provides us with a set of tools and techniques that we can use to organize the chaos, describe the patterns, quantify the uncertainty, and make more informed decisions. While most people are inherently averse to DOING statistics themselves to accomplish this, we all place tremendous weight on statistics that others DO and present to us. People tend to think of statistics as though they are undisputed facts that exist completely independent of people. This would all be well and good if statistics worked like general math and could just spit out an answer like a calculator (e.g., $2+2=4$). In reality, people are involved in almost every step of statistical analyses:

- Someone must decide what data to collect, how to collect them, and where to collect them from.
- Someone must determine what questions to ask, how to ask them, and what analyses to run.
- Someone has to determine what other factors should also be considered to understand complex interactions among relationships.
- Someone has to interpret and then communicate those results.

Notice that there are a lot of "someones" in this list, and "someones" are very rarely perfect. People can incorrectly design analyses, make mistakes in the data they collect, make incorrect assumptions, or ignore key factors. They can misinterpret results or omit context essential to understanding the implications. At their worst, statistics can be twisted and distorted to change the interpretation entirely. That doesn't mean that data presented to us are always wrong; we just need to know which specific questions to ask before we accept any conclusions. A thorough critique would include a broad understanding of the statistical methods used and assumptions made. But even without a PhD in statistics, you can still ask some basic questions that will help you sniff out bogus statistics.

Your first reaction whenever you encounter data (or their interpretation) should be to start asking questions. Consider this a good list to start with.

- **Are the methods sound?** Any statistical analysis is only as good as the data that went into it. If the application of treatments, collection of measurements, analysis of samples, or input of data are flawed in any way, the results will be misleading. If treatments were applied, we can ask if those methods match what could be expected in nature. If data were collected across a landscape, we can ask how the sampling locations were selected. If plots or subplots were used, we can ask how data were aggregated for analysis. If laboratory analyses were necessary, we can assess the methods used in the analyses. The goal is to ensure that sound scientific methods were followed so that the data are of high quality, errors were minimized, and the sampling and statistical methods were appropriate to address the specific research objective.

- **Is the sample representative of the population that conclusions refer to?** We perform statistical analyses so that we can understand the world around us without having to collect data for every single organism or location in the world. Therefore, data collection must be designed so that it reflects the general characteristics of the population we are interested in. For example, if I wanted to examine the impact of arsenic contamination on human health in India, but I collected data only from homes serviced by municipal water systems in urban areas, I would not be capturing the potential impacts on rural residents or impoverished populations whose exposure may be higher and symptoms worse. This doesn't mean the study is useless; it just limits the population about which we can draw conclusions. Typically, a random selection of observations from the larger population will be representative if the sample contains a sufficient number of observations.

- **How many observations are included?** Too few, and the population of interest may not be adequately represented. Too many, and the power of the test may be so large that significant results will be found even when no meaningful relationships exist.

- **Are there other FACTORS that should be considered?** Even when we are given proper context, often assumptions are made that may not be relevant because other important factors have been ignored. For example, did you know that since 1970, the number of young adults who live at home has increased 48%? Does this mean that your generation is a bunch of freeloading slackers (Fig. 1.2.2)? While the number living at home has gone up 48%, the general population has also increased 36% over the same time period. But the number of young Americans living at home is likely related to more than just the size of the general population. For example, how does the health of the economy compare between the two time periods? Has the unemployment rate risen, making it harder for young adults to find jobs? Has the cost of living gone up, essentially eliminating the possibility for many to afford their own home? Are more young people going to college, essentially extending their entry into the workforce? We also need to consider historical differences. For example, in 1970, many young men from this age group had been drafted to fight the war in Vietnam. Is it possible that the number of youths living at home was unusually low in 1970 because so many were in the military? This may seem like a trivial example, but these confounding factors are even more problematic in ecosystems where a myriad of variables are at work.

FIGURE 1.2.2
Is your generation a bunch of couch potatoes? Before you trust the numbers, be sure to investigate potential confounding factors. *Photo by Lookcatalog (CC by-SA-2.0) via Flickr.*

- **What INFORMATION are you missing?** The statistics that are presented to you are important, but sometimes equally important are the statistics that are left out. Consider a political candidate touting their support for environmental issues. If this candidate says that she voted for green legislation 20 times during her last term, we might be inclined to think that she has demonstrated strong support for environmental legislation. To know for sure, we need some context, including how many times she voted against environmental legislation. But we also need some additional information. We should ask how she defines "green legislation." For example, is voting to provide laptops for schools considered "proenvironment" because it reduces paper usage? Or is it "antienvironment" because manufacturing, distributing, powering, and disposing of laptops have a cumulatively large environmental impact? Is voting for laptops in schools even an environmental issue at all?

- **What is the CONTEXT?** While a statistic presented to you may seem staggering, without perspective, the numbers are meaningless. For example, New Smyrna Beach in Florida is the shark attack capital of the world. It is estimated that anyone who has swum there has been within 3 m of a shark at least once. In 2014 alone, there were 79 shark attacks worldwide, 28 of which were in Florida (https://www.flmnh.ufl.edu/fish/sharks/statistics/statsw.htm). However, before you decide to never set foot in the ocean again, you have to put these numbers into perspective. You have a 1 in 63 chance of dying when you get the flu and a 1 in 218 chance of dying from an accidental fall, compared to a 1 in 3,700,000 chance of being killed by a shark during your lifetime. The human mind is naturally drawn to the "fantastical." We must always be aware of this bias in ourselves and seek to put numbers presented to us into context before drawing any conclusions.

- **Is the level of SIGNIFICANCE in results presented?** Significance thresholds are set prior to analysis, based on the expected power of the test. Yet many scientists try to describe nonsignificant results as something more interesting, using terms like "borderline significant," "trend toward significance," "marginally significant," or "robust, distinct, or marked trends." *P*-values, confidence intervals, and margins of error should always be presented so that you can determine if results are truly significant. Perhaps even more problematic in our age of "big data" are significant results that arise simply because of the power of the test. Results should include a discussion of how ecologically meaningful or relevant these significant results are.

- **Do they imply causation without sufficient evidence?** "Correlation DOES NOT EQUAL causation." This statement is considered the golden rule of statistics by many. Just because a relationship exists between two variables does not mean that changes in one are driving the response of the other. Sometimes significant relationships arise by random chance. Other times, a relationship between two variables exists because of their common relationship with another variable. However, there are many cases where causation can be inferred from statistical analyses. Perhaps one of my favorite quotes comes from Randall Monroe:

  > Correlation does not imply causation, but it does waggle its eyebrows suggestively and gesture furtively while mouthing 'look over there'.

In designed, randomized experiments where other potentially confounding factors are held constant, one may infer a causal relationship. In such studies, potentially confounding factors are carefully controlled, with the treatment applied randomly across many replicates. Cause and effect relationships can also be inferred when a "confluence of evidence" is presented. This occurs when the same pattern presents under varying conditions across a number of studies. For both randomized experiments and confluence of evidence justifications, there must be a reasonable theoretical explanation for the causal relationship.

An example of this in practice is how the scientific community has come to the conclusion that carbon emissions are driving the many changes in climate patterns witnessed across the globe. We know that atmospheric CO_2 levels have been rising and that they are strongly correlated with a simultaneous increase in mean global air and sea surface temperatures (Fig. 1.2.3). Many climate change naysayers will cite the rule that correlation does not equal causation. They will also show you many pieces of anecdotal data to point out how cold or snowy particular locations or dates are. But what they do not consider is that hundreds of observational studies from across the globe comparing greenhouse gas (GHG) concentrations to a variety of environmental climate variables all find the same pattern in the relationship between CO_2 concentration and temperature.

These observations are backed up by experimental studies in which concentrations of GHGs are manipulated in closed systems while all other variables are held constant. In addition, theoretical models that consider hundreds of

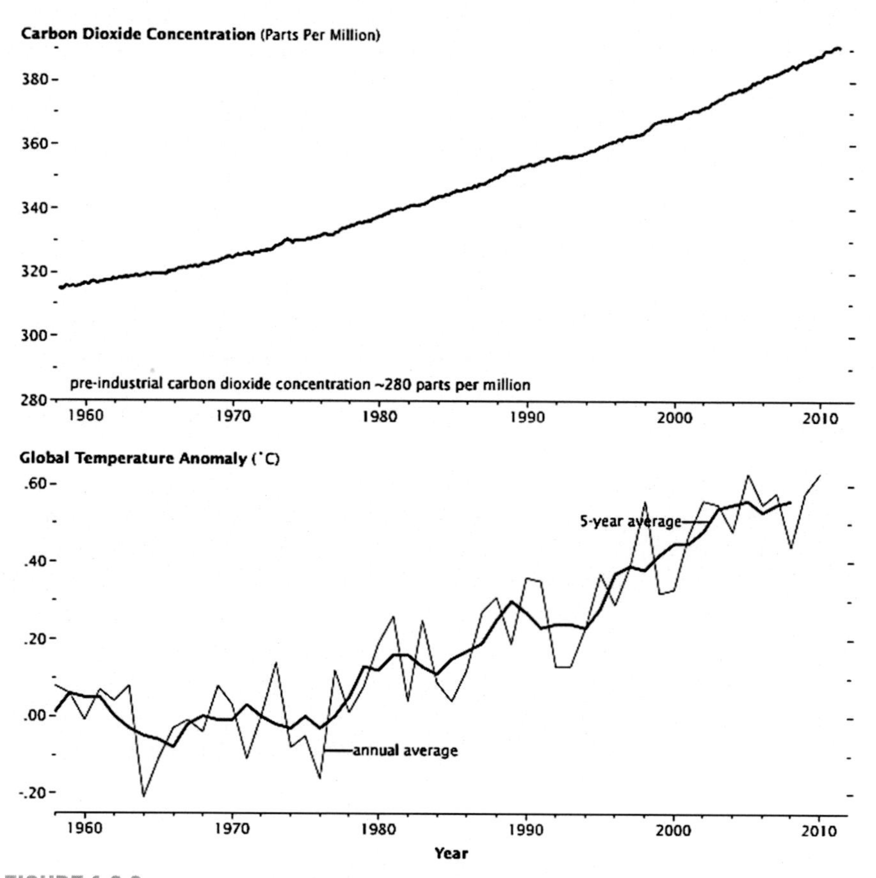

FIGURE 1.2.3

Atmospheric CO_2 concentrations are strongly correlated with a simultaneous increase in global temperature anomalies. *By Bvelevski (CC BY-SA 4.0) via Wikimedia Commons.*

variables that impact climate, as well as their interactions and feedback mechanisms, all report the same relationship. This is why even the most conservative scientists, professionals trained to methodically evaluate data before drawing conclusions, agree that increases in GHGs in the Earth's atmosphere are causing dramatic changes in our climate system.

The Take-Home Message

Data and statistics help us make sense of the world around us. We all, in some way, are beholden to those who collect and interpret these data. From the medications we take, food we eat, clothes we wear, cars we drive, and actions we take, statistics have played a major role in our lifestyle. Perhaps because of this, there is a vast industry of statisticians, researchers, engineers, lobbyists, and marketers constantly presenting us with data and their interpretations of those data. The take-home message is to always ask questions. This is easy to do when we disagree with some conclusion. But we must also be asking these questions when we desperately want to agree with the conclusions. By asking the questions, you can strengthen the argument, preparing for any possible rebuttals or attack. Good science and data interpretation are the keys to making people aware of the many environmental issues that currently face us. Each of us should play a role in raising awareness, either by sharing information with our families, commenting online, or creating our own outreach materials. In all cases, we must present well-vetted scientific evidence, rather than anecdotes, opinion, or fear mongering.

Practice

1. In the realm of statistics, we are trained to carefully evaluate data, consider nuances, and use probability to inform our decisions. But we need more than a mechanical understanding of the statistical method to interpret our results successfully or use those statistics to inform decision-making (the ultimate purpose).

   We each have a different approach to making decisions. Take the quiz found at the following website (and linked on your Resource Page) to see how you typically approach decision-making (https://www.mindtools.com/pages/article/newTED_79.htm).

   Briefly (one sentence each) answer the following:
   - What do your responses to this quiz tell you about your own tendencies in evaluating evidence and drawing conclusions?
   - How might this impact your interpretation of statistics you encounter?
   - What steps might you take (or considerations might you make) to improve your decision-making (including how you interpret and communicate statistics)?

2. Read the Washington Post article "Beer Tax Lowers Clap" found on your Resource Page, and offer your own **one-paragraph assessment of the statistics presented, including your current opinion about the efficacy of using a tax on alcohol to reduce rates of sexually transmitted diseases.**

3. Now read the rebuttal (located on your Resource Page) to the alcohol tax article written by a statistician. Make note of the points he makes in his critique of the research.

   - What are the key statistical points he makes to disprove the conclusions of this study?
   - Did you catch anything that he did not include in his assessment?

Chapter 1.3

Experimental Design and Sampling

GETTING WHAT YOU WANT FROM YOUR DATA

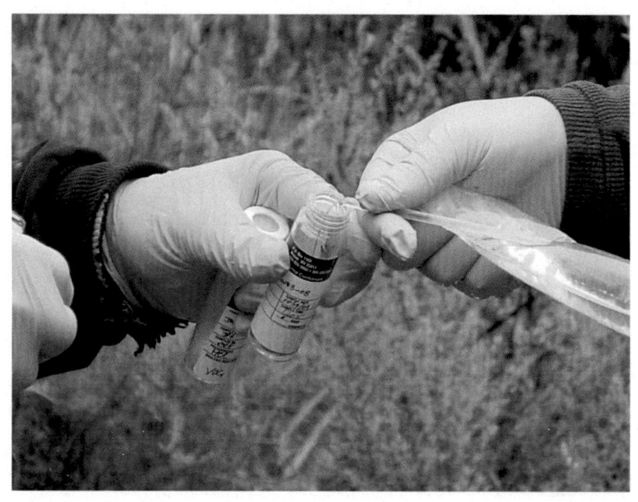

FIGURE 1.3.1
While many of us are trained in basic field and laboratory methods, where and how to collect those samples require more careful consideration in order to successfully meet our study objectives. *Photo by LouisvilleUSACE (CC by-SA-2.0) via Flickr.*

In a Nutshell

Hopefully, many of you will have opportunities to collect data in your professional careers (Fig. 1.3.1). This might be to monitor the conditions of a given ecosystem, test the efficacy of various treatments, or describe temporal or spatial patterns in key ecosystem characteristics. Most students who have made it through an environmental program have been trained in proper field and laboratory techniques to collect measurements. But our training in how to select observations for which to collect those measurements typically falls short. As a result, there are many practicing environmental professionals who are really good at collecting data but not as good at ensuring that the data they collect can tell them what they want to know. This is where a little knowledge of experimental design can go a long way.

Experimental Design is the creation of a detailed experimental plan that allows you to obtain the maximum amount of information specific to your objectives.

The critical parts of any experimental design include:

1. **State the study objective or research hypothesis.** This may sound like stating the obvious, but often we rush to the field to collect samples without really considering ***exactly*** what it is that we want to know. Narrowing this down to a specific set of research objectives will help inform the rest of your design. This should be a set of clear, concise statements that identify the exact information you are looking for or questions you want to ask.

2. **Identify the population of interest.** Once you collect, analyze, and interpret your data, you should be able to draw conclusions about a larger entity than just the observations you collected. This is your population, the set of ALL possible observations from which you select your samples for measurements. You have to be sure that the sample you have collected included the full range of characteristics you would expect to see in the entire population.

3. **State the variables of interest.**

    a. What is the **response variable** that you are interested in? How will it be quantified?
    b. Is there a specific **treatment** that you are hoping to test or specific factor you are hoping to isolate the effect of? These are your independent variables, the ones that influence your response variable.
    c. Are there **controls** that could be included to better isolate the relationships you are interested in? These are factors that you may not be directly interested in but that you know you must hold constant across your samples so as to minimize their influence.
    d. Are there other variables you will measure to relate to this response variable? For example, are there **covariates** that are likely to influence your response variable but cannot be controlled?

4. **Specify a sampling design.** A sampling design lays out exactly ***how you will select the observations*** on which you make your measurements. Typically, these observations (which comprise your "sample") will be representative of the population of interest you have identified above. A good sampling design considers many details of data collection such as:

    a. What is the **unit of observation**? What represents one data point or one observation for you? Again this may seem simplistic, but imagine you are collecting measurements of the growth of all trees on a plot, one of a set of several plots along an elevational transect. Is the unit of observation a tree, a plot, or transect? Really, it comes down to independence. For most statistical analyses, each unit of observation should be independent of the others, meaning that your chance of picking one observation does not influence your chance of picking any other observation.

In this example, because I am measuring all trees on each plot, the trees are NOT independent. Instead, they are replicates of tree growth on a given plot. My plots are located along a transect at set intervals (a systematic design described below), which ensures an unbiased selection of plot location. Thus, the plot is my unit of observation.

b. **How many observations should you collect?** This is a trick question because usually you will want to collect as many observations as possible so as to most accurately represent the larger population and to increase the power of your statistical analysis (statistical power is your ability to identify a significant result when it truly exists). But, in reality, environmental scientists are often limited by time and money. The key is to make sure that you have sufficient resources to collect enough observations to make the study worthwhile. Power analyses can help with this, but they are beyond the scope of this text.

c. **How will you select observations to measure?** Typically, in order to collect a sample that is representative of the larger population (Fig. 1.3.2), we want a large set of randomly selected observations. However, there are times when other sampling protocols are more useful. Here are some examples of common ecological sampling techniques:

   i. **Probability-based (random) sampling**: Every member of the population has an equal chance of being selected. It is important to verify that enough observations are collected to capture the range of conditions in the population and that all observations are positioned to achieve good spatial dispersion. This provides a statistically unbiased estimate of the larger population and is necessary for most experimental or modeling studies.

   ii. **Stratified random sampling**: When the target population is separated into distinct strata or subpopulations of interest exist within the larger population, it is often useful to conduct separate random sampling *within* each of these distinct groups. This ensures more accurate representation of subgroups of special interest.

   iii. **Systematic (grid) sampling**: When large areas are of interest, it is often useful to choose an initial starting location randomly and then identify remaining sampling locations across a spatial grid of regular intervals. This approach ensures uniform coverage across larger landscapes and is particularly useful for examining spatial patterns.

iv. **Adaptive cluster sampling**: When looking for rare characteristics or "hotspots," it is often useful to start with a random sample but then intensify sampling around observations that meet some set criteria. Several additional rounds of sampling may be used to identify the location of interest or delineate the boundaries of hotspots.

v. **Judgmental sampling**: The selection of observations is based on professional knowledge of the feature or condition under investigation. While inferential analyses are not possible based on judgmental sampling, this is particularly useful for descriptive studies when looking for rare or specific characteristics, or for pilot studies where time is of the essence and larger samples cannot be collected.

d. **What is the timing or frequency of measurements?** Is there a particular seasonality that should be captured in your data? Do you expect conditions to vary over time? Do you need to capture this variability? Are you hoping to repeat measurements at the same locations over time to test for temporal trends?

FIGURE 1.3.2
NASA scientists study melt ponds forming atop Arctic sea ice. It is impossible to sample every pond across this vast system, so scientists must carefully construct an experimental design to ensure that the samples they do collect are representative of the larger set of freshwater ponds now forming in the Arctic. *Source: NASA Goddard Space Flight Center.*

5. **Select the appropriate statistical test**. There are thousands of statistical tests, and figuring out the right one to use is not always straightforward. Sometimes there are multiple tests that can help to address your study objectives. But armed with the information from your experimental design, you should be able to identify the right statistical test to use to answer your specific research question or study

objective. I find it helpful to walk through a set of specific questions that can then be used to guide you through a sort of statistical dichotomous key (Fig. 1.3.3).

i. What **type of response data** do you have? Is it continuous (measurements could take any value), categorical (observations can be described as classes or groups), or frequencies (counts of observations)?
   - **Frequency response**: When you are counting the number of observations that fall into a particular group, you can typically use the simple Chi-square analysis described in Chapter 1.8.
   - **Categorical response**: Sometimes you are interested in determining the probability that a given observation falls into a specific categorical response class based on a set of input measurements or in quantifying the influence of various factors in determining class assignment. There are several approaches for this type of analysis, including logistic regression and discriminant function analysis, but these techniques are less common across the disciplines and beyond the scope of this text, so we will save them for your more advanced statistics coursework.
   - **Continuous response**: Most often environmental professionals are measuring continuous response variables on their observations. In this case, there are additional steps necessary to determine the appropriate test.

ii. **What is the nature of your research question**? If you are just looking to describe data, you can simply work through the standard descriptive techniques outlined in Chapter 1.4: Visualizing and Describing Data. But if you hope to make inferences about larger populations, you need to consider whether you are looking primarily for DIFFERENCES between groups or for RELATIONSHIPS among variables, or if you are trying to MODEL the response based on various inputs.
   - If you are looking for **DIFFERENCES in a CONTINUOUS response variable**, you need to consider:
     - **How many treatment variables** (or factors) are you examining in relation to your response?
     - **How many groups** within these variables are you hoping to compare?
     - Are the observations in each of your groups selected **independently**, or are they **purposely paired** (a common design used to control for many possibly confounding factors).
     - **Are your data normally distributed?** Most inferential tests are based on a normal distribution curve, where the mean, median, and mode are all similar (and other characteristics of a standard

"When to Do What" Basic Stats Analysis Flow Chart

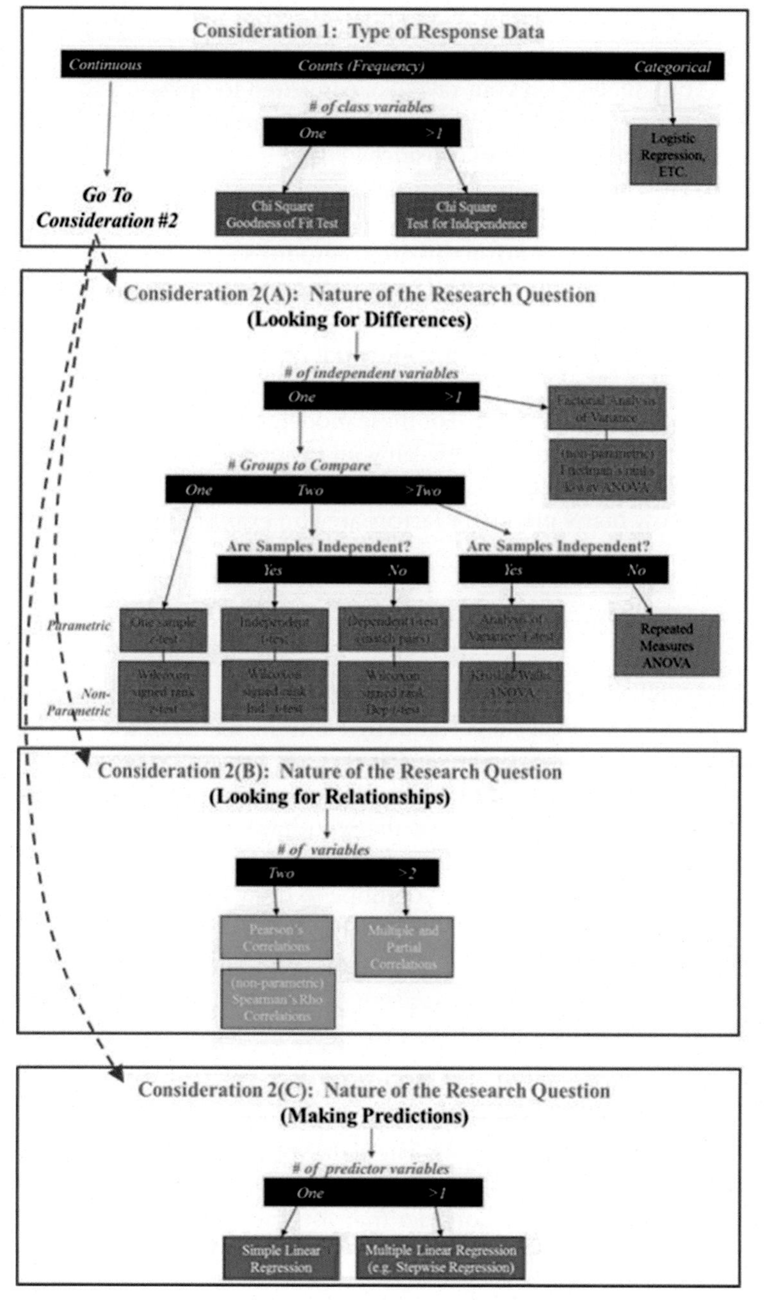

FIGURE 1.3.3

This is an example of a statistical "dichotomous key" that I use to help students in my applied statistics course identify which statistical analysis is appropriate based on a set of simple experimental design questions. Other examples of such keys are available on your Resource Page.

bell curve are approximated). If your data do not approximate a normal distribution, you will need to switch to a nonparametric version of your test. In many cases, this simply involves running your analysis on your raw data converted to ranks.

- If you are looking for **RELATIONSHIPS among CONTINUOUS variables**, you need to consider:
  - **How many treatment variables** (or factors) are you examining in relation to your response?
  - **Are your data normally distributed?** (See above.)
- If you are hoping to **MODEL a continuous response variable**, you are moving into the regression family. There are many regression methods out there to use, all with the intent of creating a mathematical equation that can be used to predict the response variable based on a set of input parameters. For these models, you need to consider:
  - Is the relationship between your model predictors and the response **linear?**
  - **How many** predictive factors are you examining in relation to your response?

Important Things to Remember About Experimental Design

There is no ONE correct way to design an experiment, and often you must balance the need for an ideal statistical design and the reality of the time, money, and access to observations you have. The key is to make sure that before your data collection even starts, you have carefully considered how to address your specific research or study objectives, that you have documented and justified your choices for experimental design, and that you have considered the potential limitations to the conclusions you hope to draw from your efforts.

Practice

A water-sampling example illustrates the level of detail required to design a solid experiment.

Consider that you have been tasked with monitoring *E. coli* levels at a municipal swimming beach (Fig. 1.3.4). The data you collect will serve to inform beach closures to protect public safety, but you also want to examine temporal patterns in *E. coli* outbreaks. Consider that you specifically want to:

- Monitor daily *E. coli* concentrations to inform beach closure.
- Determine if *E. coli* concentrations differ over the summer season (compare June, July, and August).

FIGURE 1.3.4
Your task is to spearhead the monitoring and assessment of *Escherichia coli* outbreaks at a large municipal beach. *Photo by dcwriterdawn (CC by-SA-2.0) via Flickr.*

- Determine if *E. coli* concentrations differ over the course of the day (morning versus afternoon).
- Determine if *E. coli* concentrations differ based on location (picnic area versus reserve area).

Consider that while you have lifeguards on duty to collect water samples at your whim, you only have funding to test 1,000 samples over the course of the season. Lay out an experimental design to address these study objectives, including the following:

1. population of interest;
2. sampling unit (consider the need for any subsamples for quality assurance/quality control (QA/QC) purposes);
3. sampling locations (feel free to include a hypothetical map to help describe your sampling design and specify the type of sampling design you'd use);
4. sampling frequency and duration of the study;

5. type of statistical analysis for each objective listed above;
6. QA/QC protocol to verify proper collection and handling of samples;
7. any other considerations before you embark on your data collection; and
8. potential limitations of this study (what conclusions do you expect you will be able to draw about patterns of *E. coli* contamination?).

Chapter 1.4

Describing and Visualizing Data

In a Nutshell: Once you have data in hand, the next step is to describe those data. Just like going on a blind date, there is some key information that you may want to know. Armed with this information, you can then make decisions about whether it is even worth going on the date, and if so, where to go or what to do on the date.

In a similar way, we want to know some basic information about our data before we delve into a long, in-depth analysis. This information will help us decide if the data are appropriate for our study, what type of analyses we can conduct, and what initial patterns we should look for. Describing and visualizing the data are important first steps in ANY statistical analysis.

Visualizing Your Data: There are as many different ways to plot and graph data as there are types of plots and graphs. When doing basic data exploration, you often want to use whatever is the simplest. Typically, our main goals in visualizing the data are to (1) assess the basic distribution of the data (are they normally distributed?) and look for outliers that might skew our data or bias our analysis, (2) examine the basic nature of the relationships or differences we are ultimately looking for, or (3) let our data "tell a story."

EXAMINING THE DISTRIBUTION OF YOUR DATA

When you have collected continuous data, you will want to see how your observations are distributed across the range of values collected. This should be a precursor to any statistical analysis, as determining if your data are normally or not normally distributed will inform what type of statistical analysis to run. Your goal in examining these graphs is to determine if:

- Your data approximate a normal distribution (this allows you to use parametric statistical tests).
- There are any gaps in your data collection (did you cover the full range you intended to in collecting your data?).

- There are any interesting phenomena in your observations (for example, is there a high frequency of observations at two different data ranges? If so, this might suggest that you are collecting data from two distinct populations).
- You have potential outliers (extreme observations) that you should examine more closely (or at least keep an eye on during your analyses).

To see how your observations fall across the full range of data values, histograms or box-and-whisker plots are your best options (Fig. 1.4.1). Histograms simply create bins of data values and create a bar chart with the count of observations that fall into each bin. If your data are normally distributed, you should see something that approximates a bell-shaped curve. If your data do not approximate a normal curve, you likely do not have normally distributed data.

Box-and whisker-plots give you the same basic information about the spread of observed values, but also include several specific markers for important descriptive metrics in your data.

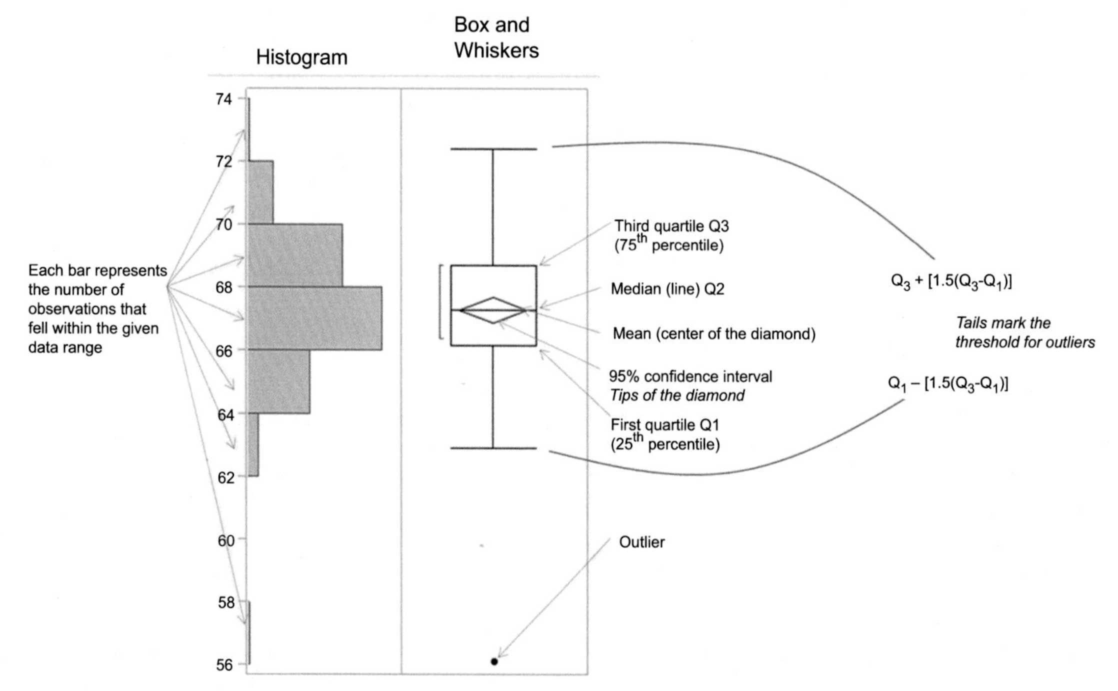

FIGURE 1.4.1
Histograms and box-and-whisker plots provide useful information about the general distribution of your data. This set of figures was made using the SAS Software JMP Version 12.0.

LOOKING FOR RELATIONSHIPS AND DIFFERENCES

Determine how variables are related *(common precursor for correlation and regression)*:

When you have two continuous variables and you want to see how they are related, you will want to use a scatter plot. Scatter plots allow you to see how much common variance exists between the two variables of interest. Typically, when examining scatter plots, you are looking for several pieces of information:

1. Are the variables related to each other (i.e., how much variance do they have in common?)? For example, do observations fall in a random pattern, or do you see points falling along a common line? The tighter your observations group along the horizontal (1:1) line, the stronger your relationship.

2. What is the nature of any relationship present (Fig. 1.4.2)?

   a. When one goes up, the other does as well. This is likely a positive (direct) relationship.
   b. When one goes up, the other goes down. This is likely a negative (indirect) relationship (correlation).

3. Is the relationship linear (consistent in its nature across the full range of data), or does it change across the range of variables?

4. Are there any gaps in your data collection (i.e., regions of the range of possible values where you cannot see the nature of the relationship because of missing data)?

5. Are there any interesting patterns or thresholds that may be ecologically meaningful?

6. Are there any outliers that don't fit the general pattern of the relationship? You will have to do some digging to see if these are erroneous data points that should be removed from the analysis or if they are really interesting observations that warrant further investigation (Fig. 1.4.3).

Determine how data are divided into parts *(common precursor for Chi-square tests)*:

Sometimes you will want to describe how your data fall into different categories. The goal is to understand how large each group is relative to the others. Pie charts can be useful when there are many categories to consider. But when there are groups with low counts that will be difficult to see in the "pie," a simple bar chart may be more helpful (Fig. 1.4.4). Remember that fancier isn't

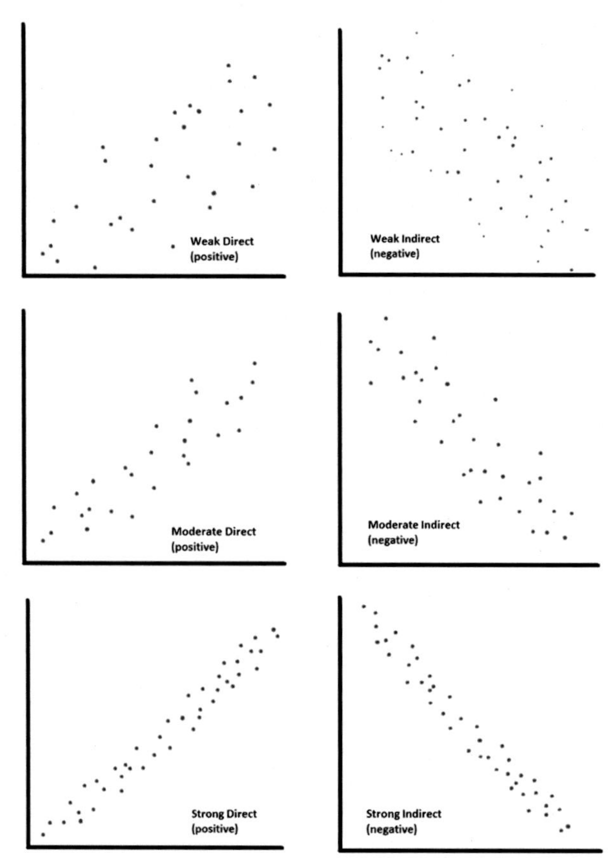

FIGURE 1.4.2

Relationships between variables can be described by the strength (weak, moderate, or strong) and direction (direct or indirect) of the relationship.

always better. Your goal is to find the way to best display the information that you need, and this is often best accomplished using a simple figure.

Look for differences between groups *[common precursor for t-tests and analyses of variance (ANOVAs)]:*

When you are interested in comparing measured values across different groups, bar charts are the easiest way to show how mean values for each group differ. However, it is critical to remember that if you want to make any sort of inference to the larger population (this is what inferential statistics are all about), the bar chart must include error bars that allow you to see how much variability there is within each group around that mean. If variability is high within a group, it is difficult to have confidence in the mean. Therefore, even if the means of the groups appear different for your sample of collected data, the same may not be true of the larger population. The general

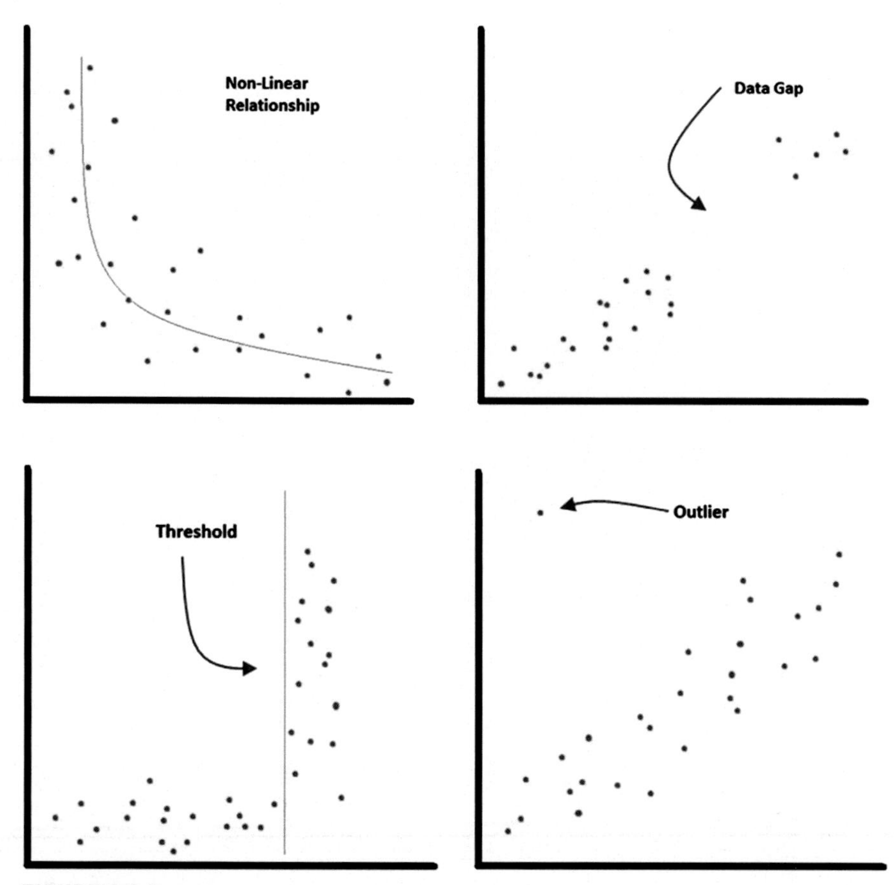

FIGURE 1.4.3

Visualizing your data before you run any analyses is critical to understanding the nature of your data and the story they have to tell. For example, if your data are nonlinear, a standard regression analysis is likely to conclude that there is no relationship between the variables when, in fact, there is a relationship, just not one that fits a straight line. Similarly, you can build a regression based on data with large gaps, but the applicability of that regression is valid only across the data ranges that are well represented.

rule of thumb for data exploration is that if the standard error bars around the means do not overlap, you likely do have significant differences between your groups.

In Figure 1.4.5, we are looking to see how the relative importance value (RIV) of various tree species differs in this forest inventory and also how RIV has changed over time for each species. The error bars allow us to see that while some species appear to have significantly increased in importance (BE = American beech and EH = eastern hemlock), some have decreased significantly (PB = paper birch and YB = yellow birch). Still others may have increased (BF) or decreased (SM) in RIV over the study period, but the overlap of error bars about the mean indicates that this may not be a significant change.

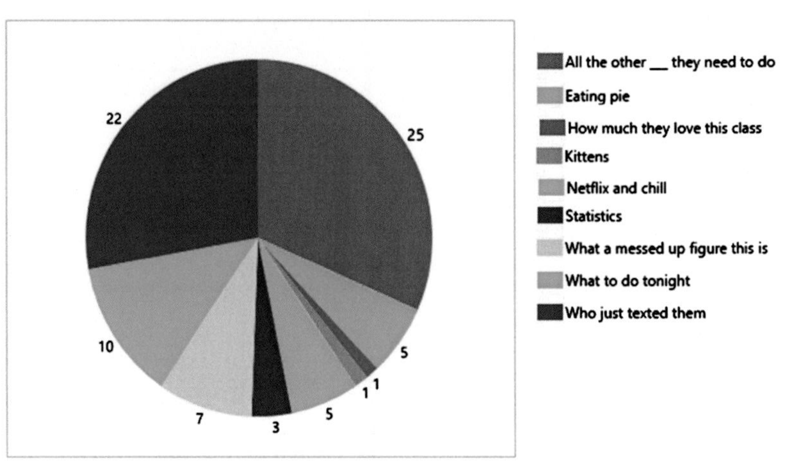

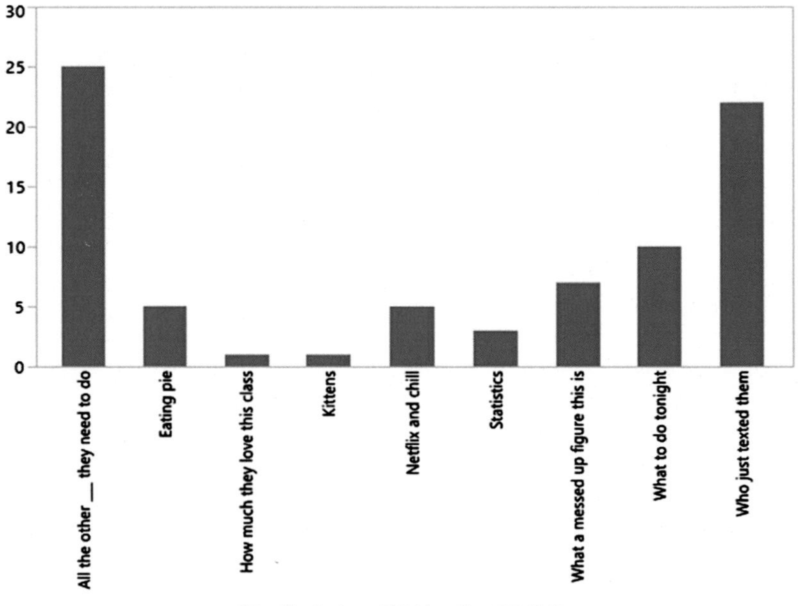

FIGURE 1.4.4

Pie charts can be appealing when general information is to be conveyed, but if you want to more directly compare various groups, a bar chart is often better.

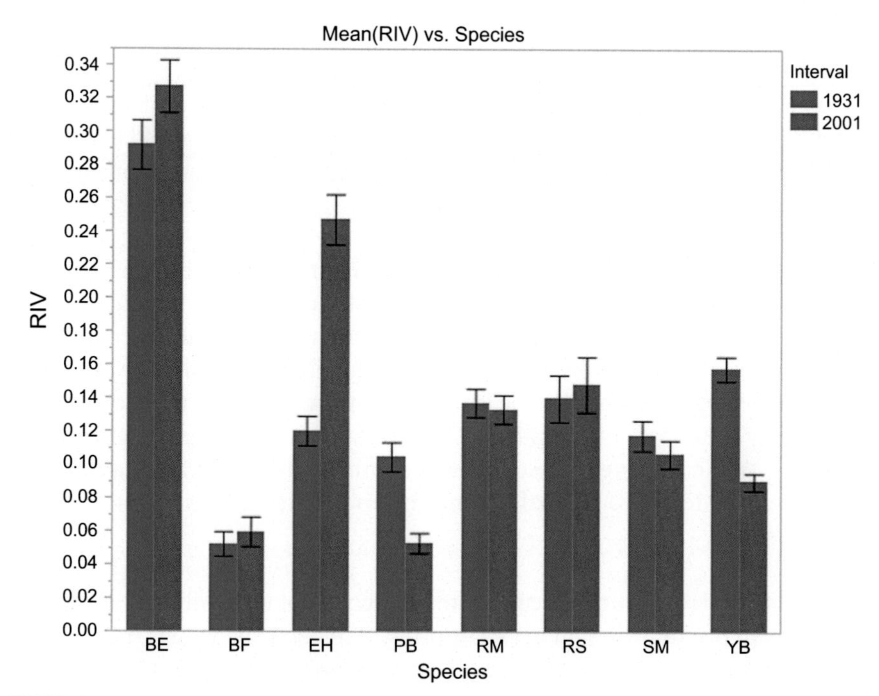

FIGURE 1.4.5
Measurement comparisons across groups should always include error bars to aid in interpreting the potential significance of differences seen in the means.

Metrics to Describe Your Data

1. **Central Tendency**: Almost all data in nature will cluster around a central or "typical" value. The goal is to identify one value that best represents an entire group. The most common metrics are the **mean** (average), **median**, and **mode**, but which to use depends on your data. Nature loves norms, and even though we all like to think that we (or our data) are special, in fact, the majority of us fall somewhere in the middle. You may think that you are the smartest person in the world, but chances are you fall somewhere in the middle.

   Mean: The mean **($\overline{X}$)**, or average, is the most common measure of central tendency. It is simply the sum **($\sum$)** of all the values **(X)** in your data divided by the number of observations **(n)**. While simple to calculate, the mean is very sensitive to outliers and is therefore *not appropriate to report when data are significantly skewed* (see normality below).

$$\overline{X} = \frac{\sum X}{n}$$

Median: The median (*Md*) is simply the value that falls in the middle of all your data, with 50% of all observations falling above and 50% falling below. After ranking your data from low to high, simply find the middle value, or, for even sample sizes, take the average of the two middle values. Median is appropriate to report instead of the mean for continuous data that are skewed by extreme values.

Mode: The mode (*Mo*) is the value that occurs most frequently in your data. The mode is not appropriate to report for continuous data unless it is binned into groups. This is because having the same value occur for measurements with many significant digits is highly unlikely. For ordinal or other categorical data, mode is the only appropriate metric of central tendency to report. It is possible to have data with no mode or multiple modes. While helping to describe how common different values are, it can also be used to assess normality, since normal distributions are unimodal (one mode).

2. **Variability**: How much spread there is in your data can tell you a lot about the population you are examining. Are "typical" values truly typical (i.e., most observations will report similar values) or simply in the middle of a wide range of equally likely values? Measures of variability allow us to set confidence intervals around our metrics of central tendency and are an essential part of determining the significance of inferential tests. Common metrics of variability include the **range**, **standard deviation**, and **standard error.**

Range: The range (R) is simply the maximum minus the minimum value in your data set. Because it uses only two observations to compute, range is not very useful for understanding the spread in your full dataset. However, it is very important when modeling or making predictions, since your algorithms are valid only over the range of values used to calibrate your predictive model.

Standard deviation: The standard deviation (s) quantifies how far, on average, each observation is from the mean. The larger the standard deviation, the more highly variable your data.

$$s = \sqrt{\frac{\sum (X - \overline{X})^2}{n - 1}}$$

- First, you determine how far each observation (X) is from the mean ($\overline{X}$),
- then square each of these differences;
- we sum these all up ($\sum$),
- and then divide by the number of observations (n) minus 1
- to get back to the original units of our data, we take the square root.

Standard error of the mean: Standard error of the mean estimates the amount of variability you would see in the mean of your sample if you collected multiple samples from the population. In this way, it serves as a sort of standard deviation for your statistical mean. Lower standard error values indicate higher confidence in the mean value that you report for your target population. This metric becomes important because it is used to compute other measures like confidence intervals and margins of error and is also the standard metric for error bars in bar charts displaying the means of various groups. The calculation itself is quite easy once you know the standard deviation **(s)** of your sample. Simply divide your standard deviation by the square root of your sample size **(n)**.

$$\frac{s}{\sqrt{n}}$$

3. **Normality**: Most traditional statistical analyses use a standard normal distribution to quantify the probability of outcomes and thus determine the significance of statistical analyses. As such, they require that your data meet the assumption of normality. For a population to be normally distributed, the following criteria must be met:

   - A histogram of the distribution approximates a bell curve, with no significant **kurtosis**. Kurtosis is simply a measure of how pointy or flat the peak of your distribution curve is. Any deviation from a bell shape, with the peak either too flat (platykurtic) or too peaked (leptokurtic), suggests that your data are not normally distributed.
   - The histogram is symmetrical with similar central values for the mean, median, and mode with no significant **skew**. Skew is a measure of how symmetrical your distribution curve is on either side of the peak. If your curve looks less like a bell and more like a slide on a playground, you likely have nonnormal data.
   - The histogram is unimodal (i.e., only one mode or peak in your distribution curve).
   - The tails are asymptotic (a fancy word indicating that infinitely higher and lower values are always theoretically possible, and the tail ends of your distribution curve never reach zero).

Many environmental data sets follow a normal distribution; nature loves norms. However, you should always check to make sure your data are normal before running any of these parametric tests. If you don't have access to software that can test your data for normality, you can go with what I like to call a quickie check:

1. Plot out your data in a histogram, and check to see if the distribution resembles a "bell" shape.

2. Compute the mean and median. These should be relatively close to each other (compared to the range of the data).
3. Calculate the skewness and kurtosis of your data. This will give you an idea of how bell-shaped your distribution is. Comparing these values to the standard error for each (see below) will allow you to estimate whether or not skew or kurtosis is significant in your data. The general rule of thumb is that if your skewness or kurtosis is less than two times the standard error, you do not have a significant departure from normality (Fig. 1.4.6)

Skewness: This metric quantifies how balanced (symmetrical) your distribution curve is. A normal distribution will have its mean and median

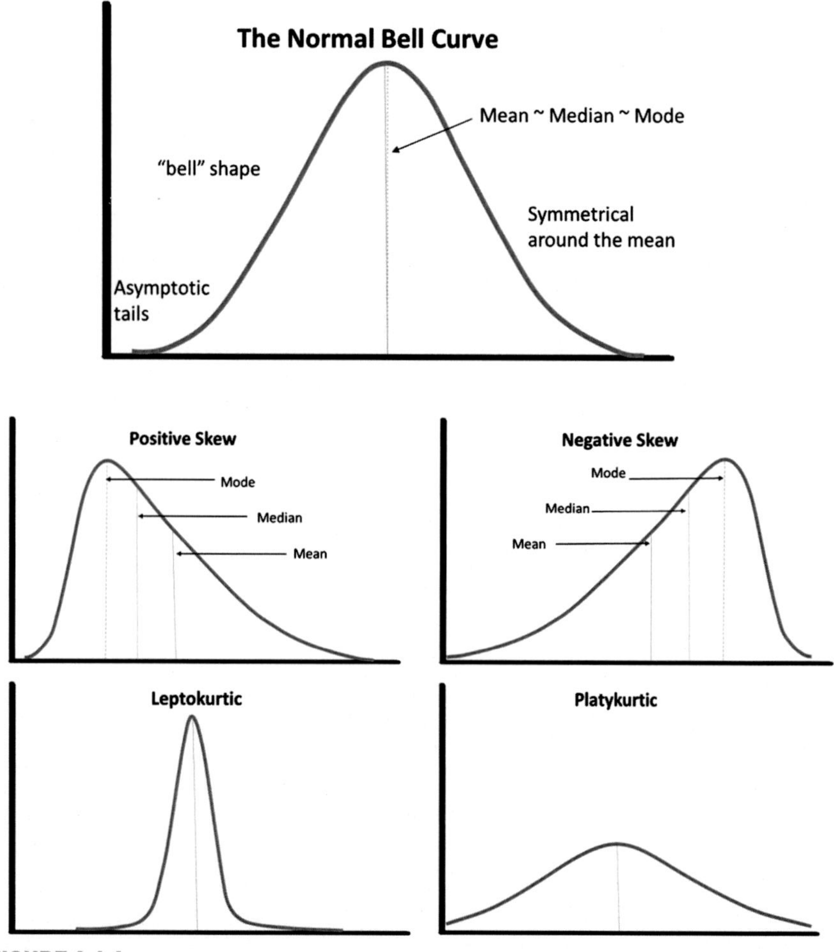

FIGURE 1.4.6
While "goodness of fit" tests are ideal, there are several metrics that can be used to help us assess the normality of our data.

values located somewhere near the center of its range. Skew of this peak away from center is common when extreme values pull the median away from the mean.

There are different formulas to approximate skewness, but the simplest is a measure of how far apart the mean and median are from each other relative to the amount of variability (quantified with the standard deviation) in the data.

$$skew = \frac{3(M - \bar{X})}{s}$$

To calculate skew:

- Take the difference between the median and mean of your data set,
- multiply it by 3, and then
- divide by the standard deviation.

Values near zero indicate no skew, while positive values indicate positive skew and negative values indicate negative skew.

To determine if this deviation from zero in the skew statistic is likely a significant departure from normality, compare it to the **standard error of skew (ses)**. If the skew you have calculated is more than two times the **ses**, then you likely have significant skew, which means you have nonnormal data and should consider a nonparametric test for your statistical analyses.

$$ses = \sqrt{\frac{6}{n}}$$

**If your calculated skew is more than 2 time this ses, then you likely have significant skew*

Kurtosis: In addition to skew, distributions can deviate from normality in how much the peak of the distribution deviates from a bell shape. If most observations fall around the mean value, you might expect a very narrow pointed curve (leptokurtic distribution). If observations are spread more equally across the full range of values, the curve may be relatively flat (platykurtic distribution). Kurtosis allows us to quantify how pointed, flat, or just right (bell-shaped) our distribution curve is.

$$K = \frac{\Sigma\left[\left(\frac{X - \bar{X}}{s}\right)^4 - 3\right]}{n}$$

This is another mathematically simple formula that can really mess you up if you get the order of operations wrong. Breaking down the steps from inside the parentheses out, you need to:

- Take every data value (X), subtract the mean ($\overline{X}$), and then divide by the standard deviation (s) of the full data set.
- Next, raise every one of these newly calculated values to the fourth power and subtract 3.
- Now, sum all of these up across all of your observation values.
- Divide by the sample size (n), and you have a metric for kurtosis.

Values near zero indicate no kurtosis and a close approximation of a normal bell curve. Positive values indicate peaked, leptokurtic distribution, while negative values indicate flat, platykurtic distribution.

To determine if this deviation from zero in the kurtosis statistic is likely a significant departure from normality, compare it to the **standard error of kurtosis (sek)**. If the kurtosis you have calculated is more than twice the sek, you likely have nonnormal data and should consider a nonparametric test for your statistical analyses.

$$sek = \sqrt{\frac{24}{n}}$$

**If your calculated kurtosis is more than twice this sek, then you likely have significant kurtosis*

Outliers: Outliers are extreme observations that don't fit the distribution of your sample. Sometimes outliers are used as a sort of quality control step. Even the most meticulous studies can include errors in the data, either from the measurements themselves or from data recording or entry. When outliers are true, correct measurements, they often can be the most interesting part of an analysis, but they do tend to skew our data distribution. Either way, these observations require additional data investigation, but they cannot be removed from our data unless they are proven to be:

- false (incorrect with error in either measurement, collection, or recording);
- outside of our target range; or
- unrelated to our population of interest (i.e., from a different population that we are not representing in our sample).

Descriptive statistics allow us to identify extreme or anomalous observations that may be wrong or that may just be really weird observations (every population does have its own set of weirdos). There are several common approaches for identifying potential outliers. One uses percentiles to identify any observations below the 2.5th percentile or above the 97.5th percentile. In a similar way, some use z-scores to identify the data value beyond which only 5% of observations would fall. A limitation is that both of these approaches depend on the assumption that the data are normally distributed. It is therefore safer to use an interquartile range (IQR) approach when working with data for which you cannot assume normality. This approach calculates upper and lower thresholds based on the range of values between the first and third quartiles. Any observations that fall beyond these thresholds are considered statistical outliers (Fig. 1.4.1).

Outlier identification with the IQR:

$$IQR = (Q_3 - Q_1)$$

$$Lower\ Threshold = Q_1 - [IQR * 1.5]$$

$$Upper\ Threshold = Q_3 + [IQR * 1.5]$$

- Find the median of your data set to split the data into an upper and lower half.
- Now, find the median of each of these halves (upper median and lower median).
- The median of your lower half of data is the first quartile (Q_1), below which 25% of your observations fall, and
- The median of the upper half is your third quartile (Q_3), above which 25% of your observations fall.
- Take the difference between the third and first quartile values to calculate the interquartile range (**IQR**).
- Multiply this IQR by 1.5.
- Add this value to the third quartile to establish an upper threshold.
- Subtract this value from the first quartile to establish a lower threshold.
- Any values that fall below or above these thresholds can be considered statistical outliers.

An Easier Way

Most spreadsheet software and all statistical software have built in functions to calculate basic descriptive statistical metrics. For products like Excel, you simply have to call up the right function (=function_name), and then, in parentheses, highlight the array of data for which you want the calculations made. A list of the most common descriptive statistical functions in Excel and similar spreadsheet platforms can be found on your Resource Page.

Important Things to Remember About Descriptive Statistics

Always report the basics: A table that describes the data included should be part of any study. In addition to allowing the reader to better assess the quality and nature of the data, it also provides critical information for others to determine how widely applicable the findings are.

Visualize your data first: The beginning of any statistical inquiry should always start with a visual examination of the data. The type of figure to use depends on the research questions you are asking. Relationships are usually examined with scatter plots, differences with bar charts, and temporal trends with line charts. Any of these visualization techniques must also include some measure of variability in addition to central values. For scatter plots, this might include a 95% confidence ellipsoid; for bar charts, error bars are common; and, for line charts, confidence borders should be included. Any figures without this additional information are difficult to interpret at best and misleading at worst.

Outliers: can strongly impact normality by skewing the mean. It is always smart to graph out your data first, and look for obvious data points that deviate from the general pattern. Remember that you cannot remove data points unless you suspect they are incorrect or not from your target population.

Meeting assumptions: One of the basic assumptions of many statistical analyses is that your data are normally distributed. Testing for normality is easy in statistical software packages, but harder if you are left to your own devices. If testing for normality is not an option, you can simply conduct nonparametric tests.

Practice

It's time to practice. On your Resource Page, you will find the "Tools example 1.4.txt" data file. These data provide the yearly mean snow depth (from measurements taken daily) at the summit of Mount Mansfield in northern Vermont between 1954 and 2014. Use the following guide to complete a descriptive statistical summary for these data.

- Create a histogram to visualize the distribution of your data. What are your initial assumptions about this distribution?
- Calculate the MEAN.
- Identify the MEDIAN.

- If you rounded each snow depth to the nearest integer (whole number), what is the MODE of the Yearly Mean Snow Depth Data? (Note that this will require first making a new column with the rounded value.)
- Calculate the STANDARD DEVIATION.
- What is the INTERQUARTILE RANGE for Yearly Mean Snow Depth?
- Using the interquartile range technique, how many OUTLIERS are there in your Yearly Mean Snow Depth Data? List any outlier years.
- Now calculate SKEW for the Yearly Mean Snow Depth Data.
- Calculate the ses for these data so we can determine the significance of our skew.
- Based on the ses technique, is this skew likely SIGNIFICANT?
- Now, determine the KURTOSIS for the Yearly Mean Snow Depth Data.
- Is this kurtosis likely SIGNIFICANT?
- Based on your statistical summary values, do you think that your data are likely NORMALLY DISTRIBUTED? Justify your response.

Chapter 1.5

Examining Relationships: Correlations

IN A NUTSHELL

Often in nature, different variables display similar patterns of variability. As one changes, the other is likely to change in some predictable way (they are said to be related to each other). In the environmental sciences, it is often useful to quantify the strength, nature, and significance of these relationships. It helps us to understand how the various components of an ecosystem are interconnected and informs management decisions designed to help mitigate environmental problems. The most common test of relationships is the Pearson's correlation (r), which provides a simple metric for us to quantify the proportion of variance in common between two continuous variables.

WHEN TO RUN A PEARSON'S CORRELATION (R)

- You have two continuous variables, and you want to test to see if there is a relationship between them.
- You assume that there is a linear relationship between the two measurements (i.e., the nature of the relationship between the two variables is similar across the range of possible values).
- Your data are normally distributed (See Chapter 1.4.). This allows you to use a normal probability table to assess the significance of any relationships you find.
- Your samples are independent and unbiased (i.e., you've selected an unbiased, representative set of observations on which to take measurements of the variables of interest).

WHAT THE PEARSON'S CORRELATION TELLS YOU

The Strength of Relationships Between Variables

The calculated Pearson's correlation test statistic **(r)** gives you a standardized metric with which to quantify the strength of the relationship between two variables.

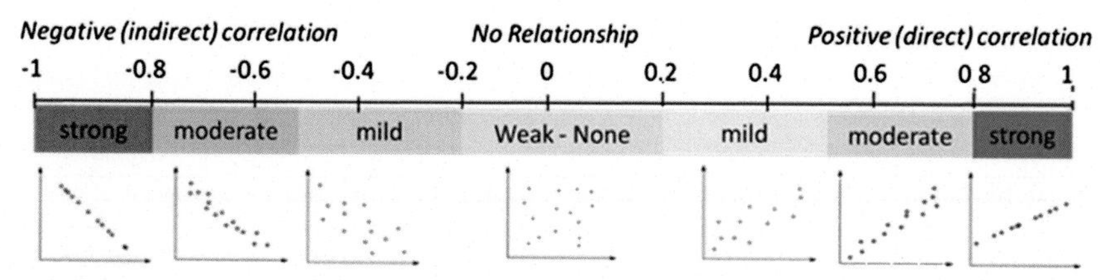

FIGURE 1.5.1
Visualizing a scatter plot of your variables of interest provides useful information about the strength and nature of any relationships before even calculating the Pearson's correlation (r) test statistic.

This metric ranges between −1 and 1. The closer to zero, the weaker the relationship is and the more a scatter plot of the two variables will look like, well, a plot of scattered points. The closer to −1 or 1, the stronger the relationship is and the more a scatter plot of the two variables will resemble a tight diagonal line (Fig. 1.5.1).

The Nature of the Relationship

Correlations are described as positive (**direct**) or negative (**indirect**). For a direct correlation, as one variable changes, the other is likely to change in the same direction (e.g., as the pH of precipitation increases, the pH of stream water also increases). In an indirect correlation, as the value of one variable changes, the other will change in the opposite direction (e.g., as the temperature of water increases, the amount of dissolved oxygen it can hold decreases).

The Significance of the Relationship

We can simply use the Pearson correlation coefficient (r) to describe relationships, but we can also use it to test hypotheses. No matter what the analysis, hypothesis testing follows a standard set of steps.

Steps for Inferential Statistical Analyses

1. **Specify your hypothesis**: Any analysis begins by stating the hypothesis to be tested. This is important to know because it will help you determine if you are interested in a directional (one-tailed) or nondirectional (two-tailed) test. Which test you are running will determine which probability to use in order to interpret the significance of your results.
2. **Set your significance threshold (α)**: To keep yourself honest, it is also useful to specify a significance threshold to use right up front. This is the threshold for your calculated probability, below which you will conclude a significant result. Most common is a $p<0.05$ threshold. This would signify that there is less than a 5% chance that the result you got occurred simply due to random chance and a greater than 95% chance that it occurred because the relationship truly exists.
3. **Test for normality**: Run some basic descriptive statistics on your data to see if they follow a normal distribution (see Chapter 1.4). The tests that we go over in this text all have assumptions for normally distributed data.

4. **Calculate the appropriate test statistic**: This will be a specific formula that quantifies whatever it is that you are trying to test (for correlations, this test statistic quantifies the amount of co-variability between two variables, while for *t*-tests, this test statistic quantifies how different the means of two groups are, considering their inherent variability).
5. **Determine the probability of getting this calculated result**: The test statistic that you calculate has an associated distribution curve that can be used to determine the probability of calculating what you did (or more extreme values). The lower this probability, the more likely it is that your calculated Pearson correlation coefficient (r) resulted from a true relationship between the two variables.

   Most statistical software packages will simply give you the exact probability associated with your test statistic. There are also functions in most spreadsheet programs and online calculators that will return the appropriate probability. The good old-fashioned way involves a look-up table, where as long as your calculated test statistic was more extreme (absolute value is greater) than the critical value in the look-up table, you knew you had a significant result (although you wouldn't know the exact probability associated with that test statistic).

   You will find links to both online calculators and look-up tables on your Resource Page. No matter which approach you use to determine the probability associated with your test statistic, you are going to need some key pieces of information:
   - the directionality of the test (one- or two-tailed, see #1 above)
   - significance threshold for the test (see #2 above)
   - how many degrees of freedom are associated with this test
6. **Conclude**:
   - If the probability associated with your calculated test statistic is less than your significance threshold, you HAVE a significant relationship.
   - If the probability associated with your calculated test statistic is greater than your significance threshold, you do NOT HAVE a significant relationship.
7. **How meaningful is this relationship?** While determining if you have a significant result is your first step, significance tests can sometimes be misleading. If you have a very small number of observations, it is possible that a truly significant result exists, but the power of your test is too low to be able to detect that significance. Similarly, if you have a very large number of observations (more and more common in our world of big data), it is possible to get a significant result that isn't at all meaningful in an ecological sense. Each test should have additional statistical metrics to help you assess how meaningful results are.
8. **Summarize your results**: Any statistical analysis should be able to be communicated in a complete but concise paragraph. To avoid writing a bunch of text interspersed with a bunch of numbers and symbols to summarize your statistical metrics, most analyses have a shorthand format to summarize your results. In addition to this shorthand, you should include the following in any summary of your results:
   a. State your original hypothesis or purpose of the study.
   b. Specify the type of test you ran and any pertinent information about the distribution of your data.
   c. Clearly state your results in their appropriate shorthand format and with text to describe the nature of any significant results and any follow-up tests that were conducted.
   d. Interpret your results. Are these meaningful? How do they relate back to your original objectives? What is the big picture? What do these results tell us and why do we care?

STEPS FOR THE INFERENTIAL PEARSON'S CORRELATION TEST

It is easiest to see how this process for inferential analysis works by doing an example. Let's assume that the following data represent the number of genetic mutations in amphibians and their exposure to ultraviolet radiation (UVR) during development. We want to know if there is a relationship between these two variables.

UV radiation	Mutations
3	61
1	54
14	111
15	125
11	53
7	33
3	47
13	101
12	95
11	72

1. **Hypothesis**: Do you assume a specific direction in your relationship? If so, this is a one-tailed test. If you don't hypothesize a specific direction for the relationship, this is a two-tailed test. Based on the literature and increasing evidence that exposure to UVR can damage cellular DNA, let's assume that there should be a direct relationship (more exposure = more mutations).

2. **Significance threshold**: We'll just stick with the standard threshold of $\alpha = 0.05$.

3. **Normality**: I can use descriptive statistics to estimate normality for this data set:

Statistic	UV radiation	Mutations
Mean	9	75.2
Median	11	66.5
Standard deviation	5.10	30.82
Skew	1.18	−0.85
2*ses	1.55	1.55
Significant skew?	|1.17| < 1.55 = No	|−0.85| < 1.55 = No
Kurtosis	−1.67	−1.59
2*sek	3.10	3.10
Significant kurtosis?	|−1.67| < 3.1 = No	|−1.59| < 3.10 = No
Outliers?	None	None

Based on the fact that I have no significant skew, no significant kurtosis, and no outliers, let's assume that both of our variables are normally distributed.

4. **Pearson's test statistic**: The Pearson's correlation test simply compares how much variability the two measurements have in common, relative to how much variation differs between them. The formula might look complex, but, breaking it down, you see that you really need only five pieces of information, and it mostly involves summing up columns of data:

$$r_{XY} = \frac{n\sum XY - \sum X \sum Y}{\sqrt{[n\sum X^2 - (\sum X)^2] * [n\sum Y^2 - (\sum Y)^2]}}$$

n = the number of observations (paired measurements)
$\sum X$ = all of your values for one of the variables summed
$\sum Y$ = all of your values for the other variable summed
$\sum XY$ = each of your observations multiplied together and then summed
$\sum X^2$ = the square of each value for one variable, then sum them
$\sum Y^2$ = the square of each value for the other variable, then sum them

If I do this for my example data, I get:

	UV radiation (X)	Mutations (Y)	XY	X^2	Y^2
	1	54	54	1	2916
	3	61	183	9	3721
	3	47	141	9	2209
	7	33	231	49	1089
	11	53	583	121	2809
	11	72	792	121	5184
	12	95	1140	144	9025
	13	101	1313	169	10,201
	14	111	1554	196	12321
	15	125	875	225	15,625
SUM	**90**	**752**	**7866**	**1044**	**65,100**
n	**10**				

$$= \frac{(10 * 7866) - (90 * 752)}{\sqrt{[(10 * 1044) - (90)^2] * [(10 * 65{,}100) - (752)^2]}}$$

$$r = 0.776$$

This is our Pearson's correlation coefficient, which tells us that we have a strong, direct relationship between UVR and mutations. But is it significant? We need the probability associated with our correlation to know that.

5. **Probability**: We can find the exact probability associated with this calculated test statistic by using any of a host of online statistical calculators. But note that you need some additional information in order to determine the probability for this test:

- the directionality of the test (we specified one-tailed for a test of a direct relationship)
- significance threshold for the test (we selected 0.05)
- how many degrees of freedom are associated with this test? For the Pearson's correlation, the degrees of freedom is a simple calculation of the number of observations in your data set (n) minus 2.

$$df = n - 2$$

Google "Pearson probability calculator," and select one. These calculators should ask for the calculated r value, the df (or sample size), and the significance threshold. They should also provide probability results for both a one- and two-tailed test. If not, try another calculator.

For this example, I am returned a probability for $r = 0.776$ of 0.00832 for the two-tailed test and 0.00416 for the one-tailed test.

$$p = 0.00416$$

6. **Conclusion**: Because I was conducting a one-tailed test, I can conclude that I have a significant relationship between UVR and mutations because my probability (or *p*-value) of 0.00416 is less than my significance threshold of 0.05. This means that there is only about four-tenths of a percent chance that the relationship we have just described occurred simply because of random chance. Instead, it is highly probable that there is a true, significant relationship between UVR exposure and the number of mutations in amphibians.

7. **Meaningful**? While we have a significant relationship, that doesn't mean that it is necessarily meaningful. We have a secondary metric we can use for that.

If you square the Pearson correlation coefficient (r), you get the coefficient of determination (r^2), a measure of *how much of the total variability in the data can be explained by the covariance (related variability) of the two*. The closer to zero, the less common variability there is. The closer to 1, the more common variability is shared between the two variables. How you interpret this value depends on how much of the variability you would expect to see in common before considering it to be ecologically meaningful. Most students are really uncomfortable with this ambiguity because it relies on expert opinion and that isn't something most students feel they can provide. But the key is to be able to justify your conclusion.

For this example, I am returned:

$$r^2 = 0.6026$$

In this example, an $r^2 = 0.6026$ indicates that approximately 60% of the total variability in these data can be explained by the relationship between the two variables. Considering how many other environmental factors might be impacting the number of genetic mutations (i.e., other conditions during development such as pollution or natural variability in cell division and replication), 60% may seem like a lot to you. Or it may not. You should be able to use ecological principles to back up your interpretation of this metric.

8. **Summarize your results**: Any statistical analysis should be able to be communicated in a complete but concise paragraph. To avoid writing a bunch of text interspersed with a bunch of numbers and symbols to summarize your statistical metrics, most analyses have a shorthand format to summarize your results. For the Pearson's correlation, this uses the following format:

$$r_{(df)} = \textit{calculated r value},\ \ \textit{probability},\ \ r^2$$

For our example, that would look like this:

$$r_{(8)} = 0.776,\ \ p = 0.00416,\ \ r^2 = 0.6026$$

We can then try to summarize it all in a succinct paragraph:

> We tested the relationship between the UV radiation exposure (UVR) and the number of genetic mutations in 10 amphibians. The Pearson's correlation indicated a significant and meaningful direct relationship between the two variables ($r_{[10]} = 0.776$, $p = 0.00416$, $r^2 = 0.6026$). The strength of this relationship in amphibians indicates that populations in high UV regions may be at risk of increased mutations.

IMPORTANT THINGS TO REMEMBER ABOUT CORRELATIONS

Visualize your data first: The beginning of any statistical inquiry should always start with an examination of the data. With correlations, it is usually easiest to explore your data using a scatter plot. This will help you avoid two common pitfalls:

Outliers can strongly impact correlation values: It is always smart to graph your data first and look for obvious data points that deviate from the general pattern. Remember that you cannot remove data points unless you suspect they are incorrect or not from your target population. But careful inspection will help you interpret your results, particularly if you find a surprisingly low correlation. One outlier may be hiding a significant pattern in the remainder of the data.

Linearity: Pearson's correlation tests only for linear relationships between variables. If a nonlinear relationship exists, your Pearson's test will not pick this up, but examining your scatter plot will tell you if you need to explore nonlinear options.

Meeting Assumptions: One of the basic assumptions of the Pearson's correlation is that your data are normally distributed. As discussed earlier, testing for normality is easy in statistical software packages but harder if you are left to your own devices. If testing for normality is not an option, you can simply conduct a nonparametric correlation test. There are many available that do not retain the assumption of normally distributed data. The simplest and most common is Spearman's Rho (ϱ). This test is essentially identical to Pearson's, but instead of being run on the raw data values, the data are first converted to ranks and these ranks are the values used in the calculations.

Interpretation: Often, correlations are interpreted to assign causation. However, you have to be very careful in assigning causation just because a significant relationship exists. It is quite common for logically unrelated factors to demonstrate what is called a spurious correlation (a mathematical relationship exists, but it is because each measurement is related to another common variable). For example, it is true that there is a significant positive correlation between the number of crimes committed in a municipality and the number of police officers. Does this mean that crime causes politicians to pour money into hiring additional officers? Possibly, but it is more likely that both the number of crimes and the number of police officers are tied to the population size for a given municipality. Small towns have both fewer police officers and fewer crimes simply because they have fewer people.

Significant relationships can also exist just by chance (remember that a significance level of 0.05 still means you have a 5% chance of saying there is a relationship when, in fact, the pattern you see is simply due to random chance). For example, over the past decade, there has been a strong correlation between divorce rates in Maine and the per capita consumption of margarine, but that doesn't mean that eating margarine leads to marital strife.

For many, "correlation does not equal causation" is considered the golden rule of correlations. However, scientists can use correlations, along with other pieces of evidence, to infer possible cause and effect relationships. We must always be careful in basing an entire argument on just one test, but, when considered as a part of the larger weights of evidence, backed up by scientific theory and experimental studies, it is fair to discuss the possibility of a cause and effect relationship using correlations.

Practice

Now you know just enough about correlations to be dangerous. It's time to practice. On your Resource Page, you will find the "Tools example 1.5.txt" data file. These data show decadal mean temperatures and atmospheric CO_2 concentrations.

1. Test the hypothesis that there is a positive correlation between temperature and CO_2 concentrations, BUT do this for two different time periods: 1880–1930 and 1962–2012. Report the results of these two Pearson correlation tests in a succinct one-paragraph summary and discuss how or why the relationship does or does not differ between the two 50-year intervals.

2. Now try the same test, BUT this time, use the full data set. Report the results of the analysis of the full-study duration. How does this differ from the results of your analysis of the early and later decadal averages? Statistically speaking, why might this be?

Chapter 1.6

Testing for Differences Between Means: *t*-Tests and Analysis of Variance

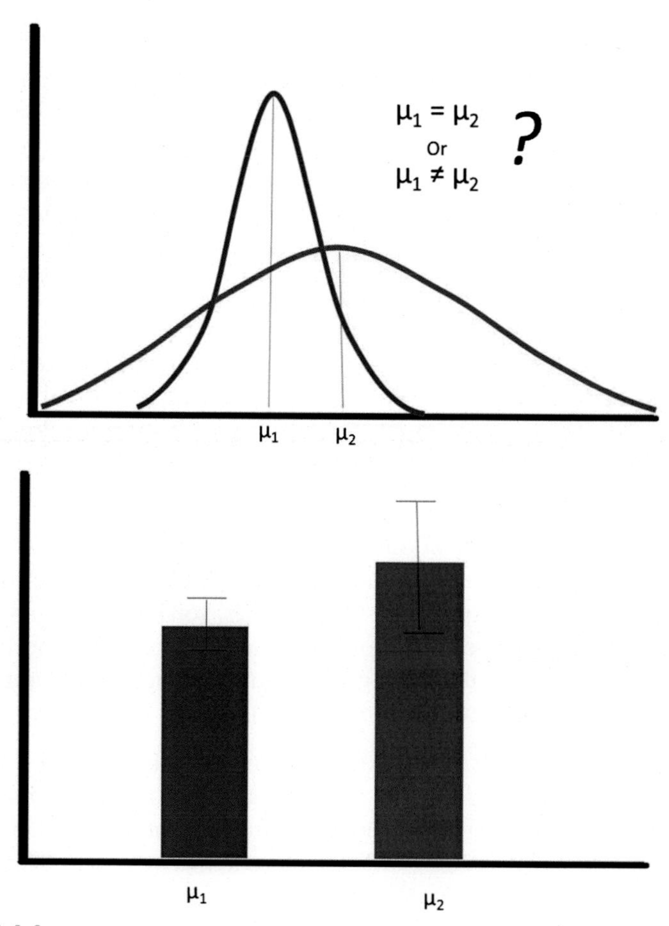

FIGURE 1.6.1
While measured differences in the means of different sample groups may look different, statistical analysis is necessary to determine if those differences (while considering the variability within each population) are significant.

TWO GROUPS TO COMPARE?

t-Tests in a Nutshell

Often, we are interested in comparing measurements from different groups (Fig. 1.6.1). For example, we might want to know if there is a different response for a treatment group compared to a control group that did not receive any treatment. To know if ecosystem restoration efforts are effective, we have to compare environmental metrics before and after restoration activities. To know if two species respond to environmental forcings in similar ways, we have to compare their response.

Because it is not feasible to collect and measure every drop of water or every organism present in an ecosystem, we collect data from a sample of observations that represent the larger population. We can then use inferential statistics to test for differences in the means of these groups. This test will tell us how likely it is that the measured means for each sample group are from the same distribution, and this allows us to draw conclusions about the larger population. When there are only two groups to compare, the most common test in environmental sciences is the *t*-test.

Independent or dependent?: *t*-tests can get a little tricky in that there are two different tests, depending on how your observations were selected for each group. One test compares the means of two independent groups, while the other compares differences in dependent (paired) observations. This is often decided not so much by your research question but by your experimental design. It all comes down to how you selected observations for your sample groups.

If observations are selected randomly, then it is safe to assume that your two groups are independent (the selection of one observation did not influence the selection of any other observations). In such cases, you would be hard-pressed to figure out which exact observation from one group should be linked to which exact observation in the other group. You simply took measurements on a bunch of random observations from each group.

Dependent tests dictate that for each observation you pick within a given group, there is a specific match, or pair, in the other group. So when you select one observation, it is very clear which observation from the other population group you should match it with. Before and after measurements are a perfect example of this pairing. You have taken measurements at the same location or on the same individual, just at two different times. These two measurements at different times are matched by their common location. Therefore, the selection of observations is dependent upon what other observations are selected. In these paired experimental designs, the careful pairing of observations eliminates many other sources of variability that might confound the test of your treatment (what it is that defines the two groups as different groups?). In this way, paired *t*-tests can be much more powerful in isolating the influence of the grouping factor you are interested in.

When to Run a *t*-Test

- You have a continuous response variable measured for two different groups.
- You want to test for a difference in this response variable between those groups.
- Your data are normally distributed with approximately equal variance in each group.
- IF: your samples are independent and unbiased (i.e., you've randomly selected a representative set of observations) = **independent *t*-test**.
- IF: your samples are paired (i.e., the selection of one observation determines which observation to pick from the other group) = **dependent *t*-test (matched pairs)**.

What the *t*-Test Tells You

A *t*-test allows you to calculate the probability that the two groups you are comparing come from the same population or if the means are sufficiently different to indicate that they are from two distinct populations. *t*-tests do this while considering how much variability there is in both groups (Fig. 1.6.2). For example, you may have group means that appear very different, but if both samples are highly variable (large variance), then it would be difficult to say with confidence that the means of their respective populations would be as different if you were to sample over and over again.

Steps for a *t*-test (refer back to the Steps for an Inferential Analysis in Chapter 1.5 for more general details on each of these steps):

1. **Specify your hypothesis**: The null hypothesis for a *t*-test is simply that the means of the two groups (or the mean of the differences between the paired observations) are equal: $\mathbf{H_0}$: $\boldsymbol{\mu_1 = \mu_2}$. But it is also important to know if you are just looking for a difference (nondirectional = two-tailed test), or if you expect one group to be higher than the other (specific direction to the difference = one-tailed test).

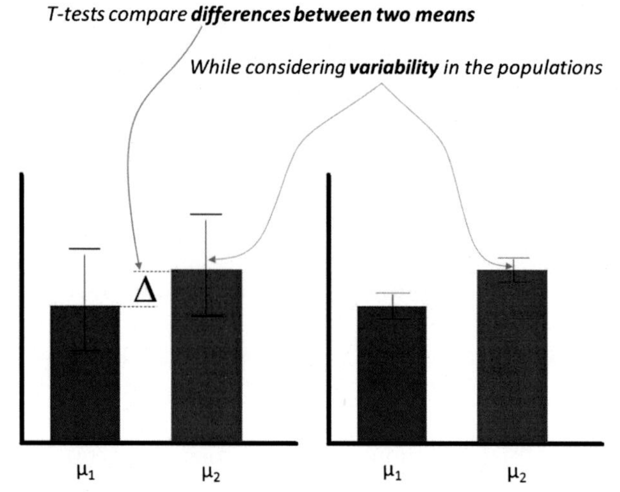

FIGURE 1.6.2
t-tests will quantify the probability that the means of two groups are different considering the amount of variability in each. In the bar chart on the left, variability is sufficiently high that we can't say with confidence that the means of the two populations are truly different. But in the bar chart on the right, variability is relatively low, indicating that this same measured difference in the means is likely different for the two populations.

2. **Set your significance threshold (α)**: $\alpha = 0.05$ is usually fine. Select a lower threshold (i.e., 0.01) if you have a very large sample size or want a more conservative test. Select a higher threshold (i.e., 0.10) if you have a very small sample size and want to reduce your risk of missing a significant result.

3. **Test for normality**: Run some basic descriptive statistics on your data to see if they follow a normal distribution (see Chapter 1.4).

4. **Calculate the appropriate test statistic**:

   - **Independent *t*-test** (quantifying the difference between two group means while accounting for the variability inherent in both):

$$t = \frac{\overline{X}_1 - \overline{X}_2}{\sqrt{\left[\frac{(n_1 - 1)s_1^2 + (n_2 - 1)s_2^2}{n_1 + n_2 - 2}\right]\left[\frac{n_1 + n_2}{n_1 n_2}\right]}}$$

   where $\overline{X}_1$, mean of sample group 1; $\overline{X}_2$, mean of sample group 2; n_1, # observations in sample group 1; n_2, # observations in sample group 2; s_1^2, variance of sample group 1; and s_2^2, variance of sample group 2.

- **Dependent (matched pairs) *t*-test** (quantifying the mean difference between observations while accounting for the variability in these differences)

$$t = \frac{\Sigma D}{\sqrt{\frac{n\Sigma D^2 - (\Sigma D)^2}{n-1}}}$$

where ΣD, sum of the paired differences; n, # of paired observations.

Note that to compute this dependent test statistic, you need to create a new column that calculates the difference between EACH pair of observations. In this way, the dependent *t*-test is not comparing the means of your paired groups, but the magnitude and variability in the differences between your paired observations.

5. **Determine the probability of getting this calculated result**: Once you have your calculated test statistic quantifying the difference you want to test, you need a few other pieces of information to either look up (**students *t*-distribution look-up table** found on your Resource Page), or calculate (using spreadsheet functions or online **probability calculators** also listed on your Resource Page) the probability associated with this test statistic.

   - the type of test (one- or two-tailed) (see step 1)
   - your significance threshold (see step 2)
   - degrees of freedom based on your sample size
     - independent *t*-tests **$df = (n_1 + n_2) - 2$**
     - dependent *t*-tests **$df = n - 1$**

   (Remember that n for the dependent t-test represents the number of PAIRS.)

6. **Conclude**:

   - If the probability associated with your calculated test statistic is less than your significance threshold, you HAVE a significant relationship.
   - If the probability associated with your calculated test statistic is greater than your significance threshold, you do NOT HAVE a significant relationship.

7. **How meaningful is any significant result?** Because significance tests may return significant results that aren't ecologically meaningful when sample size (i.e., power of the test) is large, or return nonsignificant results that still may be of interest when sample size (i.e., power of the test) is small, it is helpful to include a secondary metric that quantifies the magnitude of the difference without consideration of sample size when interpreting your results.

The most common assessment for this is the **Cohen's effect size**, which takes the difference between your group means divided by the mean standard deviations of the two groups.

$$Cohen's\ ES\ (d) = \frac{\bar{X}_1 - \bar{X}_2}{\bar{s}}$$

where small effect = 0.0–0.20; medium effect = 0.21–0.50; and large effect = 0.51 and above.

8. **Summarize your results**: For more detail on the various pieces of information that should be included in any statistical summary, refer back to Chapter 1.5: Steps for an Inferential Analysis. Don't forget to specify which type of *t*-test was used (independent or dependent, one- or two-tailed). Also specific to *t*-tests, you should include the proper shorthand to make it easier to convey key statistics metrics without rambling on with a bunch of numbers interspersed in the text.

   $t_{(df)}$ = calculated test statistic, *p*-value, effect size)
   For example: $t_{(29)} = 3.24$; $p = 0.003$; $d = 0.26$

IMPORTANT THINGS TO REMEMBER ABOUT *T*-TESTS

Visualize your data first: The beginning of any statistical inquiry should always start with an examination of the data. With *t*-tests, it is usually easiest to explore your data using a bar chart of the means with error bars. This will help you anticipate (and quality check) what your statistical results might tell you. Remember that if you are working with two dependent groups, you also want to graph out the mean difference between pairs (along with the standard deviation error bars).

Meeting assumptions: There are several assumptions that must be met to use a student's *t*-distribution in our *t*-tests. One of the basic assumptions is that your data are normally distributed. If you suspect that your data are not normally distributed, you can simply conduct a nonparametric *t*-test. There are many available, but perhaps the simplest and most common is the Wilcoxon *t*-test. This test is essentially identical to the student's *t*-test, but instead of being run on the raw data values, the data are first converted to ranks and then used in the calculations.

Interpretation: *t*-tests provide a probability-based assessment of how likely it is that the means of the populations represented by our two group samples differ. But remember, significant relationships can also exist just by chance (remember that a significance level of 0.05 still means you have a 5% chance of saying there is a relationship when, in fact, the pattern you see is simply due to random chance). In addition, significance tests are highly sensitive to the "power of the test." Particularly large samples will often return significant results even if those results are not particularly meaningful in an environmental sense. Because of this, it is wise to include a second follow-up test, such as the Cohen's effect size, that quantifies the magnitude of the difference between your groups without considering the sample size.

THREE OR MORE GROUPS TO COMPARE?

Analysis of Variance F-test

ANOVA in a Nutshell

Similar to *t*-tests, ANOVA is designed to examine the differences in the means of groups, but ANOVA is reserved for cases where you have three or more groups to compare. This is a particularly common test, considering that if we go through the trouble of setting up an experiment, we typically want to include more than two types or levels of treatments. For example, if I want to compare the impact of various restoration techniques for reducing contamination on a given population, I'd likely want to account for more than just a "treated" and "untreated" control group. Because there are many different restoration techniques that may impact contamination at different rates or in different amounts, I would likely want to compare several of these to inform which might be best in similar situations. This is when a well-constructed ANOVA comes in handy.

When to Run an ANOVA

- You have a continuous response variable measured in three or more different groups.
- You want to test for a difference in this response variable among those groups.
- Your data are normally distributed with approximately equal variance in each group. *(A quick check of equal variance compares the ratio of largest to smallest group standard deviation. If this ratio is less than 2:1, you can assume approximately equal variance.)*
- Your samples are independent and unbiased (i.e., you've randomly selected a representative set of observations)

What the ANOVA Tells You

One of the other differences to be aware of with the ANOVA test is that the initial inferential analysis only tells you whether or not there is *some* significant difference among your groups, but it does not tell you where that difference is. Follow-up tests are required to identify which of your possible groups are different from each other. As such, when running an ANOVA, you actually get to run the initial ANOVA and follow-up *t*-tests (lucky you)!

Steps for an ANOVA *(Refer back to the Steps for an Inferential Analysis in Chapter 1.5 for more general details on each of these steps.)*

1. **Specify your hypothesis**: The null hypothesis for an ANOVA is simply that the means of all groups are equal: $\mathbf{H_0}$: $\boldsymbol{\mu_1 = \mu_2 = \mu_3}$.

2. **Set your significance threshold (α)**: $\alpha = 0.05$ is usually fine. Select a lower threshold (i.e., 0.01) if you have a very large sample size or want a more conservative test. Select a higher threshold (i.e., 0.10) if you have a very small sample size or want to reduce your risk of missing a significant result.

3. **Test for normality**: Run some basic descriptive statistics on your data to see if they follow a normal distribution (see Chapter 1.4), and compare the ratio of standard deviations across your groups. The ratio between the group with the largest standard deviation and the smallest standard deviation should be less than 2:1 to approximate equal variance. If you violate this assumption, there are alternative tests that can be used with adjustments for unequal variance.

4. **Calculate the appropriate test statistic**: The ANOVA (F) test statistic essentially quantifies how much of the total variability in your data set can be attributed to differences among groups against how much of this total variability can be attributed to natural variability within each of the groups. To get at this requires more than one simple formula. There are several pieces of information that go into its calculation. Because of this, it is common to construct an **ANOVA table** to organize all of the pieces necessary for its calculation.

Source of variation	Degrees of freedom (df)	Sum of squares (SS)	Mean squares (MS)	F-statistic
Grouping factor (between)	$df_B = c - 1$	SS_B *(see formula)*	$MS_B = SS_B/df_B$	$F = MS_B/MS_W$
Error (within)	$df_W = n - c$	SS_w *(see formula)*	$MS_W = SS_W/df_W$	
Total	$df_T = n - 1$	$SS_T = SS_B + BB_W$		

c, the number of groups you are comparing; n, the number of observations in your data set.

Don't be overwhelmed by the table; there is nothing more complicated than adding, subtracting, and squaring involved in these calculations. You just need to take it step by step:

- Calculating the degrees of freedom is straightforward. Just take your number of groups minus one for the df between, and take the total number of observations minus the number of groups for the df within.
- Calculating the sum of squares gets a little more complicated.
  - **Sum of Squares Between:**

$$SSB = \sum_{j=1}^{c} n\,(\bar{X}_j - \bar{\bar{X}})^2$$

    where c, # of groups (and j is a given group); n, # of observations, $\bar{X}_j$, the mean of a given group (j); and $\bar{\bar{X}}$, the grand mean (mean of ALL observations).
    Essentially all you have to do is:
    — Calculate how far each of your group means is from the overall mean of the whole data set.
    — Square this value for each group.
    — Multiply this for each group by the number of observations.
    — Then, sum it up for all of your groups.
  - **Sum of squares within:**

$$SSW = \sum_{j=1}^{c}\sum_{i=1}^{n}(X_i - \bar{X}_j)^2$$

    where c, # of groups (and j is a given group); n, # of observations (and i is a given observation); X_i, a given observation value; and $\bar{X}_j$, the mean of a given group (j).
    Essentially all you have to do is:
    — Calculate how far each observation is from ITS group mean.
    — Square this value for each observation.
    — Add them ALL up (all observations in all groups).
- Calculating the mean squares is easy once you have your sum of squares figured out. Simply divide each by its appropriate degrees of freedom.
- Finally, you are ready to calculate what you ultimately need to determine the significance for your test, the F-statistic. This one is simply the ratio of the mean square between and the mean square within.

5. **Determine the probability of getting this calculated result**: Once you have your ANOVA table completed and the resulting calculated F-test statistic, you need a few other pieces of information to either look

up (**students F-distribution look-up table** found on your Resource Page), or calculate (using **spreadsheet functions** or online **probability calculators** also listed on your Resource Page) the probability associated with this test statistic.

- your significance threshold (see step 2)
- degrees of freedom based on the number of groups you are comparing and your sample size (note that this information is embedded right in your ANOVA table)
  - within groups: $\mathbf{df_W = n - c}$
  - between groups: $\mathbf{df_B = c - 1}$

6. **Conclude**:

- If the probability associated with your calculated test statistic is less than your significance threshold, you HAVE a significant relationship.
- If the probability associated with your calculated test statistic is greater than your significance threshold, you do NOT HAVE a significant relationship.

7. **How meaningful is any significant result?** Because significance tests may return significant results that aren't ecologically meaningful, it is helpful to include a secondary metric that quantifies the magnitude of the difference.

The most common assessment for this in an ANOVA is the coefficient of determination (r^2), which essentially tells you how much of the total variability in your data set is attributable to the grouping factor you were testing. It's easy to calculate once you have your ANOVA table filled out. You simply divide your sum of squares between by the sum of squares total. Values closer to one indicate that a larger proportion of the variability can be attributed to your grouping factor, making the results more likely to be ecologically meaningful.

$$r^2 = \frac{SS_B}{SS_T}$$

8. **Posthoc tests**: Because the ANOVA F-test simply tells you if there are differences but not exactly which groups are different, you will need to conduct some follow-up *t*-tests to see which of your multiple groups are different from each other. While you can conduct a simple *t*-test (described above) for each of the possible group comparisons, you have to apply a modification to the significance threshold of your test before interpreting

those results. This is because the initial ANOVA was designed to compare multiple groups simultaneously, but when you break down those groups into many individual comparisons, your chance of falsely claiming a significant result (set by your significance level) is compounded with each additional comparison. The most common solution is to apply a **Bonferroni adjustment** to find a new significance threshold to use when interpreting the individual *t*-tests.

$$\propto_{Bonferroni} = \propto_{original} * \frac{c!}{2! * (c-2)!}$$

where *c*, the total number of groups; !, remember that this is the factorial symbol, which just means to take any given number (x) and multiply it by ($x-1$), then multiply that by ($x-2$) and so on and so forth until you are down to multiplying by the number 1. For example $4! = 4 * 3 * 2 * 1 = 24$.

9. **Summarize your results**: For more detail on the various pieces of information that should be included in any statistical summary, refer back to Chapter 1.5: Steps for an Inferential Analysis. You should include the proper shorthand specific to the ANOVA test, and also be sure to discuss the nature of any significant differences between groups you find. Which groups were different from each other? How were they different (which was higher or lower)? Are these differences likely meaningful? Justify your conclusions with ecological theory.

$t(df_B, df_W)$ = calculated F-test statistic, *p*-value, r^2)

For example: ($F_{(2,12)} = 18.56$, $p = 0.0002$, $r^2 = 0.76$)

IMPORTANT THINGS TO REMEMBER ABOUT ANALYSES OF VARIANCE

ANOVA is a common analysis that is really quite easy to do by hand once you get used to the ANOVA tables (and even easier to do with software). But you still need to be careful that you are meeting all of the assumptions of the test. We usually remember to test for normality, but don't often think about equal variance. If either is violated, a simple nonparametric option (**Kruskal–Wallis**) can use the exact same formulas described above but run on the ranks. Remember that because the ANOVA is an omnibus test, it only tells you if there is a significant difference. To figure out which of your groups are different from which others, posthoc *t*-tests (with a Bonferroni adjustment to control for compounding error) are required.

Practice

Now that you know the basics of *t*-tests and ANOVAs, it's time to practice. On your Resource Page, you will find the "Tools example 1.6.txt" data file. These data include biological metrics (weight) taken on banded birds from three different species that were collected before and after their seasonal migration.

1. Start by testing to see if there is a difference in the initial (preweight) size of the three species of interest.

2. Is there a difference between the preweights of males (M) and females (F)?

3. Now test to see if there is a significant loss of weight for all species (all observations pooled). This involves testing for differences in the pre- and postweight data.

4. What if the change in weight during migration differs by species (i.e., some species lose more weight than others)? Create a new column that calculates the difference between your pre- and postweight data. With this new "change" data column, test to see if there is a difference among the species.

For more practice using *t*-tests and ANOVAs, try your hand at the exercises presented in the case studies that follow.

Chapter 1.7

Making Predictions: Regression Analyses

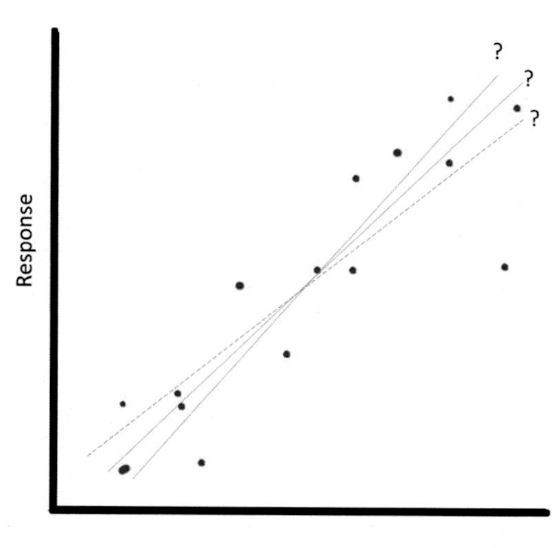

FIGURE 1.7.1
Least squares regression simply tries to find the equation for the line that best fits our data (by minimizing the squared distances of each point from that line).

REGRESSION IN A NUTSHELL

Regression analysis is a way to create a model (equation) to predict your response of interest based on a set of input predictor variables (Fig. 1.7.1). In its most basic form (a simple linear regression with one continuous input variable used to predict one continuous response variable), this process is as easy as finding the equation for a line that best fits a scatter plot of the data (Fig. 1.7.2). But more complex regressions, with multiple predictor variables, nonlinear relationships, or noncontinuous variables, are also common.

The idea is that if you can build an accurate and robust (stable when new observations are used) model to predict the response, you should be able to use that prediction as a substitute for making additional measurements. This is particularly useful when the input variables are relatively easy to measure and collect, but the response is more difficult to measure directly or is unknown (i.e., the outcome hasn't happened yet).

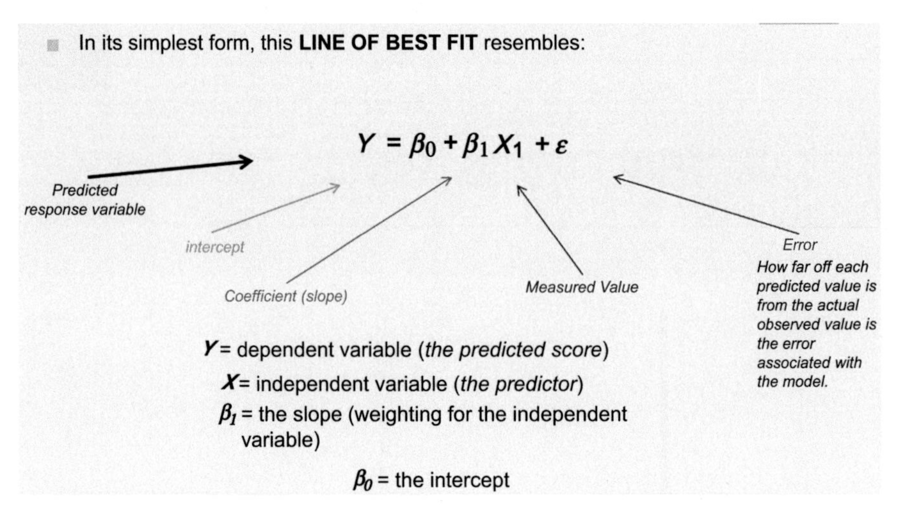

FIGURE 1.7.2
The simple linear regression model only requires that we calculate the slope and intercept for the line that best fits our data.

WHEN TO CREATE A LINEAR REGRESSION MODEL

- You have a continuous response variable.
- You believe there is a linear relationship (correlation) between this response variable and one or more continuous predictor variables (note that regressions can also be built with categorical predictor variables, but this will not be discussed in this text).
- You want to create an equation (model) to be able to predict this response variable based on those input predictor variables.
- Your samples are independent, unbiased and representative of the population you want to model (i.e., you've randomly selected a representative set of observations).

Note that normality of the data is not required, but normality of the residuals is.

STEPS FOR A LINEAR REGRESSION

Regression is different from the inferential tests we have covered so far in that our goal is not to test some hypothesis, but instead to build a predictive model based on significant correlations. This means that while we do need to run a correlation test (see Chapter 1.5) to determine if there is a significant relationship from which we can build a predictive model, there are several other steps that are necessary to both build and test the utility of any regression model.

1. **Visualize your data**: Linear regression only works for, well, linear relationships. Therefore, it is useful to start with an examination of a scatter plot between your input and response variables.

   a. Look to make sure that the general relationship is linear.
   b. Just like correlations, regression is sensitive to outliers. Outliers in your scatter plot that don't conform to the general relationship typically have "high leverage," meaning that they will be heavily weighted in the calculation of the slope and intercept of your equation.
      To remove or not to remove? Take outliers out of your data if:
      - There's something erroneous about this observation.
      - It doesn't fit the working range of the model.
      - It makes the model unstable (does the model change significantly when you remove it?).
   c. Examine the range that both your input and response variables cover. You cannot make any predictions outside of this range, so be sure that your calibration data set covers the range you would expect to see in the larger population. Also verify that there are no gaps in the distribution of your data points.

2. **Do you have a significant relationship? Is this relationship meaningful?** Before you can run a linear regression, you have to know that there is a significant correlation upon which to base the equation. See Chapter 1.5 for more details on correlation tests. By calculating the correlation (r), you can also compute the coefficient of determination (r^2) to help with the interpretation of how meaningful your predictive model might be based on this relationship.

3. **Calculate the slope and intercept for your equation**:

   - **Slope**:

$$\beta_1 = \frac{\sum XY - [(\sum X \sum Y)/n]}{\sum X^2 - [(\sum X)^2/n]}$$

   where X input (predictor) values, Y response (output) values, and n, number of observations; note that similar to computing r for a correlation coefficient, you need to create a new column to calculate the $X * Y$ for each observation and a new column to calculate X^2 for each observation.

   - **Intercept**:

$$\beta_0 = \frac{\sum Y - \beta_1 \sum X}{n}$$

where X input (predictor) values, Y response (output) values, n, number of observations, and β_1, slope of your line; note that you need to calculate the slope first to determine the intercept.

4. **Calculate the accuracy of your model**: You may have a significant model (based on your correlation analysis), and it may even be meaningful (based on your interpretation of the r^2 value), but any model without reported accuracy is nothing more than a pretty equation. Luckily, we have an easy way to quantify the typical error for our model using the **root mean square error (RMSE)**.

$$RMSE = \sqrt{\frac{\Sigma(Y - Y')^2}{n - 1}}$$

where Y, actual (measured) response values, Y', predicted response values, and n, number of observations.

I also prefer to include an interpretation of the relative accuracy of my model by converting the RMSE (in the same units as the response variable) to percent accuracy by simply dividing by the mean of the response. This gives me an idea of how far off (relative to the magnitude of the response) my predictions are.

$$\%\, error = \frac{RMSE}{\bar{X}}$$

5. **Summarize your results**: When you develop a predictive model, people will want all the information you can give them about the model itself and your interpretation of its potential utility. This means you have a lot of information to pack into your summary. At a minimum, this includes:

- the purpose (intended application) of the model;
- the nature of your input (predictor) and response variables (what range do they cover and what population do they represent?);
- the actual equation (with slope and intercept coefficients);
- the significance (p-value of the correlation) of the model;
- how meaningful this might be (in terms of how much of the total variability in the model can be accounted for with this model, interpreted from the r^2);
- how accurate the model is (RMSE at the least, but percent accuracy puts it all into perspective for potential users); and
- your general conclusions about how and when this model would be useful and any caveats that might limit its utility.

IMPORTANT THINGS TO REMEMBER ABOUT LINEAR REGRESSION

We have only covered the basics of linear regression, but modeling is a powerful tool that has tremendous potential to help us not only understand the drivers of environmental processes we study, but also test how changes to those drivers might impact the response of ecosystems. When you dabble in designing predictive models, be sure to carefully consider the significance, accuracy, and stability of the model.

While we have covered significance and accuracy here, the true test of whether or not a predictive model is stable (unlikely to change coefficients, decrease in accuracy, or fall apart entirely) when a new set of observations is introduced is to try out your model on a new set of data for which you have measured response variables (**independent validation**). If your accuracy (RMSE) with this new set of independent observations is similar to your original calibration RMSE, then you likely have a stable model. If you see a big difference between the validation RMSE and original calibration RMSE (e.g., 50% higher than the calibration RMSE), then it is likely that your model was simply fit to those calibration points rather than capturing a "true" relationship between variables.

Practice

Now you know just enough about regression to be dangerous. It's time to practice. On your Resource Page, you will find two data files: "Tools example 1.7.1.txt" and "Tools example 1.7.2.txt". The data in 1.7.1 were collected from an *in situ* fish farm relating the age of fish to their length. The goal is to develop a model so that scientists in the field can measure the length of fish and infer the number of days since they hatched to better understand reproductive success over the course of a season. The data in 1.7.2 show mean hourly fish counts through a fish ladder related to the mean number of fish retrieved by shock-and-net efforts upstream. The goal is to develop a model to estimate population size without having to shock the poor buggers.

1. Using the fish hatchery data in 1.7.1, develop a model to predict age based on length for assessments in local streams so that you can estimate the date of hatch for the fish you net. **Develop a simple linear equation**, and summarize your conclusions about its **quality for use in the field**.

2. Using the Fish Ladder data in 1.7.2, develop a model to predict upstream fish numbers based on fish ladder counts. Be sure to **summarize your final model** along with your conclusions about its **quality for use in the field**.

Chapter 1.8

Chi-Square (χ^2) Test

MAKING IT COUNT: WORKING WITH FREQUENCIES

In a Nutshell

Sometimes (most commonly in wildlife or vegetation inventories), the response measured is actually a count of how many observations are taken, categorized into distinct categories. For example, data collected at a newly installed fish ladder may include the count of fish in various species. At the end of the sampling interval, the observer would know how many of each species of interest had passed through the gates. Because counts are not truly continuous data (rather integer), we need a different tool designed to test hypotheses about how many observations we think should fall into each of the predetermined groups.
For example, you might expect that there are 50% males and 50% females in a given population. You can test to see if this is true by quantifying the difference between the observed and expected counts of each gender in a Chi-square test. More commonly, we just want to know if there is any statistical difference in the frequency with which observations fall into the various groups. A Chi-square test is a nonparametric tool that can be used to see if observed counts of observations differ from how they are expected to occur across several classes.

When to Use a Chi-Square Test

Anytime you have one or more categorical variables with **COUNTS** of the number of observations that fall into each class, this is the test for you. Because the Chi-square test is technically a nonparametric test, there are no assumptions for normality, but your observations must still be independent and randomly sampled to provide an unbiased estimate of the population.

There are two types of Chi-square analysis:

- When you are examining only the counts of observations that fall into the different categories of **ONE categorical variable,** use the **Chi-square goodness of fit** (Fig. 1.8.1 and Fig. 1.8.2A). This allows you to test to see if there is any difference in the frequency with which your observations fall into the classes or to test if your counts differ from some specific expected distribution.

- When you are examining the counts of observations that fall into the different categories across **several categorical variables**, use the **Chi-square test for independence**. This allows you to determine if the category into which an observation falls influences the category for the other variables. In other words, you are testing to see if these two categorical variables are associated with one another.

Goodness of Fit

Highest Education Level	Count
High School	382
Associates Degree	107
Bachelor's Degree	209
Graduate Degree	125
Total Column Count	823

Test for Independence

Highest Education Level	Views on Climate Change: Does not exist	Exists but is not related to GHG emission	Exists and is related to human GHG emissions	Total Row Count
High School	136	94	152	382
Associates Degree	22	20	65	107
Bachelor's Degree	57	42	110	209
Graduate Degree	15	29	81	125
Total Column Count	230	185	408	823

FIGURE 1.8.1
Chi-square analyses examine the observed counts of observations that fall into different groups or categories. The Chi-square goodness of fit test considers only one variable (the number of people with different levels of educational attainment) (left). The Chi-square test of independence (right) considers two or more variables (the same number of people with different levels of educational attainment AND their opinions on climate change) to see if there is an association between the two.

WHAT A CHI-SQUARE TEST TELLS YOU

Chi-square analyses tell you the likelihood that the distribution of observations you see matches either a random (or some hypothesized) distribution. In plain English, did you see as many observations as you thought you would in each of the categories you are testing? This information can be used to determine if there is some difference between the groups for which you are collecting observations or if there is some association between your variables of interest.

Steps to Conduct a Chi-Square Analysis

Whether you are conducting a Chi-square goodness of fit or independence test, the required steps are essentially the same and consistent with the steps that we use for any inferential statistical analysis *(refer back to the Steps for an Inferential Analysis in Chapter 1.5 for more general details on each of these steps)*.

1. **Specify your hypothesis**: The null hypothesis for a Chi-square test is that the frequency of observations you actually see in your data is the same as the frequency you expected to see. Typically, you are interested in testing a random distribution of observations that fall into each category (indicating that there is no difference between the categories), but Chi

square also allows you to test specific expected frequencies that are informed by the goals of the study.

2. **Set your significance threshold (α)**: $\alpha = 0.05$ is usually fine. Select a lower threshold (i.e., 0.01) if you have a very large sample size or want a more conservative test. Select a higher threshold (i.e., 0.10) if you have a very small sample size and want to reduce your risk of missing a significant result.

3. **Calculate the appropriate test statistic**: The Chi-square test statistic is one of the easiest of our inferential tests to calculate. This formula quantifies how far off your observed and expected frequencies are relative to the total expected count. You calculate this for each cell (each category or possible combination of categories) in your data. The final Chi-square test statistic for your inferential test is the sum of all of these individual cell Chi-square values.

$$\chi^2 = \sum \frac{(F_O - F_E)^2}{F_E}$$

where: F_O, the observed count (frequency) of observations; F_E, the expected/hypothesized count (frequency) of observations.

Knowing what your **observed frequency** is comes straight from your data, but how you calculate your **expected frequency** depends on which type of Chi-square analysis you want to do. If you are running a simple Chi-square goodness of fit test with only one variable to work with, the expected frequency for each category is a straightforward proportion of the total count you expect to see. If you are interested only in testing a random distribution, the proportion for each category should be equal (Fig. 1.8.2A), but if you have a specific, nonrandom proportion you want to test, you can do that too. For each category for your variable of interest, find the expected value with:

$$F_E = n * \propto$$

where n, the total number of observations, and $\propto$, the expected/hypothesized proportion. Note that if you are testing a random distribution of counts, the proportion is simply 1 divided by the number of categories present.

For the Chi-square test for independence (Fig. 1.8.2), when you have two or more variables to compare, each with its own set of categories, the formula to determine the expected frequency for each possible combination of categories is proportional to the total number of observations in each of the individual categories:

$$\frac{F_R * F_C}{n}$$

where n, the total number of observations, F_R, total frequency for a given row, and F_C, total frequency for a given column (Fig. 1.8.2B).

A smart way to check your calculated expected frequencies is to add them up. The expected count should always equal your observed count.

4. **Determine the probability of getting this calculated result**: Once you have your calculated test statistic quantifying the difference between your observed and expected frequencies, you need a few other pieces of information to either look up (**Chi-square distribution look-up table** found on your Resource Page) or calculate (using spreadsheet functions or online **probability calculators** also listed on your Resource Page) the probability associated with this test statistic.

- your significance threshold (see step 2)
- degrees of freedom based on the number of categories you are considering

For the goodness of fit test, you have only one column of data, so the degrees of freedom is simply:

$$\mathbf{df = (R - 1)}$$

where R, # rows in your data table (i.e., how many categories you are considering).

For the test for independence, you have both columns and rows in your data, so the degrees of freedom is:

$$\mathbf{df = (R - 1) * (C - 1)}$$

where R, # rows in your data table; C, # columns in your data table.

5. **Conclude**:

- If the probability associated with your calculated test statistic is less than your significance threshold, you HAVE a significant relationship.
- If the probability associated with your calculated test statistic is greater than your significance threshold, you do NOT HAVE a significant relationship.

(A) **Goodness of Fit Chi-Square Test**

Highest Education Level	Observed Count	Expected Count	Difference	Cell Chi-square Value
High School	382	205.75	176.25	150.98
Associates	107	205.75	-98.75	47.40
Bachelor's	209	205.75	3.25	0.05
Graduate	125	205.75	-80.75	31.69
Column Total	*823*	*823*	*0*	***230.12***

Where

expected value for each cell = total count / # categories

Difference = observed - expected

Cell Chi-square = = (observed - expected)2 / expected

Final Chi-square test statistic = sum of all cell Chi-square values

(B) **Test for Independence**

Observed Values

Highest Education	Views on Climate Change: Does not exist	Exists but not human driven	Exists and human driven	*Row Total*
High School	136	94	152	*382*
Associate	22	20	65	*107*
Bachelor	57	42	110	*209*
Graduate	15	29	81	*125*
Column Total	*230*	*185*	*408*	*823*

Expected Values

Highest Education	Does not exist	Exists but not human driven	Exists and human driven
High School	106.76	85.87	189.38
Associate	29.90	24.05	53.04
Bachelor	58.41	46.98	103.61
Graduate	34.93	28.10	61.97

*Where the expected value for each cell = (row total * column total) / total count*

Difference

Highest Education	Does not exist	Exists but not human driven	Exists and human driven
High School	29.24	8.13	-37.38
Associate	-7.90	-4.05	11.96
Bachelor	-1.41	-4.98	6.39
Graduate	-19.93	0.90	19.03

Where Difference = Observed - Expected

Cell Chi-Square

Highest Education	Does not exist	Exists but not human driven	Exists and human driven
High School	8.01	0.77	7.38
Associate	2.09	0.68	2.69
Bachelor	0.03	0.53	0.39
Graduate	11.37	0.03	5.84

Chi square value for each cell = (observed - expected)2 / expected

Final Chi-square test statistic = sum of all cell Chi-square values **39.83**

FIGURE 1.8.2

Using the data from Fig. 1.8.1, we walk through the steps for calculating a Chi-square value for each cell. (A) shows the calculation of the Chi-square Goodness of Fit test for one variable (highest education level), while (B) shows the calculation for the Chi-square test for Independence between two variables (highest education level and views on climate change). The sum of all cell Chi-square values is the final Chi-square test statistic for both probability tests.

6. **How meaningful is any significant result?** Because significance tests may return significant results that aren't ecologically meaningful when sample size (i.e., power of the test) is large or return nonsignificant results that still may be of interest when sample size (i.e., power of the test) is low, it is helpful to include a secondary metric that quantifies the magnitude of the difference without consideration of sample size when interpreting your results.

   The most common assessment tool for this is the **phi coefficient**, which provides a separate measure of the magnitude of the difference between groups or association between variables.

$$\phi = \sqrt{\frac{\chi^2}{n}}$$

   where $|0 \text{ to } 0.3|$ little or no association, $|0.3 \text{ to } 0.7|$ weak association, and $|0.7 \text{ to } 1.0|$ strong association.

7. **Describe significant differences**: You can look at individual class Chi-square values to see which differ the most from the expected value. The higher the cell Chi-square value, the more it deviates from what was expected. Then, examine the nature of the difference between observed and expected (was it positive or negative?) to see HOW that class count differs (higher or lower than expected). These differences should be clearly described in your summary. It isn't enough to know that there is an association, especially when you are considering many groups. Instead, we want to know which groups were the most different from what was expected and whether they were higher or lower than expected. This information is critical for interpreting the results.

   In our example in Fig. 1.8.2B, we can see that the greatest association for the test of independence between the level of terminal degrees and views on climate change comes from far fewer people with graduate degrees who believe climate change does not exist than would be expected if the two (education and climate views) were not associated with one another. We also see more people who stopped their education after high school who believe climate change does not exist than would occur randomly.

8. **Summarize your results**: For more detail on the various pieces of information that should be included in any statistical summary, refer back to Chapter 1.5: Steps for an Inferential Analysis. Specific to Chi-square-tests, you should include the proper shorthand to make it

easier to convey key statistics metrics without rambling on with a bunch of numbers interspersed in text.

$\chi^2_{(df)}$ = **(calculated test statistic, *p*-value, phi coefficient)**
For example: ($\chi^2_{(6)} = 39.83$, $p < 0.001$, $\varnothing = 0.26$)

Important Things to Remember about Chi-Square

Chi-square is an easy way to look for differences among groups or associations between variables that is particularly useful in social sciences where it might be difficult to quantify variables of interest but easier to assign observations to general categories. Hopefully, you noticed that we didn't have to test for normality when we conducted this inferential test. This is because the Chi-square is a nonparametric test with no assumptions about the distribution of your data. However, Chi-square is very sensitive to low cell counts. This means that if you have some categories you are testing that have few observations, it is possible that you either have not obtained a sufficient sample size to accurately represent your population of interest or you are including categories to test that are not relevant (i.e., so rare as to not be of interest). These low cell counts can inflate your Chi-square test statistic and lead to incorrect conclusions of significance.

Practice

Now you know have the basics of Chi-square down, let's practice before starting in on the case studies.

1. You want to determine if a certain native species of bird is still successfully nesting in small fields, given the presence of two nonnative bird species. If nesting success is random, you would expect equal representation of all three species. Test to see if this is true.

   You have roamed around small fields for days and have collected the following counts of successful nests (intact nests with live chicks). Test to see if nesting success is random or if some species is favored in this ecosystem.

Species	Successful nests
A	23
B	17
C	50

2. The US Environmental Protection Agency (EPA) has collected air quality data from locations in various neighborhoods to better understand how poor air quality may disproportionately affect populations at different income levels. Here, they report the number of days over a 3-year period where air quality did not meet US EPA standards (air quality was poor enough to be considered a threat to human health). Describe how the number of days with poor air quality differs across the various neighborhoods.

Neighborhood mean annual income	Met US EPA protection agency standards?	
	Yes	No
<$30,000	832	263
30,000–60,000	996	99
>60,000	1040	55

Chapter 1.9

Carbon Footprints

FIGURE 1.9.1

Cutout feet relative to the size of various nation's carbon footprints. *Source: allispossible.org.uk (CC-by-SA 2.0) via Flickr.*

IN A NUTSHELL

Because of global concerns about climate change and the need to reduce emissions of CO_2 and other greenhouse gases (GHGs) (CH_4 and N_2O among others) that contribute to climate change, there have been many different attempts to quantify and describe the various sources of those emissions. Life cycle assessments (LCAs) (see Chapter 1.10) are used to quantify how much environmental impact is associated with the production, transport, and use of various items. A carbon footprint (Fig. 1.9.1) is an important component of an LCA since it estimates the amount of carbon released from the production, transport, and use of a product.

But carbon footprints can also be calculated for individuals, organizations, or events. These assessments can allow us to look at the cumulative contribution of our lifestyle to carbon emissions and evaluate which of our actions contributes the most. Perhaps what is most interesting about carbon footprints is that they allow us to quantify our personal contribution to climate change. As a result of recent global agreements to slow the progress of climate change, it is likely that ever-increasing attention will be paid to where GHGs are coming from. Individuals skilled in "carbon accounting" should have excellent professional opportunities in the future, as businesses gear up to meet the carbon-cutting challenge.

WHAT IS A CARBON FOOTPRINT AND HOW IS IT USED?

There have been a number of definitions of "carbon footprint" in recent years. A commonly used version includes the phrase "**the total set of GHGs caused directly and indirectly by an individual, organization, event, or product**" (Carbon Trust, 2007). Identifying and quantifying emissions of GHGs can serve a number of purposes. Government agencies charged with reducing GHG emissions from large point sources like power plants need good estimates of emissions so that they can detect changes over time. Industries facing regulations designed to reduce or eliminate GHG emissions need accurate numbers to determine whether or not their control strategies are working. As global efforts to reduce carbon emissions increase, makers of all consumer products will likely be seeking ways to lower their carbon footprints. As individuals, it is useful for us to see how different components of our daily lives contribute to GHG emissions. Of course, the challenge is to come up with accurate numbers.

Carbon footprints have even made their way into the marketplace. Some companies use "carbon labeling" to identify the specific amount of carbon released during the lifespan of their products. For example, some companies now attach tags to their products to show customers exactly how many carbon emissions resulted from their purchase, thus appealing to individuals seeking to make the "greenest" choice (Fig. 1.9.2). Unfortunately, since there are different methods used to calculate total CO_2 emissions, it can be difficult for a consumer to really know which the best choice is.

Because there are additional GHGs to consider besides CO_2, we need to express footprints in **carbon equivalents** (CO_2-e), the term used to express the GHG potency when there is a mixture of gases present. If a product's life cycle gives off methane and nitrous oxides, both GHGs, their contributions can be expressed as an equivalent amount of CO_2 by converting their potency as a GHG to that of CO_2 based on the **global warming potential** (GWP) of each gas. For example, CO_2 has a GWP of 1, while methane's value is 25. So a given amount of CH_4 would be valued at 25 times what the same amount of CO_2 would.

FIGURE 1.9.2
An example of a carbon footprint tag from Japan's Ecoleaf program. *Source: Kevin Dooley (CC-by-SA 2.0) via Flickr.*

Remember that carbon footprints are inclusive. They deal with every aspect of a product or service, including procurement of the raw materials for a product, manufacture of the product, its transportation and storage, and its use and disposal by the consumer. Carbon footprints for individuals, organizations, or events must consider every activity and production of the items required for that activity. This can be daunting to quantify, but luckily, many models exist to estimate carbon footprints for common products or to predict individual carbon footprints based on information about our lifestyle.

Some Examples of Carbon Footprints

Some carbon footprints have been calculated at large scales. A 2006 study by the Global Action Plan estimated the carbon footprint for all the schools in the UK. Their report indicated that in 2001, the schools accounted for about 1.3% of the UK's total carbon emissions, with 26% of the schools' emissions resulting from heating, with the remaining coming from such indirect sources as furniture (5%) and paper (4%).

Other models exist to estimate individual carbon footprints. Many websites are available that will walk you through a series of questions about your lifestyle (where you live, how you get around, what you eat, etc.) to provide an estimate of your carbon footprint and show you how you compare to others. The US EPA carbon calculator linked on your Resource Page provides a rough estimate of your carbon footprint by using US average values for home heating, transportation, and waste. For a more accurate estimate, you can gather your utility bills (electricity, natural gas, fuel oil, propane) to calculate your average use over a year.

Other carbon footprint calculations have focused on individual consumer products. One of the largest product-specific carbon footprints comes from our vehicles. Ball (2008) looked at carbon footprints of several types of cars. At the time of the evaluation, the average midsize car made in the US emitted an average of 63 tons of CO_2 over its 120,000-mile lifespan. This figure included emissions from the entire life cycle, from procuring the raw materials to shredding the junked vehicle. Most of the emissions, about 86%, were attributed to the car's fuel use, so the models with the best fuel economy had the lowest carbon footprints per miles driven. For instance, Toyota's hybrid Prius had a lifetime carbon footprint of 44 metric tons, while its SUV 4Runner came in at 118 tons.

The footprints of everyday items like food and apparel are now commonly calculated. An assessment of running shoes gives us a good look at how the global economy impacts carbon emissions associated with what may seem like an inconsequential product. A group of scientists from MIT's Materials Systems Laboratory (Cheah et al., 2013) measured the carbon footprint of a pair of running shoes (Fig. 1.9.3). While perhaps each individual pair of shoes has a relatively small carbon footprint, running shoe sales each year amount to 25 billion pairs globally, so the cumulative footprint is much larger.

The research team assessed the materials, manufacturing, and transportation, noting that most of the shoes are made in China. The manufacturing assessment alone required GHG data from 26 distinct materials used in making the upper shoe, the sole, and the packaging material. In fact, a single shoe can contain 65 discrete parts requiring 360 assembly steps.

The results of the analysis showed that total GHG emissions over the shoes' lifespan were 14 ± 2.7 kg $CO_{2\text{-e}}$. The greatest contributor was the manufacturing process (Fig. 1.9.4). An important contributor to the carbon footprint was the source of energy used in manufacturing. China relies heavily on coal, a significant GHG contributor.

Important components of analyses like this are suggestions about how to reduce the output of GHG. In this case, important steps recommended to reduce the carbon footprint for running shoes included (1) consolidating shoe components to reduce the manufacturing steps and energy use required; (2) increasing the efficiency of the machinery being used, thus saving electricity; and (3) using recycled materials whenever possible. Obviously, switching from coal to a less carbon-intensive energy source would reduce emissions as well.

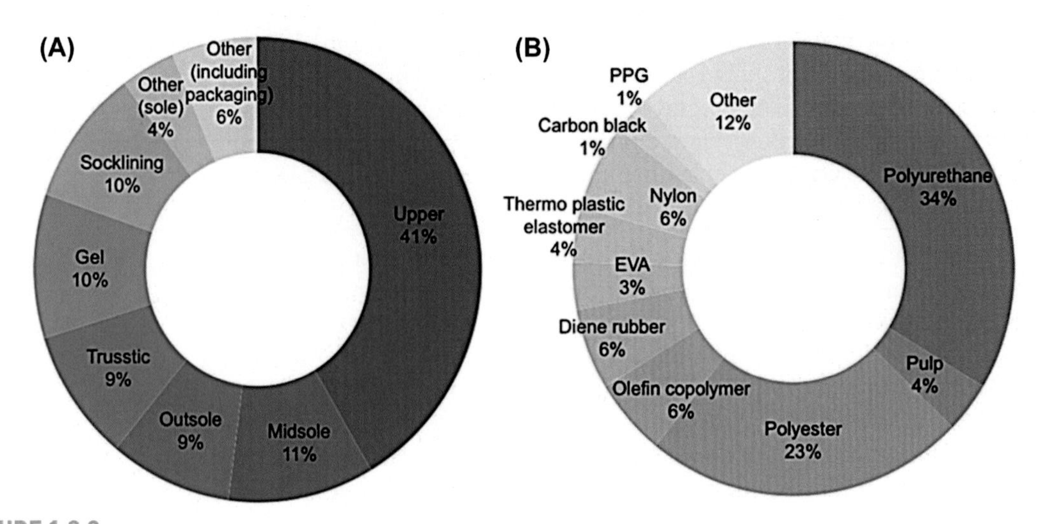

FIGURE 1.9.3

Global Warming Potential (GWP) (kg CO_2-eq) of materials by percent within a pair of running shoes including (A) breakdown by part and (B) breakdown by material (Cheah et al., 2013).

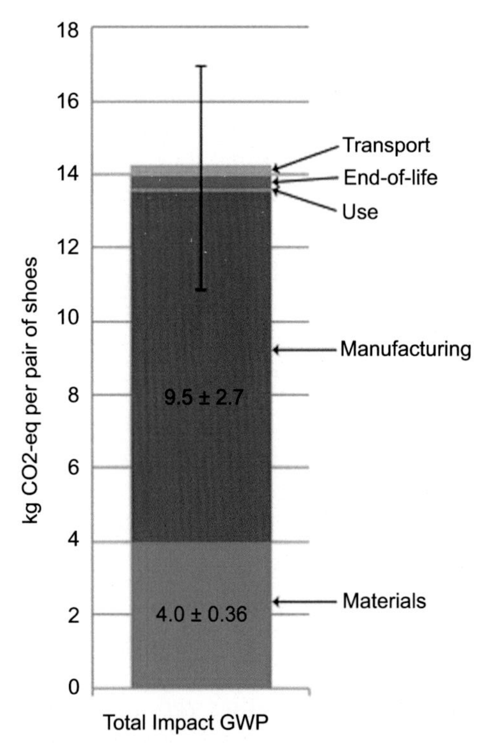

FIGURE 1.9.4

Breakdown of total GWP of running shoes by life cycle phase (Cheah et al., 2013).

An Important Thing to Remember About Carbon Footprints

It's complicated. We noted that many different pieces of material were needed to assemble a running shoe in the example above. Manufacture of electronics can be even more complex. With this many different materials and processes, data collection and verification can be a very time-consuming task. Errors in your estimate of each component compound to elevate the error associated with the final product. Precision and accuracy of data on GHG emissions are particular concerns, as are variabilities inherent in the manufacturing process (e.g., the amount of raw material actually ending up in a shoe as opposed to ending up as scrap). As you might suspect, the availability and quality of data on GHG emissions from all the various steps in the process may also be a problem. These numbers should serve as a good reference, but may not be exact.

Practice

1. Several different tools are available online to help you estimate your personal (or household) carbon footprint. Several of these are listed on your Resource Page.

   - Chose two of these models, and complete the questionnaire to help you estimate your carbon footprint.
   - How did these two models agree or differ in their calculation of your carbon footprint? What might account for those differences? Which do you think is more accurate?
   - What component of your daily life contributes most to your carbon footprint?
   - What are some ways that you could make changes to reduce these contributions?

2. In terms of their carbon footprints, everyday items can surprise you. Consider these three objects:

   - a quart of orange juice,
   - a cotton T-shirt, and
   - a 6-pack of soft drinks

   Assume that the GHG release for each of these products occurs at four stages: raw material procurement, manufacture, transportation and storage, and consumer use and disposal.

   - Do some research and, for each of the three products, identify which of the four stages in the lifespan accounts for the greatest proportion

of the object's carbon footprint. If you can, identify the specific step responsible for releasing the greatest percentage of the total.
- Come up with at least one suggestion for each product that could shrink its carbon footprint.

REFERENCES

Ball, J., 2008. Six products, six carbon footprints. Wall Str. J. October 6 edition. http://www.wsj.com/news/articles/SB122304950601802565.

Carbon Trust, 2007. Carbon Footprinting Guide. https://www.carbontrust.com/resources/guides/carbon-footprinting-and-reporting/carbon-footprinting.

Cheah, L., Ciceri, N.D., Olivetti, E., Matsumura, S., Forterre, D., Roth, R., Kirchain, R., 2013. Manufacturing-focused emissions reductions in footwear production. J. Clean. Prod. 44, 18–29.

Global Action Plan, Stockholm Environment Institute, and Eco-Logica Ltd., March 2006. UK Schools Carbon Footprint Scoping Study Report for the Sustainable Development Commission, London. http://www.sd-commission.org.uk/publications.php?id=389.

Chapter 1.10

Ecological Footprints

FIGURE 1.10.1
How many people the Earth can sustainably support is tied to how many resources each of us requires to support our standard of living. *Source: NASA.*

IN A NUTSHELL

Among the tools for assessing human impact on the planet is the concept of the ecological footprint (EF). An EF differs from LCAs and carbon footprints in that it quantifies the amount of global resources, both land and water, needed to support an individual, household, city, or larger population (Fig. 1.10.1). This includes both our resource needs (think food, water, and consumables) as well as the amount of land required to dispose of our solid waste and absorb the CO_2 we emit.

Experts estimate that the Earth has about 12 billion hectares (ha) of biologically productive land, or about 2 ha for each person on Earth. To put things into perspective, the EF for the average US citizen is about 7 ha, meaning that it takes 7 ha to supply what we currently consume to maintain our standard of living in the US.

Our EF is closely tied to the carrying capacity of the Earth, which is the total population that the Earth can sustainably support. Because the base of productive land is relatively constant (or decreasing, considering stressors like development and desertification), how many people the Earth can support is dependent upon the size of our EFs (Fig. 1.10.2). The obvious challenge is to manage global resources sustainably so that everyone can enjoy a decent standard of living now and in the future without destroying the Earth's natural resource base. If human populations exceed the Earth's carrying capacity or our cumulative EFs exceed the land base available, there could be widespread shortages of basic supplies like food and water.

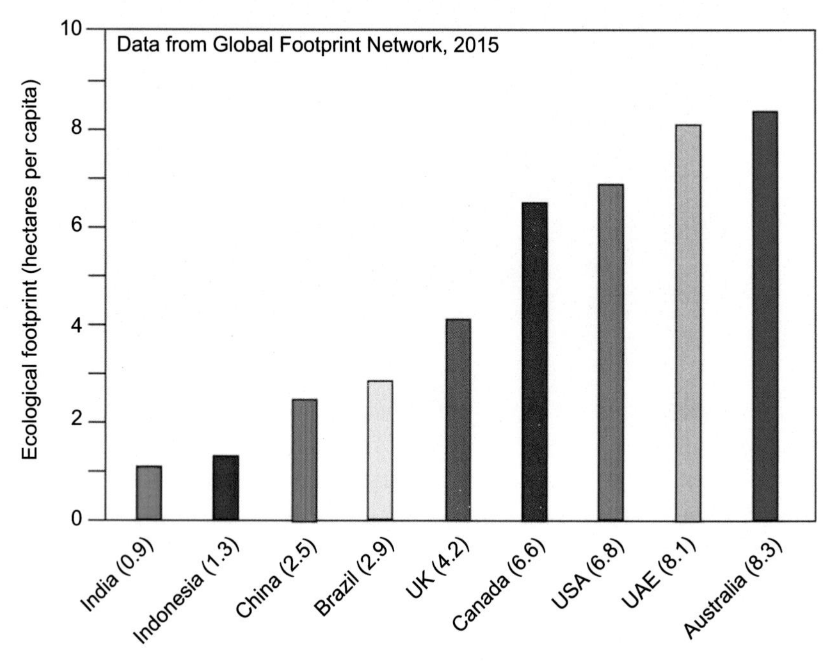

FIGURE 1.10.2

The current ecological footprint of various countries expressed as the number of hectares of land required to sustain the current standard of living in that country. *Data from the Global Footprint Network (2015) Elsevier.*

HOW IS IT USED?

When calculating an EF, we consider both the land and water required to not only produce our goods and services, but also to process the wastes we've generated and absorb the CO_2 we've emitted. So, if, for example, an individual composts and recycles most of his or her waste, this component of their EF would be lower than for someone whose garbage ends up in a landfill.

Similarly, because it takes more land to produce meat than a similar caloric quantity of vegetables, eating less meat can reduce an individual's EF.

EFs can be used in several ways. First, in the big picture, they can serve as a measure of sustainability for a community or a nation. The EF can be used to calculate **ecological overshoot**, which occurs when a population's demand on an ecosystem exceeds the capacity of that ecosystem to regenerate those resources. By definition, overshoot leads to a degradation of the ecosystem and depletion of the planet's life supporting biological capital.

This information can be used to guide the conservation and management of critical resources. For instance, if a nation finds that its EF far exceeds available resources, it may need to take steps to reduce resource demand. We already have many examples of water shortages, reduced agricultural production, deforestation, and fisheries collapse around the globe. Using an EF can help countries begin to address some of these issues and identify ways to reduce demands on resources before more strain is placed on these ecosystems with potentially catastrophic impacts on local populations.

Another use of EF calculations is at the individual level. Similar to carbon footprints, calculators available online allow individuals and households to plug in data about such activities as energy use, food consumption, driving habits, and waste generation to get an estimate of their EF. Quantifying how much land area it takes to support one's lifestyle helps us identify our biggest areas of resource consumption and learn what we can do to lower our impact on natural resources and the environment. The environmental footprint calculator provided by the World Wildlife Fund (found on your Resource Page) includes impacts from the consumption of food, transportation, use and disposal of consumer products, and characteristics of your home.

HOW IS AN ECOLOGICAL FOOTPRINT CALCULATED?

EFs can be calculated for individuals, whole populations (nations), or activities (such as manufacturing a product). The total global hectares needed to provide the necessary resources, treat waste produced, and absorb GHG emissions equal an individual's EF. The more one consumes and wastes, the higher will be their EF.

EF calculations start by estimating all the natural resources consumed and the total GHG emissions generated by that person, population, or activity annually. These data for resources used are then converted into the equivalent number of hectares needed to meet those needs. This conversion is specific to the yield of each particular land type (e.g., forest, agriculture) needed to provide each need. Because this yield can also vary from location to location, the estimate of land needed is based on a "**global hectare**" unit, which represents the average productivity of all productive global lands and seas. Biologically productive lands include cropland, forests, and fishing grounds, but exclude deserts, glaciers, and the open ocean. In this way, calculations of EFs are comparable from person to person and region to region.

Because the estimates of global hectare yield are just that, estimates, a set of international EF standards was adopted in 2006 to ensure that footprint studies are credible and consistent. These standards have helped clarify sticking points such as how sea area should be counted, how to account for fossil fuels, which data sources should be used, and when average global numbers or local numbers should be used when making estimates for a specific area.

IMPORTANT THINGS TO REMEMBER ABOUT ECOLOGICAL FOOTPRINTS

Calculating your EF is not a "back-of-the-envelope" calculation, but there are many online tools to help estimate it. When using these tools, it is important to make sure that they are aligned with the international EF standards. Sometimes you may see EFs converted into a "planetary equivalent," or how many planets it would take to support the human population if everyone had the same EF as you. While this is a broad extrapolation, it is a widely used measure, as it can be useful to take a small amount of information and provide perspective across a wider population. In this way, EFs can be a useful tool to educate people about how their lifestyle compares to the planet's ability to provide and renew the resources used.

Practice

The Global Footprint Network, tracking EFs for more than 200 nations, tabulates more than 600 data points annually for each country.

1. Following the link listed on your Resource Page, click on "Science Overview" and then "Data and Results," and follow the link to "Explore Interactive Maps." Click on any country to view the most recent summary of its EF. Note that the data are graphed from 1961 through 2012 and that both EF and "biocapacity" (the ability of the country's resources to meet its population's needs) are included in the figures. The difference between these two figures is shown as a credit or deficit. Examine the Interactive EF Map and explain the following:

    a. Despite its declining Biocapacity Per Capita, Australia still shows a credit. Why?
    b. China has a much smaller EF per capita than Australia, but shows a deficit. Why?
    c. How do the data for the US compare to those for China and Australia? Why is the US deficit the highest of the three nations?

2. Go to the Global Footprint Network's 2010 Atlas linked on your Resource Page, and examine Figures 7 and 8, which are bar graphs displaying the categories of impact for each nation. Figure 7 shows the per capita EF for each country broken down by category, including grazing land, cropland, built-up land, forest land, carbon footprint, and fishing grounds. Figure 8 shows the per capita "biocapacity" for each country based on the five types of land use identified in Figure 7. Examine these figures, and answer the following:

   a. Why are the United Arab Emirates and Qatar at the top of the list in Figure 7?
   b. If you live in a large US city, you may be surprised by the contribution of "Built-up Land" to the EF of the US. Discuss why this contribution appears the way it does.
   c. Explain the very different appearance of the bar shown for Uruguay.
   d. In Figure 8, Gabon shows the greatest biocapacity per person, yet it ranks as only the 55th wealthiest nation in the world. How could this be?
   e. How could it be that the US has a lower biocapacity per capita than nations like Latvia and Estonia?

Chapter 1.11

Cost–Benefit Analyses

FIGURE 1.11.1
Cost–benefit analyses estimate dollar values for ecological goods and services in an attempt to inform decision-making in the management or regulation of environmental resources.

IN A NUTSHELL

An environmental Cost-Benefit Analysis (CBA) is an important tool used to support decisions about the management or regulation of environmental resources (Fig. 1.11.1). CBA provides a framework to identify, quantify, and compare the costs and benefits (measured in dollars) resulting from a proposed action, such as a new environmental regulation. Benefits might include improved water quality, species preservation, or carbon sequestration. Costs may include the price of implementing the regulation, the loss of earnings for industries affected by the new regulation, higher prices for consumer goods, or higher taxes levied to pay for the regulation.

A quick scan of the literature reveals many examples of CBAs in use, from the proposed construction of a wastewater treatment plant in Serbia, cooking fuel alternatives in rural India, and a comparison of urban bus systems to improved land management practices in three watersheds in Ethiopia. In theory, anything environmental that involves costs and benefits can be analyzed in this manner. The challenge, of course, is how to find the necessary data and to place a monetary value on items like ecological services.

HOW IT WORKS

The first step in performing a CBA is to identify all the possible benefits and costs associated with the activity being analyzed. This may seem straightforward, but there are many complications. For example, whose costs and benefits should be counted? If a group is only marginally impacted by an activity, should this be included? What if it is unclear that they will be affected at all?

We also have to consider the timeframe over which costs and benefits are counted. Do you include only immediate impacts or costs that may be incurred over decades? Costs and benefits are rarely known with certainty so that risk (probabilistic outcomes) and uncertainty (when no probabilities are known) also have to be taken into account. Some of the data needed to assign values to various costs or benefits already exist. For instance, we know how much it costs to produce biofuels or add emission controls to automobiles. Risk analysts also apply a variety of formulae and use techniques like optimum fitting curves and models to estimate future costs and benefits. When all the identified costs and benefits have been converted to monetary values, a BCR, or benefit–cost ratio, can be calculated to indicate whether or not the project or regulation can be justified on a monetary basis.

Challenges

While a seemingly straightforward way to make decisions about projects and regulations, CBA has some serious limitations. For instance, if a project has a number of different costs and/or benefits, it may be challenging to find data that can be used to generate an accurate estimate. For instance, in your case study on the Three Gorges Dam in China, we'll ask you to complete a crude CBA. One of your challenges will be to find estimates of some of the many costs and benefits associated with the dam.

Another limitation relates to the intangibles. Many environmental projects might lead to the loss of some amount of forest, which might, in turn, increase the risk of degrading water quality in a stream. How do you put a price tag on the benefits provided by a forest? While it may be fairly easy to estimate the costs of adding scrubbers to a finite number of power plant smokestacks, it can be far more difficult to determine the value of some of the less tangible parts of the equation.

Let's consider a simple hypothetical example. Assume that the governor of a state is considering requiring a 300-MW coal-burning power plant to install an advanced technology filter to remove a larger proportion of air pollutants. Calculating the costs is fairly straightforward: it requires about $120 million to install the advanced technology on the plant and another $20 million annually to operate it. Over a 5-year period, the total cost would be about $220 million. But what about the benefits over that same period? It would be possible to put a value on reduced visits to doctors or hospitalization, but how do you quantify the value of lives ultimately saved by these changes? What price would people place on the aesthetic value of cleaner air? What is the value of improved water quality in nearby lakes and rivers? These are much more challenging numbers to come up with.

An entire new discipline has sprung up from the desire to understand how human societies and the environments that support them are linked economically. Ecological Economics is an interdisciplinary academic field that focuses on the valuation of natural resources. Issues of intergenerational equity, irreversibility of environmental change, uncertainty of long-term outcomes, and the maintenance of natural capital guide ecological economic analysis and valuation (Faber, 2008).

An Example From the Literature

Gao et al. (2016) performed a large-scale CBA on China's 2013 plan to reduce air pollution. Since coal is the primary energy source in the industrial sectors of China, the researchers focused on this fossil fuel. They ran CBAs for several scenarios, including business as usual, use of end-of-smokestack technologies, and reliance on energy saving strategies (e.g., closing smaller coal burning plants). Among the benefits, they focused on improved agricultural productivity and reduced incidence of human illness and death that would result from the reduction in the release of air pollutants.

For the entire nation, they estimated the cost to be about 118 billion yuan and the benefits, about 748 billion yuan, yielding a BCR of 6.32. The scenario that yielded the greatest potential benefit was the installation of end-of-smokestack technologies on existing plants. As you can see in Fig. 1.11.2, distribution of the costs and benefits across the nation was uneven, with the greatest benefits realized in the more densely populated eastern regions of China.

As you think about this study, several questions should be obvious. First, the benefits focused on increased agricultural productivity and better human health. Benefits to aquatic and terrestrial ecosystems, among others, were not factored into the calculations. Also, with a data set as large as this one, you can imagine that the uncertainties inherent in the analysis must have been substantial.

Important Things to Remember About Cost–Benefit Analyses

While CBAs offer an easy way for policy makers, government regulators, and others to evaluate various regulations and projects, they do suffer from substantial limitations, primarily the difficulty in assigning monetary values to ecological goods and services. While CBA has been a common tool in economic and social policy arenas, use of CBA for environmental applications

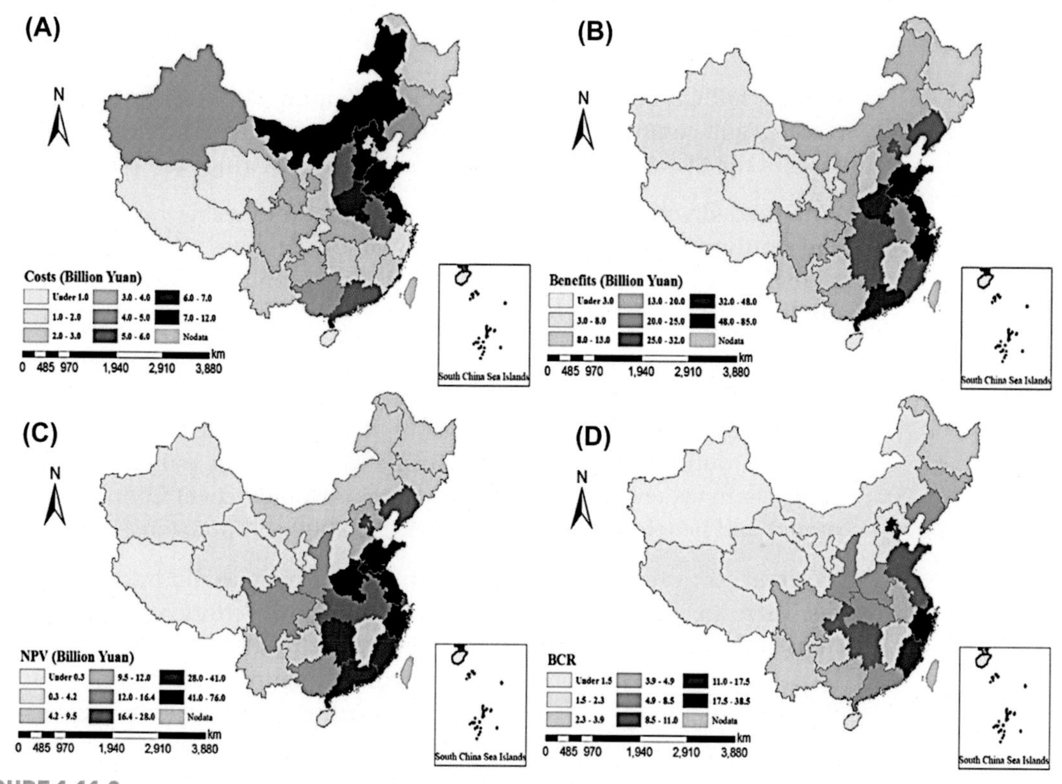

FIGURE 1.11.2

Cost–benefit analysis for different provinces and municipalities. (A) Costs under Integrated Policies (INP) scenario in different provinces. (B) Benefits under INP scenario for different provinces. (C) Net Present Value (NPV) under S3 scenario in different provinces. (D) BCR under S3 scenario in different provinces (Gao et al., 2016).

has been controversial. Dollar values for the environmental services must be inferred rather than directly observed. But how do you assign a dollar value to clean air or water? How do you set a value for biodiversity or aesthetic beauty?

In addition the ambiguity inherent in trying to assign a value to environmental services that are not actually bought or sold also raises ethical issues. Should we even try to assign dollar values to natural places, human lives saved, or species conserved? Perhaps such things are too 'priceless' to put a price on? Some argue that if environmental decisions are fundamentally framed as questions of economic welfare, the decision may more often fall with the more tangible outcomes (usually the costs). In this way, the environment would lose out to whatever policy promises more economic growth and more jobs. But without the use of CBA, it is not clear how the interests, claims, and opinions of parties affected by a proposed activity can be examined and compared. CBA is far from perfect, but it provides an important lens necessary for good decision-making.

Practice

While a full CBA can be complicated and daunting, it is possible to do some research and complete some "back-of-the-envelope" calculations to get an estimate of the BCR for many things. Consider the following example.

With bottled water all the rage, it isn't uncommon to see new water extraction and bottling facilities popping up in many rural areas. The draw of "fresh spring" water or water from a tropical paradise (brand shall remain unnamed) can bring a pretty penny on the consumer market.

Consider that you live in a rural town where an international company is hoping to site a new water extraction and bottling facility. This plant would add more than 100 mid-income permanent jobs, with another 1,000 temporary positions available during construction. Local officials expect an estimated $350,000 each year in tax revenue. Local businesses also would benefit from the influx of money and people, with an estimated $1,500,000 additional income annually.

However, not everyone is convinced that the project would bring only benefits. There are also potential costs associated with the effort. Most of the 1,000 households in your community rely on free well water that may be impacted by a lowering of the aquifer level or contamination from the new facility. Because yours is a region of moderate agricultural activity, there is also a seasonal draw on the local aquifer to support crop production. If water supplies were reduced, new wells would have to be drilled, or (worst case) a municipal water system would have to be constructed to pull from nearby surface waters.

Many residents also have strong feelings about preserving their local environment and the town's character. The idea of a big facility with an increase in truck traffic and waste generation makes many uneasy. They worry that this might reduce tourism and other recreational activities that build on the area's changes in rural character. There is also concern that local wildlife populations will be impacted by this increased traffic and changes to groundwater levels. Still others feel that bottled water is inherently anti-environmental with more widespread economic and environmental impacts.

How should the community decide whether or not to allow construction?

Perform a CBA to help this community make an informed decision.

- Identify all possible costs and benefits associated with the project. Identify both environmental and social costs and benefits in your calculation.

- Specify the timeframe to consider.

- Use the information provided above to estimate some of these costs and benefits. Also check your Resource Page for some similar studies. You can assume that economic, environmental, and social conditions in these studies were similar, and the same cost/benefit estimates can be applied here.

- For other items on your list, do some research to come up with a "best guess" of the cost or benefit from the literature or similar cost–benefit analyses.

- Once you have estimates for all of the costs and benefits you list, calculate the BCR for this project.

- What decision would you recommend the community make about siting the water extraction and bottling facility in their town?

REFERENCES

Faber, M., 2008. How to be an ecological economist. Ecol. Econ. 66 (1), 1–7.

Gao, J., Yuan, Z., Liu, X., Xia, X., Huang, X., Dong, Z., 2016. Improving air pollution control policy in China—a perspective based on cost-benefit analysis. Sci. Total Environ. 543, 307–314.

Chapter 1.12

Environmental Risk Assessments

IN A NUTSHELL

Engaging in many of your daily activities exposes you to some level of risk. Ranging from comparatively high-risk activities like smoking cigarettes or driving a car to very low risks like being struck by a falling meteorite, risks are part of daily life. The trick is to understand the risks you are exposed to because of your personal choices and to be able to make decisions fully aware of the risk involved.

There are two general types of risk: voluntary and involuntary. When you light up a cigarette or get behind the wheel of a car, you are at least vaguely aware that there is some risk involved. This is voluntary risk; you're choosing to do something that carries with it some level of risk. Involuntary risk is when you are unknowingly exposed. If there are chemical pollutants in your drinking water that you're unaware of, this is involuntary risk.

An environmental risk assessment (ERA) is a process for evaluating how likely it is that the environment may be impacted as a result of exposure to one or more environmental stressors, such as chemicals, disease, invasive species, and climate change. These ERAs then can be used to inform the public about the impact that these changes in the environment may have on human populations.

WHAT IS RISK AND HOW CAN IT BE MEASURED?

Risk analyses are routinely completed in a number of different spheres, including business and urban planning. We'll focus on ERA, which can be performed for any environmental stressor. For instance, the risk to human health from noise pollution around airports can be measured, and the risks of damage to floodplains from river flooding can be estimated.

Two terms lie at the heart of risk assessment: hazard and risk. A hazard is any situation or agent that can harm humans or cause adverse effects on the environment, while risk refers to the possibility of injury, death, or loss resulting from exposure to a hazard and to the likelihood or probability of such an occurrence. In a risk assessment, you identify the hazard(s), evaluate the risk associated with any hazards, and identify ways to minimize or eliminate the risk. Risk can be expressed qualitatively, such as high, medium, or low, or quantitatively, using a variety of techniques.

HOW IS AN ENVIRONMENTAL RISK ASSESSMENT PERFORMED?

Although the goal of any ERA is generally the same (to identify the hazard, estimate the risk posed, and consider steps to reduce risk if necessary), there are a number of different frameworks used by different groups to conduct an ERA. For instance, if the hazard you're evaluating includes economic implications (e.g., reducing noise pollution around airports by limiting use of certain types of aircraft), you'd need to include that in your risk assessment, while if you're estimating the risk posed by a new pesticide to salmon in a river, you'd be basing your assessment on a different set of measures. Regardless of the metrics used, most ERAs involve several common phases: planning and scoping, problem formulation, and risk assessment analysis and characterization (Fig. 1.12.1).

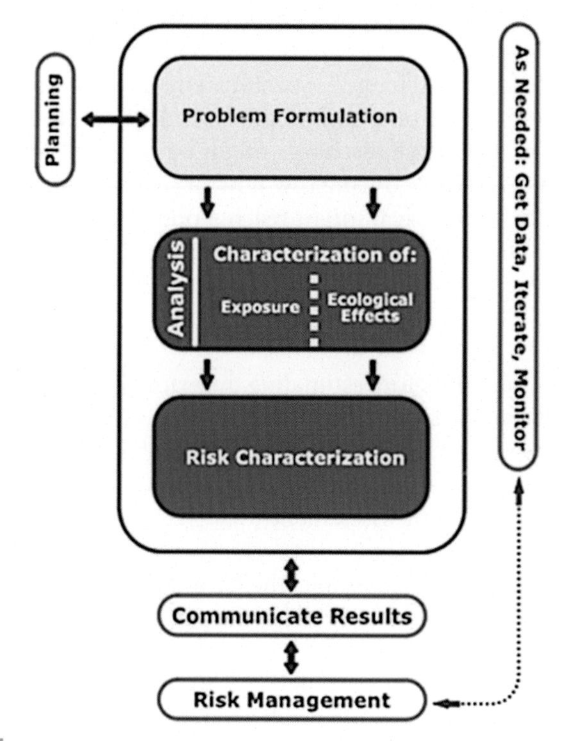

FIGURE 1.12.1
Environmental risk assessment requires several steps to characterize, model, assess, communicate, and manage risk. *Source: US EPA.*

1. **Planning and scoping**: Risk assessors will typically start by addressing a set of basic questions. Who or what is at risk? What is the environmental stressor of concern? What is the source of this stress agent? How is exposure to this stress agent propagated across environments and populations? What are the impacts of exposure to this stress agent? How long does it take for impacts to appear?

2. **Problem formulation**: The objective of the problem formulation phase is to gather all of the details needed to outline what the ERA should address. This helps to define the ecological entity you want to assess risk for and come up with an exact plan to assess this risk. This includes specifying whether the goal is to assess risk for a particular species, a functional group of species, a specific habitat, or geographic location of concern. It also needs to lay out all of the specific questions or concerns that need to be included in the assessment.

   For example, suppose a chemical plant is considering the level of waste treatment necessary to minimize the risks posed by the production of a new compound, Chemical X, that will be made at the facility. Regulators design an ERA focused on the river into which the plant effluent is discharged. Specific items to be considered in this problem formulation phase might include the following: what area downriver below the discharge might be affected, and how does the area of impact differ during low flow or high flow conditions? How much of the chemical might reach a public water supply intake located 1 km below the discharge? Which species of aquatic life might be affected? Are there other substances present in the river that might interact with Chemical X?

3. **Model development**: Before estimating the extent and likelihood of risk, it's helpful to develop a model that diagrams the relationships and pathways between the source of the hazard and the receptors at risk of harm. Such a model helps identify the key pieces of data or information necessary to complete the assessment.

   In our chemical facility example, the hazard is the chemical, the source is the discharge pipe entering the river, the pathway is movement of the chemical from the waste discharge downstream in the river, and the receptors would be fish, other aquatic organisms, or the municipal water supply of concern. This means we will need information for things such as the flow of water through the system, the solubility of Chemical X, its bioaccumulation in the species of interest, etc.

4. **Risk assessment**: Here, the actual risk to the receptors is estimated for the key components identified in the model development. Calculations are used to determine the levels of exposure that will lead to harmful effects, the expected duration or lag in the timing of this risk, which plants and animals are most at risk, and what degree of exposure is likely to have harmful effects.

   In our example, mathematical models could be used to calculate how much of Chemical X would be released into the ecosystem and accumulated in receptors over a given period of time. If data are available that show whether or not the chemical will be toxic to the species in the river, they can be used to estimate the risk of adverse effects. If not, site-specific laboratory tests may be needed to establish the risk.

   Field studies, theoretical models, or exploration of existing research could also be used to estimate the likely concentration of Chemical X entering the drinking water intake under different flow conditions and how these levels compare to federal drinking water standards. This could then be translated into the risk of harmful impacts on populations that drink the water (Fig. 1.12.2).

   An important part of the assessment process, in addition to estimating the possible risks, is to estimate the likelihood of occurrence. If, for instance, Chemical X is moderately toxic to aquatic life, but is extremely volatile, the probability of effects on aquatic life in the river would be low because the chemical would leave the water column soon after discharge. If Chemical X is persistent, the likelihood of adverse effects would be much higher.

   Typically, there will be more than one receptor to evaluate, so multiple lines of evidence will be generated. In our example, one line would be the toxicity of Chemical X to aquatic life, and a second would be human health impacts related to drinking water. If all lines of evidence lead to the same conclusion (e.g., high risk or no risk), it increases the level of confidence in the results of the assessment. Conflicting lines of evidence, of course, can make the assessment more challenging.

   Once various risks are estimated, an additional factor should be considered: level of certainty in the assessments. For example, perhaps the data on the toxicity of Chemical X are very limited, or the presence of other chemicals in the water raises the challenging prospect of synergistic effects. These would likely increase the level of uncertainty and lower the degree of confidence in the ERA.

5. **Risk management**: If an ERA identifies a significant risk, the next step is to develop a management approach to reduce or eliminate the risk. In our example, if the ERA determines that Chemical X poses a risk to aquatic life in the river, more effective removal technologies at the plant might be required, or, failing that, the plant might decide not to produce Chemical X. If the public is exposed to a risk, communication about the risk and proposed steps to manage it is also an important component of the process.

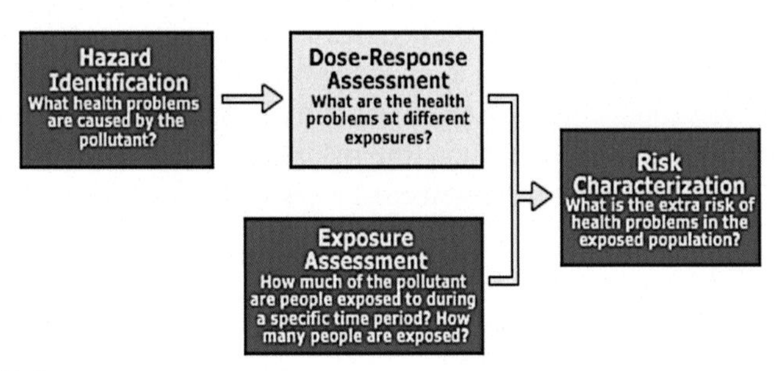

FIGURE 1.12.2
The four steps in US Environmental Protection Agency's risk assessment process used if human health is at risk. *Source: US EPA.*

An Example

Chapman et al. (2010) performed an ERA to determine the risk posed by metal contaminants in soils around two lighthouses, one in Nova Scotia and the other in New Brunswick, Canada. Because of peeling paint, levels of lead and other trace metals in soils around the two lighthouses were reported to be high.

The scientists developed five lines of evidence to estimate the risk associated with exposure to the lead contaminated soils:

a. Comparison of metal levels in those soils to Canadian soil quality guidelines;

b. Comparison of metal levels in soil leachate (water passed through soil samples) to water quality guidelines;

c. Growth of wheat shoots in soil samples (Fig. 1.12.3);

d. Growth of lettuce shoots in soil samples; and

e. Survival of the invertebrate ***Daphnia magna*** in leachate collected from soil samples. Note that in all cases, controls were used in the tests.

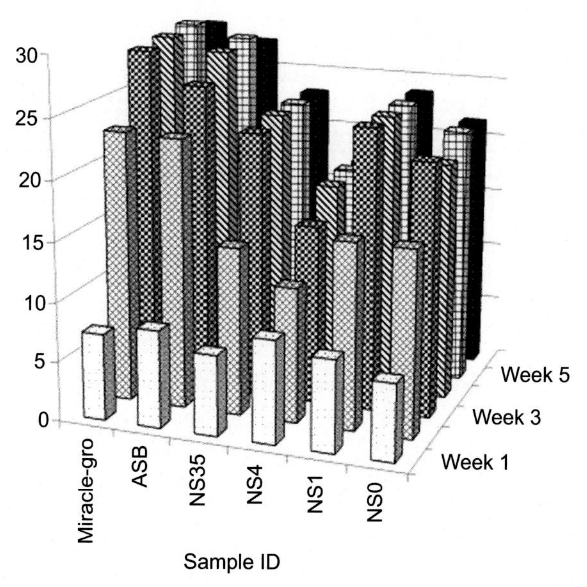

FIGURE 1.12.3
Wheat shoot growth weeks 1–6 presented as average lengths in centimeters (4 replicates) in samples from Nova Scotia. Miracle-Gro's "Seed Starting Soil" and Aurenz-Spezial-Blumenerde (ASB) Green World's "Garden Soil, Original Grower Mix" were used as control soils. Samples NS0–NS35 were collected in a transect out from the lighthouse, with NS0 collected directly in the structure's drip line and NS35 collected at a distance of 35 m from the lighthouse (Chapman et al., 2010).

While the lines of evidence from (a) and (b) above indicated higher risk (total metal levels in soils and leachates often exceeded soil quality guidelines), lines of evidence from the biological exposures (c–e) were much weaker, suggesting that the risk posed by the soils to the environment could be less than indicated by simple comparisons to available guidelines. The authors concluded that some of the metals in the soil were tightly bound and therefore biologically unavailable, resulting in reduced impacts on the test species. Note that if the scientists had simply compared metal levels in the soils to available guidelines, they would have come to a different conclusion regarding the level of risk.

Practice

A full ERA can require a long, detailed investigation. But that doesn't mean that we can't outline the basics of a risk assessment based on existing research, ecological principles, and some general frameworks. Consider the following example:

A local ski resort would like to expand their facilities to include summer activities. This will allow them to take advantage of year-round tourism.

Their plan is to build a 300,000 square foot indoor water park with additional parking and dining facilities. In addition to clearing 10 acres of forest and the creating 1 acre of impervious surface at the site, additional activities will include extraction of groundwater to maintain the water park. This resort is located in a rural, high elevational spruce fir forest, dominated by shallow soils and steep slopes. Current winter activities (downhill and cross-country skiing) at the resort bring in about 30,000 recreationists each year. It is anticipated that the expansion would bring in another 10,000 visitors each year, with all of their associated transportation, dining, and waste removal needs.

The watersheds the resort encompasses drain into nearby rivers that are listed as having high ecological sensitivity. Outline the basic steps of an ERA described above with particular attention to **how this expansion might impact the ecological integrity of the affected rivers**.

- Specify exactly what the ERA must address (what is the stress agent and system at risk).
- Work through the planning and scoping described previously in step 1 where you identify all of the questions that will need to be answered in order to perform this ERA.
- Work through the problem formulation to identify the details required to answer each of the questions described in the planning and scoping section. Some of these details will be determined by your expert opinion; other details will require a search through the literature. If information is not available, describe a study or source that could provide the necessary information. *To help you in your assessment, an article on high elevation development impacts on watershed hydrology is linked on your Resource Page.*
- Create a visual to diagram the basic connections across the ecosystem of interest and the pathways between the stress agent and the system of interest.
- Describe what would be necessary to complete this ERA.

REFERENCE

Chapman, E.E.V., Dave, G., Murimboh, J.D., 2010. Ecotoxicological risk assessment of undisturbed metal contaminated soil at two remote lighthouse sites. Ecotoxicol. Environ. Saf. 73, 961–969.

Chapter 1.13

Life Cycle Assessments

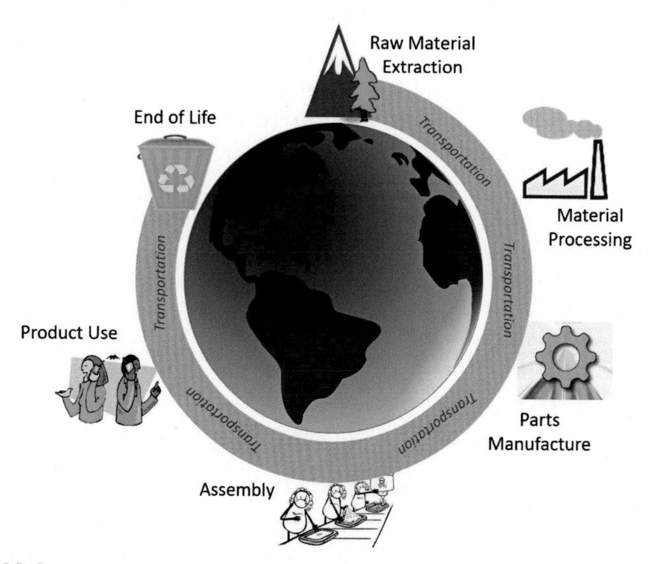

FIGURE 1.13.1
A life cycle assessment (LCA) quantifies the direct and indirect environmental impacts associated with the production, use, and disposal of a given product.

IN A NUTSHELL

A Life Cycle Assessment (LCA) is the "cradle-to-grave" analysis of the environmental costs associated with a given product. LCAs measure the environmental impacts of every step in the life cycle of a product, starting with the extraction of the raw materials, the energy needed to manufacture the product, transportation or distribution of the product to the consumer, the use of the product by the consumer, and ending with the ultimate disposal of the product at the end of its lifespan (Fig. 1.13.1).

LCA has become a common decision-support tool for both policymakers and industry in assessing the impacts of a product or process. Many organizations and businesses are moving in the direction of "life cycle accountability," the notion that an organization or business is responsible for the overall environmental impacts of their products. More and more consumers are also using LCAs to inform their purchases, further encouraging manufacturers to consider the direct and indirect impacts the production, use, and disposal of their products have on the environment.

There are several ways that LCAs can be used:

1. Consumers wishing to make the best choice for the environment when purchasing a product can use the analyses from an LCA to **compare the environmental "preferability"** of different products.

2. Manufacturers wishing to minimize the environmental impacts of their products can use LCAs to determine **where to make changes** in their products' life cycles to reduce significant environmental impacts.

3. Governments can use LCAs to pinpoint particular steps in a product's life cycle **where regulations will be the most effective** at reducing environmental degradation and/or conserving natural resources.

STEPS IN SETTING UP AND PERFORMING A LIFE CYCLE ASSESSMENT

1. Establish the purpose of the LCA (e.g., compare products, identify problem areas for improvement) and the expected outcomes from the study.

2. Life cycle inventory: identify all inputs (e.g., resources consumed, energy used) and outputs (e.g., water and air pollutants released, habitat disrupted).

3. Impact analysis: evaluate and measure environmental impacts (e.g., extent of habitat destruction, increased cancer rates from pollutants released).

4. Improvement analysis: based on your impact analysis, identify and evaluate various opportunities to reduce the environmental impacts identified in the study.

LIFE CYCLE INPUTS AND OUTPUTS

Materials extraction: Raw materials such as trees and ores are harvested/removed from the ground, transported, and processed in preparation for manufacture.
Associated impacts: release of GHGs and other air pollutants, habitat loss, soil erosion and surface runoff, energy consumption.
Manufacturing: Raw materials are converted into a product.
Associated impacts: release of air and water pollutants, energy consumption, waste generation.

Distribution: Products may be moved to distribution centers or warehouses and then to stores before purchase.
Associated impacts: release of air pollutants, including GHGs, energy consumption.
Product use: Products are operated and require cleaning or maintenance that releases pollutants.
Associated impacts: energy consumption, release of air and water pollutants.
Disposal: When products are no longer useful, where do they go?
Associated Impacts: solid waste generation or energy costs associated with recycling.

EXAMPLES OF LIFE CYCLE ASSESSMENTS

Food Transport

The US EPA (2010) conducted an LCA of three different ways to transport fresh tomatoes:

1. "Loose," minimally-packaged tomatoes that are transported in a corrugated container box with a general purpose polystyrene liner. Four tomatoes (2 lbs) are purchased at a time by the consumer in a polyethylene (PE) produce bag.

2. "PS tray," (Fig. 1.13.2), where the four tomatoes are packaged in an expanded polystyrene tray, wrapped in a PE film, and transported in bulk in corrugated container boxes.

3. "PET clamshell," where the four tomatoes are packaged in a PE terephthalate (PET) clamshell container and transported in bulk within a corrugated container.

Data on the impacts of growing, packaging, storing, and transporting 100 pounds of fresh tomatoes were collected. Impact measures for each stage of the life cycle included GWP from associated emissions, farming inputs, surface water acidity, and effects of air pollutants on human respiratory health.

As you can see in Figure 1.13.3 below, PET clamshell packaging was the worst of the three packaging types, with minimally packaged ("loose") tomatoes generally having the least environmental impact. The PET clamshell generally did the worst because it requires a greater amount of plastic, and making PET requires more energy per pound than other types of plastic.

Among the various stages of the life cycle, transport generally accounted for the greatest impact for each of the options, with growing the tomatoes having the second greatest impact in all areas measured except for smog formation.

FIGURE 1.13.2

Tomatoes packaged in PS tray. *Source: US EPA, photo by Martha Stevenson.*

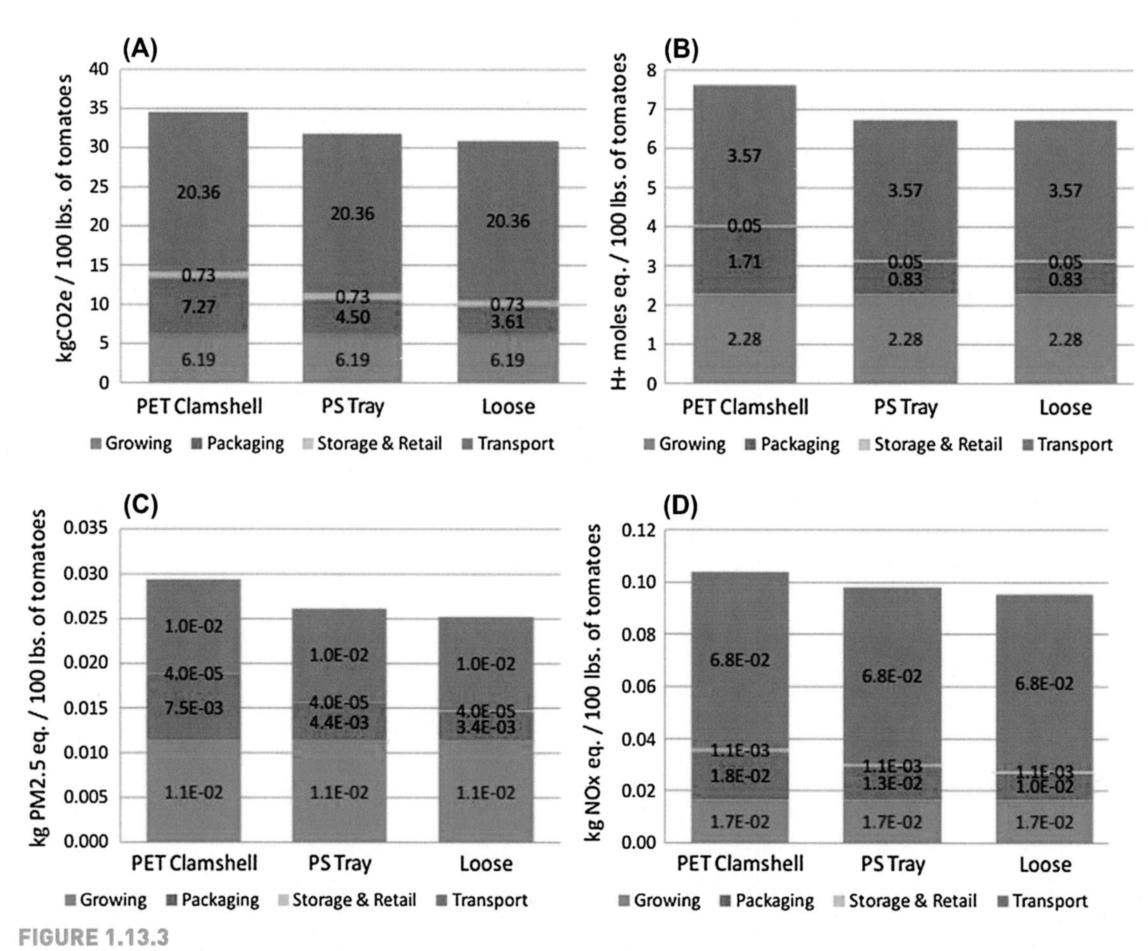

FIGURE 1.13.3

Environmental impacts by life cycle stage of three packaging options for producing and delivering 100 pounds of fresh tomatoes from San Joaquin Valley, California, to Chicago, Illinois. *Source: US EPA.*

IMPORTANT THINGS TO REMEMBER ABOUT LIFE CYCLE ASSESSMENT

Boundaries can be hard to set. You need to establish some limits on how far you want to extend your assessment. For instance, if your product requires an input of energy, do you include the environmental impact of manufacturing the tires on the trucks used to move the coal from the mine? Or only the fuel used to operate the truck?

Data can be hard to get, and sometimes what you can get is of questionable quality. If data used are from a distant part of the globe, they may not relate to local conditions. Sometimes it is necessary to collect your own data to get the best estimate of impacts.

Traditional LCAs have been criticized for not paying sufficient attention to natural capital (i.e., environmental resources and the services they provide). While most LCAs will include such things as the environmental cost of producing and applying fertilizers, they typically do not capture potential degradation to the land itself from high-intensity farming. This is particularly true for ecosystem products and services, such as the ability of a natural landscape to sequester carbon or filter water.

THINKING ECOLOGICALLY

Some LCA calculations are fairly straightforward. For example, if you know how far a product travels from Point A to Point B, calculating its GHG emissions is comparatively easy. However, many LCA estimates aren't so easy to come by. How do you measure the damage to a soil ecosystem caused by the harvesting of trees for wood to make a chair?

Because of these concerns, scientists at the Ohio State University developed Eco-LCA, a model that includes estimated values for ecosystem services that can be used to help provide a full accounting for ecological values.

Park et al. (2016) used Eco-LCA and several other tools to evaluate the impacts related to all the types of agriculture and food production in the US. They found that grain farming, dairy production, and animal production–related sectors had the greatest overall impact. They estimated that agricultural activities such as cattle ranching and farming, animal slaughter, rendering and processing, fertilizer manufacturing, grain farming, and fluid milk and butter manufacturing accounted for about 60% of the total climate change impacts.

Practice

On your Resource Page, you'll find a link to an article by Keyes et al. in the Journal of Cleaner Production that describes an LCA of organic versus conventionally grown apples in Nova Scotia, Canada. Figure 1.13.4 shows the general areas of assessment carried out in this study.

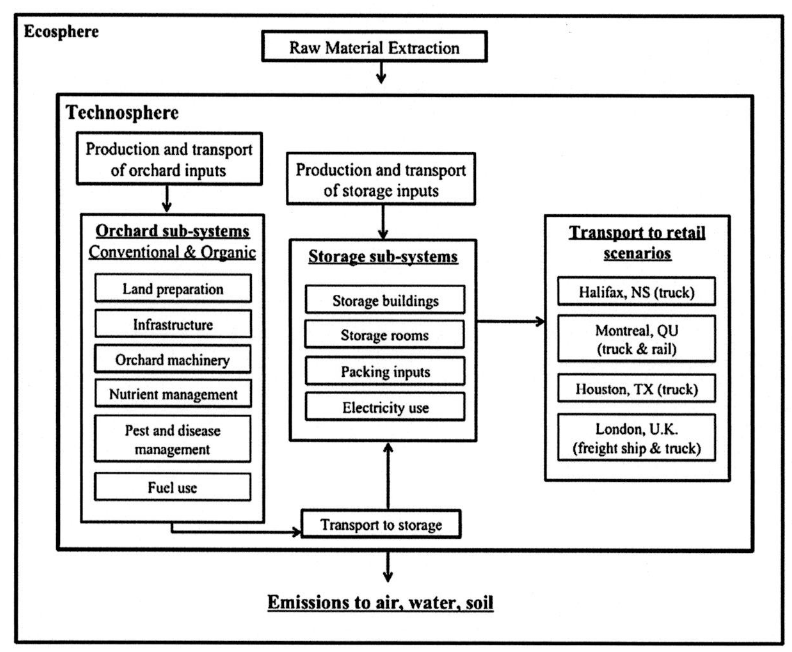

FIGURE 1.13.4
System boundaries of the apple supply chain in Nova Scotia. *Source: Keyes et al. 2015.*

Keyes et al. (2015) identified a number of general types of activities to be evaluated in their LCA, including land preparation, infrastructure, farm equipment, field use, soil amendments and fertilizers, and chemical and nonchemical crop inputs. Postproduction impacts included storage/refrigeration of the crop and transport to various markets in Canada and abroad via truck, rail, and ship.

Specific impacts assessed included GWP, terrestrial acidification potential, metal depletion potential, and freshwater and marine eutrophication potential. The authors were able to use available computer programs to examine data from 20 conventional and 8 organic orchards. One predictable finding was that the greatest areas of impact differed between the two types of orchards.

After reading through this article, answer the following about their study:

1. What were the most damaging impacts for conventional orchards? For organic orchards?
2. How were the key impacts for the two types of orchards similar? Different?
3. What is Nova Scotia's primary source of energy? How did this source affect the LCA?
4. Is there less impact from shipping Nova Scotia apples to London or to Houston? Why?
5. What is the single most important step that could be taken to reduce the environmental impact of conventional apple orchards? Organic orchards?
6. Can you tell which approach, conventional or organic, had less overall impact on the environment? Why or why not?
7. The authors acknowledge that they didn't include a full cradle-to-grave LCA in their study. What impact areas were excluded from their study? Do you suspect any of these areas would have altered the authors' major conclusions?

Challenge: Choose an item of clothing you're wearing, and do an LCA on this item. Identify the steps in the life cycle and the types of impacts associated with each step. Which types of data do you think would be the easiest to find? The most difficult? Can you find sufficient data to identify the step in the life cycle that causes the most damage to the environment? How might the impact of this product be reduced?

REFERENCES

Keyes, S., Tyedmers, P., Beazley, K., 2015. Evaluating the environmental impacts of conventional and organic apple production in Nova Scotia, Canada, through life cycle assessment. J. Clean. Prod. 104, 40–51.

Park, Y.S., Egilmex, G., Kucukvar, M., 2016. Emergy and end-point impact assessment of agricultural and food production in the United States: a supply chain-lined ecologically-based life cycle assessment. Ecol. Indic. 62, 117–137. http://ac.els-cdn.com/S1470160X15006779/1-s2.0-S1470160X15006779-main.pdf?_tid=2561cdda-1e20-11e6-b641-00000aab0f6c&acdnat=1463703589_de23623966c1663bb93914401ea97f00.

US EPA, 2010. Evaluating the Environmental Impacts of Packaging Fresh Tomatoes Using Life-cycle Thinking & Assessment: A Sustainable Materials Management Demonstration Project. Office of Resource Conservation and Recovery Report EPA530-R-11-005.

Chapter 1.14

Geospatial Analyses

READING THE LANDSCAPE: GEOSPATIAL TECHNOLOGIES

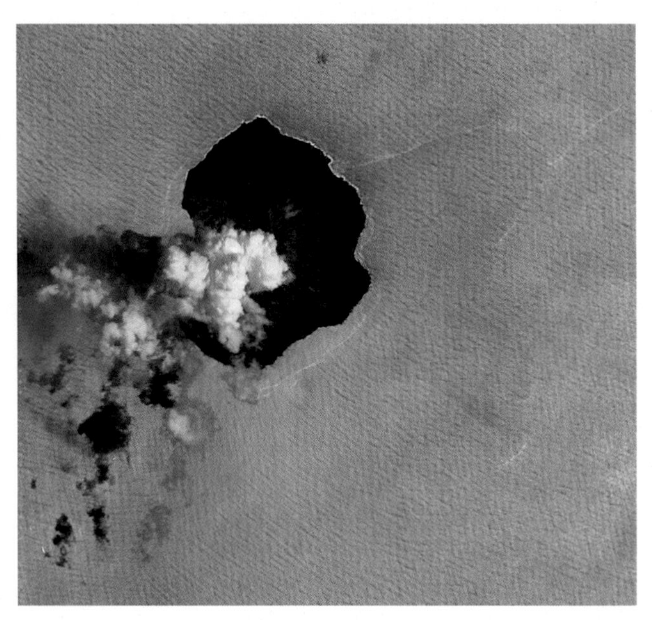

FIGURE 1.14.1
Because what happens in one ecosystem can affect other ecosystems far away, landscape context is useful to include in almost all environmental studies. Geospatial technologies help us characterize and quantify those landscapes. Here we see plumes of ash over Tinakula Island. *Source: NASA.*

In a Nutshell

Let's start off by saying that familiarity with geospatial data and tools is something that every environmental professional should have. Nothing in nature occurs in isolation, yet we are often limited in how far we can "see" or how large an area we can assess. Because of this, we are often asked to make statements about, or recommendations for, ecosystems with incomplete knowledge of the landscape around it, and how that landscape influences what we are interested in.

Luckily, over the past several decades, a growing wealth of available satellite imagery and processing tools have given us the ability to examine ecosystems on a global scale in ways that are not possible on the ground (Fig. 1.14.1). Unfortunately, to really be able to take advantage of these tools, you need coursework or training in remote sensing, geographic information systems (GIS), or GST. But, fortunately, the public's demand for a "bird's eye view" of the Earth has given us many tools that are accessible to the general user. Consider how many people use Google maps "satellite view" to get directions to their destination. This is a perfect example of how complex geospatial data sets (road vectors) and high-resolution satellite imagery (digital globe imagery) are fused into a common tool that we all but take for granted.

In the case studies that follow, we give you a chance to explore some of these online geospatial tools. But before you begin, there are some general concepts that will help you make the most of what these tools have to offer.

Key Concepts

Geospatial data can be categorized into raster or vector data. **Raster data** are images (think of a picture) where each "pixel" contains data that describe the surface below.

For **passive sensors** (like those contained on most satellites), these data are simply a measurement of the electromagnetic radiation (sunlight) reflected back from the surface and intercepted by the sensor. Different sensors are designed to measure this reflected light in different wavelengths. For example, many sensors include measurements in the red, green, and blue regions of the spectrum so that "true color" images can be rendered. But many sensors also measure wavelengths outside of what we can see with the human eye. Often these wavelengths contain key information about specific ecological processes or landscape structural characteristics. For example, near infrared wavelengths are highly reflected by vegetation, and the intensity of this reflectance can be used to quantify the amount of vegetation or the condition of that vegetation.

Because they rely on the sun's energy as the source of their measurements, these instruments are very sensitive to any conditions that might alter the amount of solar radiation reaching the Earth. This includes the composition of the atmosphere, weather conditions during image acquisition, time of day, or day of the year (different solar angles and intensities), and the variability introduced by topography (shadows, hill shade, or fore- and backscatter surfaces). These become particularly problematic when comparing images from different locations or images taken at different times. Much preprocessing goes into removing many of these "nonsurface" impacts on reflectance signals, but atmospheric and illumination anomalies can never be completely removed.

Active sensors are also used to create raster images, but they emit their own specific signal and then use the intensity of the return pulse, along with its return time, to quantify characteristics on the Earth's surface. For example, active Lidar sensors are used to assess the height of the ground surface (think of topographic maps), but they can also be used to assess the density and height of features on the surface (for example, mapping out building locations or assessing the density or biomass of forests). Active sensors with different signal types can also be used to penetrate the Earth's surface and explore geologic or hydrologic structures deep within the Earth or under the oceans.

Active sensors are not limited by illumination anomalies like passive sensors and can even collect data at night and penetrate clouds, but there are limitations to these data as well. While the rasters generated by active sensors are displayed as contiguous images, these images are typically generated from a series of point "pulses." The density of these pulses determines how much detail about the landscape is truly captured. Because the wavelengths used in active sensors are typically low energy, they can also be influenced by other radiation sources (Fig. 1.14.2).

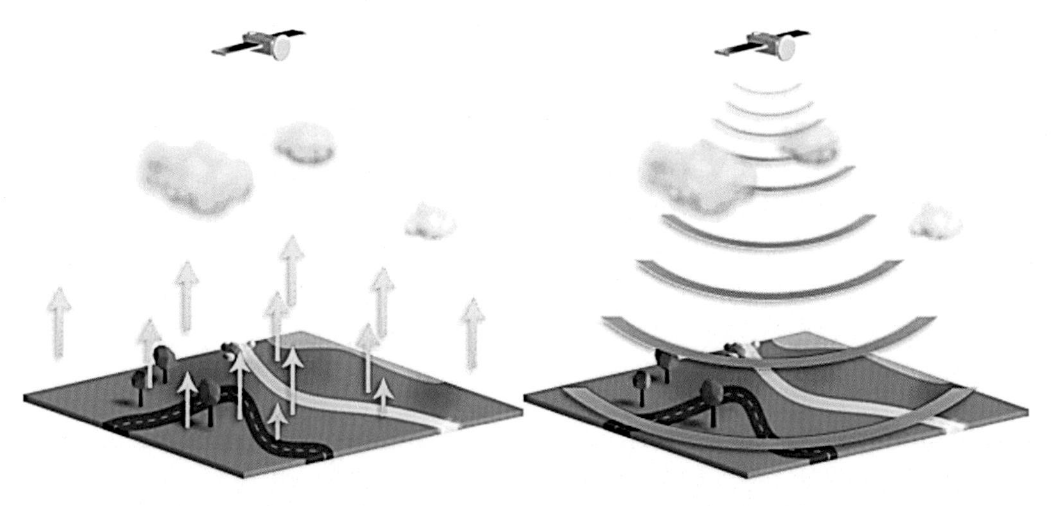

FIGURE 1.14.2
Passive and active sensor configurations. *Source: NASA.*

Vector data are the other common types of information used in GST. Vector layers usually consist of points, lines, or polygons that delineate structures of interest on the Earth's surface. These include features on many of the maps you may use, with their roads, rivers, cities, and state boundaries. These layers can be created on the ground using GPS units or can be digitized from satellite imagery. They also can be filled with information (attributes) that describe each of the features within the vector layer. In this way, you could select a given feature (for example a census block) and recover all of the information about that feature (for example, the population, mean home value, or density of buildings).

Vector data are only as good as the coordinates that were used to create them. For example, you may have noticed that sometimes when you ask to view both the satellite image and the roads in an interface such as Google Maps, there is not always a direct overlay. Similar to how some clocks are exactly calibrated to Greenwich Mean Time and some clocks are slightly off, while others may actually drift, vector data can also vary widely in their accuracy. In part, this is determined by the size of the object that the vector depicts and the scale at which you are visualizing it. While a vector of the US border may look accurate when viewing the entire US at a continental scale, zooming in to the border at the Rio Grande may not always line up (Fig. 1.14.3).

FIGURE 1.14.3
While the US–Mexico border is the Rio Grande River, the vector layers representing the border do not always align perfectly with the river location in the raster satellite imagery. *Source: Google Maps.*

Often we will use raster and vector data in combination, with each informing the other. There are many tools that have been developed to merge the data from these layers and to create new information based on the characteristics of their intersection. These tools are often useful to environmental professionals. For example, if a water quality expert has measured elevated levels of a given contaminant in certain surface waters, they could use a GIS framework to explore the land cover and land use surrounding that contaminated area and identify other potential surface waters at risk based on those characteristics.

Things to Think About When Using Geospatial Tools

The danger of geospatial data is that they have the effect of "eye candy" for many. People see a pretty picture and are instantly enthralled. Subconsciously, the information contained in the image is accepted as absolute, accurate, and informative. But sometimes a GIS framework is not much more than a pretty picture. Here are some of the things you need to consider when using geospatial data to inform the ultimate utility of the information and your interpretation of what it conveys.

Spatial resolution refers to the size of each pixel. Higher-resolution imagery will allow you to see more detail in an image, but typically it covers a smaller swath. Low-resolution imagery is useful for examining large landscapes, but it will appear blurry when zooming in. Aside from how detailed an image can look as you zoom in, spatial resolution is important to consider in terms of what it is you are trying to "see" or detect in the imagery. Some imagery is simply not able to detect smaller features or phenomena. A good rule of thumb is that to detect an object, your pixel size should be one-quarter the size of the feature you hope to see. So if I wanted to study vernal pools and be able to detect all pools greater than 10-m diameter, I would need to have access to imagery with at least 2.5-m resolution imagery. When conveying my results, I would have to include the fact that there are likely other vernal pools in this area, but that due to my spatial resolution, I cannot make any statements about their distribution or density.

Spatial accuracy refers to how far off each pixel (or vector) is from its true location on the ground. While GPS units are getting better and better, with many commercial units accurate to less than a meter, there is still some offset to geospatial data from its "true" location on the ground. This is particularly problematic in areas of extreme terrain where distances on the ground, and corresponding pixels in an image, can be skewed based on varying area on the ground captured by pixels close to the sensor (at high elevations) versus farther from the sensor (low elevations). Because of this type of variability (as well as other sources of error such as the curvature of the Earth in larger imagery), the geospatial error can vary across an image. Some locations may be more accurate than others. This is important to keep in mind when using imagery to identify specific locations to investigate on the ground or when comparing imagery over time. What may look like a change in condition for a given location could simply be a misregistration between the two images. This is particularly common along edges of surface features where being off by one pixel can alter your interpretation of the surface.

Temporal resolution refers to when (or how often) geospatial data layers are collected. Vector layers are typically updated slowly, but many satellite-based sensors are constantly orbiting the Earth with "return times" to image the same location on the order of daily to monthly. When examining your imagery, it is important to consider when it was acquired. Some imagery publicly available is outdated and thus may not represent current conditions. Seasonality in many parts of the world will also result in stark differences in surface characteristics. Sometimes this is what you want to explore (for example, studying the timing of spring phenology). But when you are hoping to capture differences from year to year (for example, studying the rate of deforestation), you would want to control for these seasonal differences by ensuring that your imagery is collected on a common date in each year interval.

READING THE LANDSCAPE

While the information you hope to glean from a geospatial analysis will vary depending on the study objective, when you are examining the broader landscape for environmental context, it is typically useful to characterize several general characteristics:

- What is the surrounding **land use**? Is the region primarily forested, developed, agricultural, desert, or barren? What are the patterns of this land use? Are there locations of higher or lower density of each land use type? Is the natural ecosystem altered or degraded in some way? Is the natural ecosystem contiguous or fragmented?
- What is the surrounding **topography**? Most physical processes are influenced by the flow of water and air. Are there any topographic features that might influence the movement of pollutants, the presence of invasive species, or other stress agents across this landscape? Are there locations more or less prone to severe weather for this ecosystem?
- What is the proximity of these land use or topographic hotspots to **sensitive environmental features**?
- Make note of any **unique features**. Does anything stand out across the landscape, perhaps a water source in an otherwise arid environment or a region of particularly high concentrations of possible pollutant sources or sensitive species?

Practice

Now that you know the basics of reading the landscape, it's time to practice. There are many mapping products freely available for public use. More and more, there are also many geospatial visualization tools that are easy to access via the Internet. We will explore several environmental geospatial visualizers throughout the case studies in this text. But let's explore a few here.

1. Start in Google Maps or some similar mapping platform. Find your home, and conduct a landscape assessment following the steps identified above in "reading the landscape."

   - First, define the extent of your home landscape.
   - How connected or fragmented is the ecosystem within your home landscape?
   - What is the mix of land uses in your home landscape?
   - What are the most sensitive ecological features in your home landscape?
   - What are the greatest risks to these features based on your landscape assessment?

2. The US Census Bureau has created an online data visualization tool that allows you to examine population size and trends across the US in a geospatial format. Visit their link on your Resource Page.

   These paired maps show the percentage change in population between 2002–03 (left map) and 2012–13 (right map). The two maps have identical data categories and use the same set of macro- and microboundaries. Clicking on an area will provide a pop-up box containing the area's title and its numeric and percentage change values in each period.

   - Compare the two maps with the swipe bar. Where are some regions of highest population growth?
   - Where has population actually decreased?
   - What is the population trend for your home landscape?

Chapter 1.15

Communicating Like a Professional

IN A NUTSHELL

There is no more important skill for an environmental professional to master than communication. The ability to speak and write effectively is critical to your success after graduation. From composing an effective job application cover letter to writing scientific articles and technical reports or speaking in front of a variety of audiences, it's hard to overstate the importance of being able to express yourself and communicate clearly. In this section, you'll practice several different types of communication. Throughout the case studies, you'll have many chances to perfect these skills.

WRITING

There are many different types of writing. We'll focus on science writing skills. Those of you who go onto graduate school or work in laboratory research may write technical reports, scientific papers, and even grant proposals. Graduates who work in the private sector, such as consulting firms, or for government agencies may be required to write project completion reports, risk assessments, or management directives. Each of these written products has unique objectives and specific formats that require different approaches. There are, however, several important keys to any writing you do:

- ***Know your story***: What are you trying to communicate? Outline this story before you start writing. You should be sure to tell your story with logical progression, using paragraphs to block the story.

- ***Be clear and precise***: It's important to say what you mean and avoid confusing wording. Avoid generalizations (for example, don't say "some" if you know of only one instance) or ambiguous antecedents (for example, don't say "this"; instead, specify what "this" refers to.)

- ***Be objective***: To be credible, a scientist must remain impartial to avoid bias and opinions in their writing. Present the facts and your interpretation of those facts without conjecture. Acknowledge the limitations in your study and avoid words like "very" or "extremely." A professional will provide quantified results rather than hyperbole.

- ***Be concise***: Scientific writing should be as straightforward as possible. Don't write three sentences when you can say what you need to in one! Use graphics or tables whenever possible to summarize results; no one wants to read a bunch of numbers in the text. If you have this information in a table, there is no need to state it again in the text. Present all the information that you need to and NO MORE!

- ***Write correctly***: Technical aspects such as grammar, spelling, word choice, and sentence structure must be mastered; in addition to spell check and grammar check, there are many online tools to help you with this. Watch your verb tense. Use the past tense to present results and present tense to discuss them. Whenever possible, use the active rather than passive voice (for example, instead of "six samples were taken" say "we took six samples"). While these may seem nitpicky, attention to detail goes a long way. In the end, it's all about practice, practice, practice!

- ***Revise***: The key to success in writing lies in smart revision, and often the best revisions come from someone else who can provide a fresh set of eyes. They may pick out misleading or confusing components that seemed very clear to you.

SPEAKING

Public speaking is probably one of the most important, yet terrifying, activities for young environmental professionals. Many of us got into this field because we love communing with nature, not necessarily people. But without professionals out there communicating their work to the public, managers, and policymakers, very little would ever change. Not everyone takes the time to read a technical report, but they may tune in to an interview or show up at a talk to hear more. When planning to present your work to just a few people, or even a large audience, the same tips will ensure that you communicate your story effectively:

- **Know your facts**: Before you communicate any information, be sure that you have the key details on the tip of your tongue. Nothing is worse than trying to "tell your story" but forgetting the actual details of the results. This doesn't require that you be an expert, but it does require that you put in the time to study the key information first.

- **Know your audience**: Whom you are talking to will inform the tone of your talk and the level of detail included. Sometimes it is important to focus on the conclusions (say for an audience of land managers), while other times, you will need to spend much more time going over the methods (say for a scientific audience who will want to vet your approach).

- **Stick to your "story"**: Remember that you have a specific story that you want to communicate. Present only the key information associated with that story so that you don't lose your audience in the details.
- **Make it a conversation**: While many people depend on slides or other visuals to communicate their story, don't let these aids get in the way of your connection to your audience. You should be making eye contact with audience members as if you were talking to each one individually.
- **Relax**: Take a breath. Be comfortable with natural pauses and silent spaces. Taking a moment to collect your thoughts before proceeding to the next point gives your audience a chance to catch up was well. This is not brain surgery, and no one expects you to be perfect.
- **Practice**: Nothing helps you to relax more than practicing the story you are going to tell. Practice out loud to yourself, or even do a dry run in front of a friendly audience. This is especially useful to help modify content. Just like having a fresh set of eyes for revisions of your writing, having a fresh set of ears listen to your talk can be valuable in making sure your story will be clear to the audience.

Practice

1. With scientific writing, it's always best to practice word conservation; if you don't need it, delete it. For the following phrases, delete extraneous words so that you convey the same information in fewer words:
   - there were several subjects who completed
   - it is suggested that a relationship may exist
   - based on the fact that
   - extremely high levels of several key pollutants were found
2. Another way to practice word conservation is to edit someone else's work. Dig through your favorite book (or one nearby) and find a particularly rambling paragraph. See if you can cut down on the number of words used without jeopardizing the flow or "story" conveyed. Change passive verbs to active and remove unnecessary words and prepositional phrases. How many total words were you able to remove without impacting the information conveyed?

Now, let's try some different types of writing:

3. Congratulations! You got the job with a consulting firm. After a year on the job, you're asked to write a report on the findings of one of your projects: a series of measurements of bacterial levels in the water of the Hudson River system done for a public interest group that paid for the monitoring program. Go to the following website: http://www.riverkeeper.org/water-quality/citizen-data/, and assume that you've collected these data. Write your report for a group of citizens very concerned about whether or not it's safe to swim in the Hudson. The report should be two to three pages long and contain a data table summarizing your findings for a nontechnical audience.

4. You've moved on to a new job at a consulting firm that specializes in fish communities in freshwater wetlands. Your first job is to do a wetlands assessment focused on the value of wetlands to important North American game fish (e.g., northern pike, walleye). First dig through the scientific literature to find three to four research papers on the role that freshwater wetlands play in supporting species of game fish. Write a literature review summarizing the conclusions stemming from this research. Have two of your fellow students peer review your paper before making final revisions.

Now let's practice some speaking and listening:

5. Create a 5-min presentation based on the report you created in #3 above. Assemble a group of five or six of your fellow students, and present your findings on bacteria counts in the Hudson River to them. Assume they're all members of the public interest group that's supported your work and will want to know whether or not it's safe to swim in the river.

6. You're working as the environmental expert for a utility company that wants to build a dam on a nearby river. Your job is to prepare a persuasive speech to present to a skeptical public audience, which includes some folks very concerned about changes in the river ecosystem that might be caused by the dam. Your classmates will make up the audience, and they will be armed with questions which you should anticipate and be prepared to answer. Your talk should be 10 min in length. Follow the link on your Resource Page for tips on how to prepare a persuasive speech.

CHAPTER 2

Global Water Resources

Science and the Global Environment. http://dx.doi.org/10.1016/B978-0-12-801712-8.00002-0

Chapter 2.1

Introduction

FIGURE 2.1.1
A view of the Pacific Ocean from space. *Source: NASA.*

Water scarcity continues to be a challenge in many parts of the globe, particularly in less developed countries. Experts estimate that as many as 2 billion of us lack access to a safe drinking water supply (Fig. 2.1.1).

Because we have a long history of disposing of our wastes in the nearest waterway and often building large cities in very dry landscapes, global concerns about water availability and quality have long been at the forefront of environmental issues.

A few interesting facts about water, a substance that makes life on Earth possible:

- It is the only natural substance that exists in three forms at ambient temperatures: solid, liquid, and gas.
- It dissolves more substances than any other liquid. Wherever it travels, water carries chemicals, minerals, and nutrients with it.

- Almost 75% of the Earth's surface is covered by water. Of this amount, only about 0.3% is readily available for human consumption.

- The average resident of the US uses between 300 and 375 L of water per day. Flushing the toilet accounts for the greatest percentage of water use.

- More freshwater is stored under the ground in aquifers than on the Earth's surface.

- The Earth is a closed system, meaning that the same water that existed on Earth millions of years ago is the same water we are using today.

- Every known living organism requires water to function. For more information on the vital role of water in sustaining life, visit http://www.nasa.gov/vision/earth/everydaylife/jamestown-water-fs.html.

Water's essential role in ecosystems and in societies stems from its unique chemical and physical properties. Water is known as the "universal solvent" because of its ability to dissolve minerals and nutrients from rocks and soils and transport them through streams and rivers into lakes and oceans. While its ability to move these substances is vitally important to the functioning of ecosystems, unfortunately, this property of flow also allows for the transport of any pollutants present to locations often far downstream.

Because of their shape, water molecules are attracted to one another, a process called hydrogen bonding (Fig. 2.1.2). When cooled to near the freezing point, these molecules form a crystal structure, ice, that is less dense than liquid water and will float atop water, allowing life to flourish in lakes in colder climates.

Its ability to exist in liquid, gas, and solid forms allows water to balance the Earth's energy budget and regulate our climate by transporting heat energy from the equator to the poles. Because it stores heat so effectively, water also influences local climatic conditions, including temperature and precipitation. A classic example is the snowbelt that lies just to the east of the Great Lakes (Fig. 2.1.3). As cold winds blow over the comparatively warmer waters of the open lakes in early winter, meters of lake-effect snow can pile up in cities like Buffalo, New York, while areas further east may receive little, if any, snow.

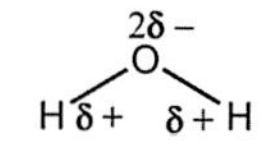

FIGURE 2.1.2
The dipole moment of a water molecule. *Created by: Marsve [Public domain], via Wikimedia Commons.*

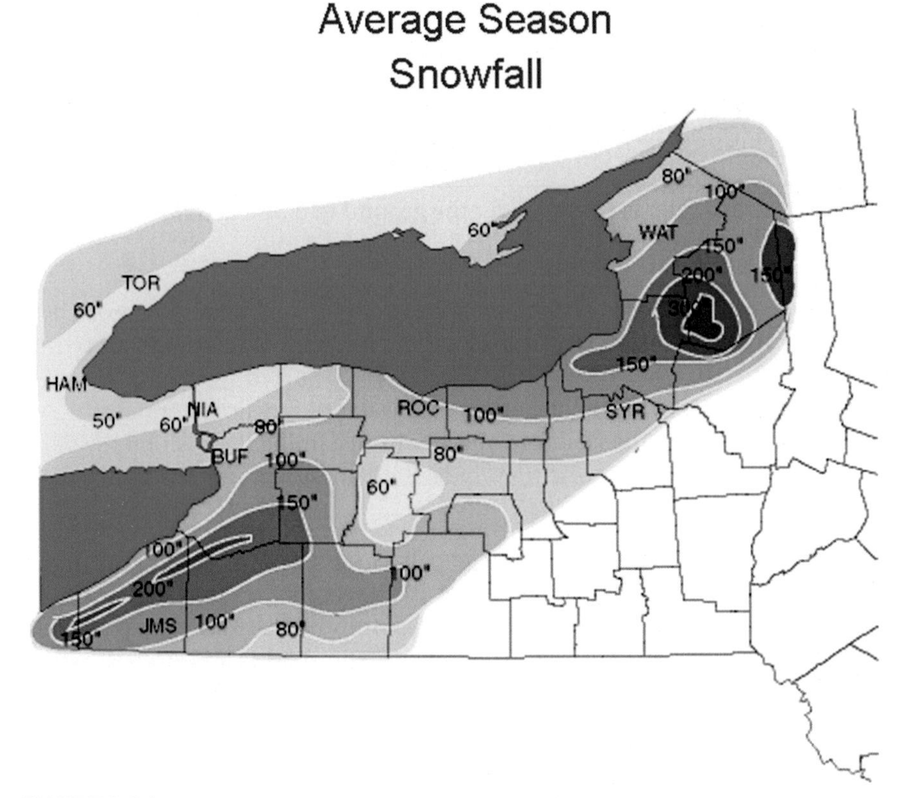

FIGURE 2.1.3
Great Lakes snowbelts. *Source: NOAA.*

Because water acts as both a solvent and delivery mechanism, our bodies use water to deliver nutrients and remove toxins. Every known organism needs water to survive. In addition to its vital role in the human body, water is important to our survival in a number of other ways. A safe and dependable water supply is critical for meeting our domestic needs and for growing crops, watering livestock, and supporting industry. We also rely on aquatic ecosystems to support fish, shellfish, and a variety of wildlife often critical to human diets across the globe.

SURFACE WATER

Humans affect water quality and the health of aquatic ecosystems in a variety of ways (Fig. 2.1.4). Some of our activities, such as the release of point and nonpoint source pollutants into rivers, impair water quality at the local or regional level, while others, like climate change, are global in their effects on water quality and quantity.

FIGURE 2.1.4
Poor water quality impacts human and environmental systems. *Photo by Alan Liefting, [CC 3.0] via Wikimedia Commons.*

Physical properties of surface waters play a major role in determining the effect of human activities on a particular system. In some environments, like the open ocean, dilution may reduce the immediate impacts of pollutants. While open areas of the ocean may remain comparatively clean, near-shore coastal areas are often degraded by activities along the shoreline, including wastewater discharges, oil spills, coastal erosion, and the presence of pollutants carried by rivers and streams.

In areas where water flow and circulation are restricted, pollutants may be present at levels that threaten both the ecosystem and human uses of the system. While rivers may be severely impacted by human activities, in some cases, they can recover quickly, assuming a clean source of water exists upstream. Conversely, lakes, reservoirs, and other water bodies with more limited water circulation tend to accumulate pollutants over time and may suffer impacts that last far beyond the initial introduction of these substances.

Threats to surface water quality can be physical, chemical, or biological. It is important to emphasize that aquatic systems are rarely exposed to just one stressor. As we'll see in our case study of the Mediterranean Sea, there may be dozens of different stressors interacting in ways that may be difficult to understand and therefore manage.

Many chemicals are potentially harmful to human health and aquatic life. These potentially toxic substances include both inorganic pollutants like mercury and synthetic organic compounds like polychlorinated biphenyls (PCBs).

But pollutants can also include nutrients, which can be too much of a good thing in aquatic ecosystems. Phosphorus and nitrogen, if present at levels above normal, can stimulate excessive growth of primary producers like algae and aquatic plants (Fig. 2.1.5).

An important consideration when addressing water quality issues is the nature of pollutant inputs into receiving waters. Point sources, such as discharge pipes from industries and wastewater treatment plants, may release waste

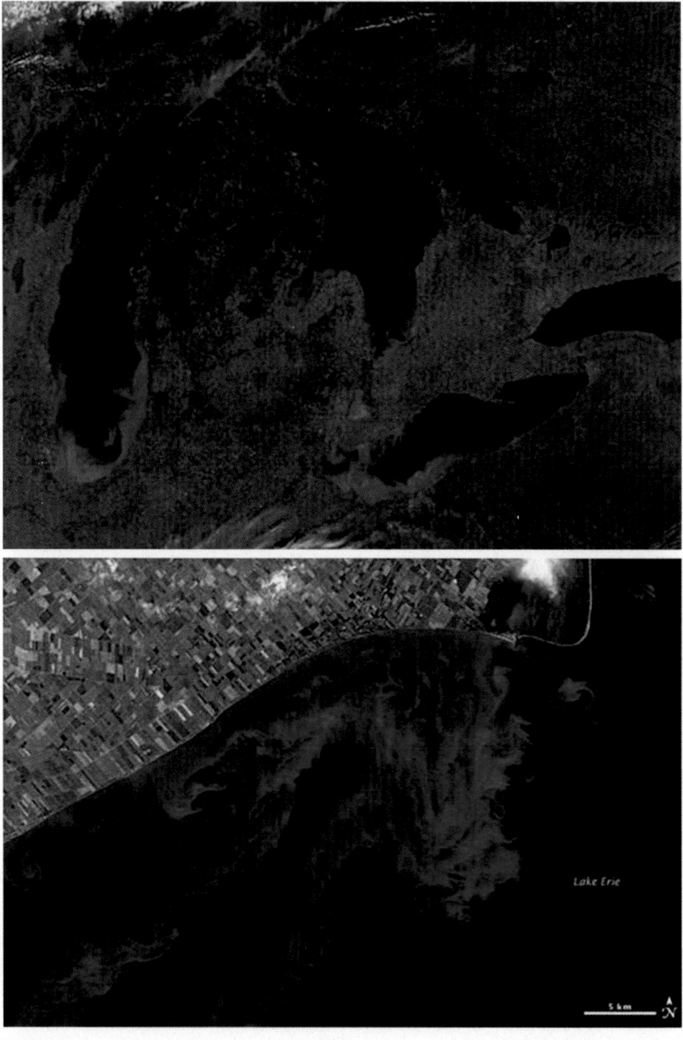

FIGURE 2.1.5
Toxic algal bloom in the Great Lakes region October 9, 2011. *Landsat images created for NASA's Earth Observatory by Jesse Allen and Robert Simmon using data provided by the USGS.*

continuously into a river. Nonpoint sources, such as storm water runoff, generate pollutants, which enter receiving waters only sporadically but often in large amounts over a short period of time. The high degree of variability of such nonpoint source pollutant episodes makes it difficult to predict the nature and extent of the impacts on receiving waters.

Physical stressors include soils carried from the exposed landscape by runoff into receiving waters where these particles can block sunlight, smother habitats, and cause other damage. Elevated temperatures in thermal discharges can push some aquatic species beyond their upper temperature tolerance limits, reduce dissolved oxygen (DO) content, and promote microbial growth.

Biological agents include bacteria and viruses, which can pose a risk to humans using surface waters for recreation or for water supply. Warming ocean waters and sea level rise associated with climate change may increase the spread of dangerous waterborne bacterial diseases like cholera.

Invasive species, such as those discussed in the Mediterranean Sea case study, can undergo explosive population growth and drive out native species. Notorious examples of alien species that have disrupted local ecosystems include the zebra mussel in freshwaters and the Asian shore crab in estuarine systems.

GROUNDWATER

While about 75% of the world's population relies on surface waters, such as rivers, lakes and reservoirs, for their water supply, the remainder tap groundwater supplies for their needs (Fig. 2.1.6). While these supplies are protected

FIGURE 2.1.6
A flowing artesian well near Valromana, Italy. *Michael Gäbler [CC 3.0], via Wikimedia Commons.*

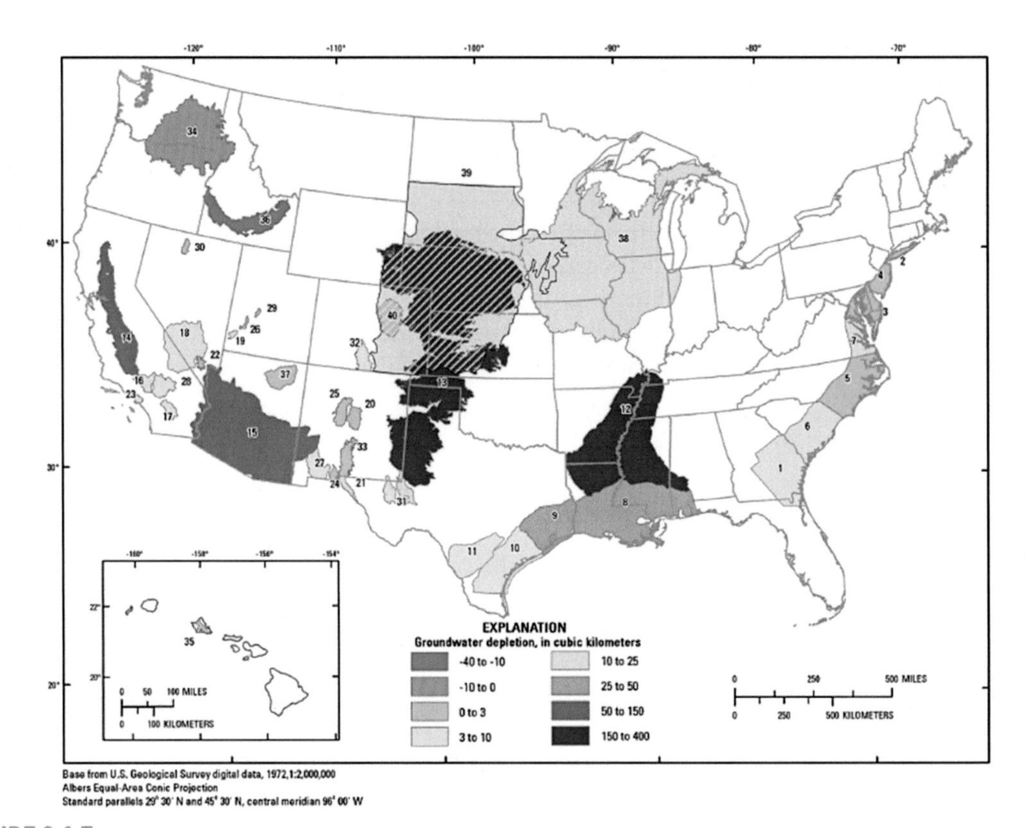

FIGURE 2.1.7

Cumulative groundwater depletion between 1900 and 2008 in 40 assessed aquifers across the continental US. *Source: USGS.*

to some extent by overlying soil formations, they are still subjected to a variety of stresses, including the following:

Overdrafts: Because recharge from the surface can be slow in areas with little precipitation, overuse of groundwater supplies can lead to depleted reserves (Fig. 2.1.7). We'll cover one of the most famous instances of groundwater depletion, the decline of the Ogallala Aquifer underlying the US High Plains, in our groundwater case study.

Contamination: Pollutants like nitrates and volatile organic compounds (VOCs) can, in some cases, move readily through the soil and reach and pollute groundwater supplies. This problem is common in areas like the Central Valley of California, where large amounts of nitrogen applied as fertilizers to agricultural lands migrate through the soil to groundwater.

Saltwater intrusion: If groundwater supplies in coastal areas are heavily used, saltwater can replace the withdrawn freshwater, contaminating the

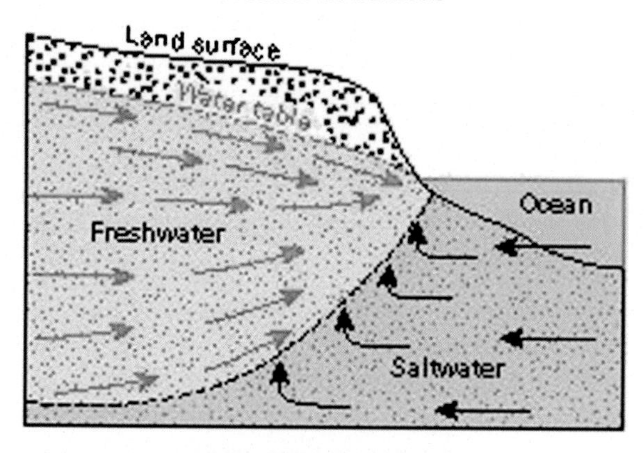

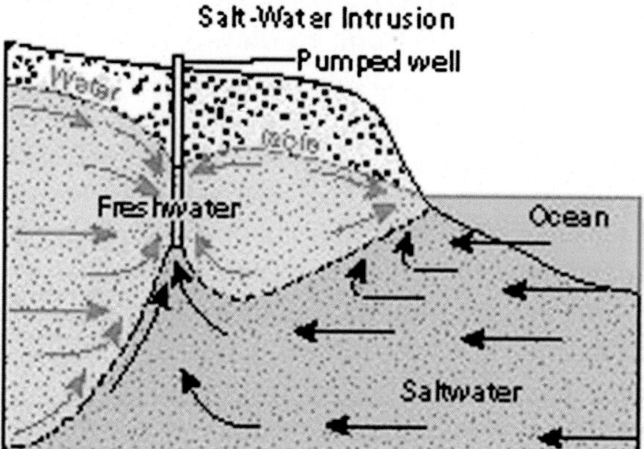

FIGURE 2.1.8

How intensive groundwater pumping can cause saltwater intrusion in coastal aquifers. *Source: USGS.*

freshwater supply (Fig. 2.1.8). About 90% of residents in South Florida rely on groundwater, primarily the Biscayne Aquifer, for their water supply. Saltwater intrusion has been an increasing threat to this important underground water supply.

WATER MANAGEMENT

Successful water management, that is, providing humans with a safe and plentiful water supply, can be both challenging and frustrating. Challenges include poor planning in developed nations like the US where too often large cities, like Los Angeles, have been built in arid regions receiving too little precipitation to support a large population. In developing nations, financial resources may not be sufficient to provide the level of treatment necessary to ensure a safe drinking water supply.

Water management can become particularly frustrating because issues related to the use and management of water resources often cross state and national boundaries. In an era of increased stress on water supplies because of overuse and our changing climate, international conflict over increasingly limited water supplies has become more likely.

Providing a complete list of issues related to water management is beyond the scope of this book; however, major challenges facing managers of water supplies include the following:

Conflicting uses: Major rivers around the globe have multiple uses, including in-stream uses like transportation, power generation, and fisheries, and off-stream uses like domestic water supply, irrigation, and industrial use. Multiple uses may not always be compatible.
A classic example of conflict over water use is the Columbia River and its hydroelectric dams in the US Pacific Northwest. Demands on the river from such uses as hydroelectric power generation, irrigation, and salmon fishing have led to long-standing disagreements among various users.

Shrinking supplies: Droughts and overuse are leading to substantial declines in the availability of both surface and groundwater (Fig. 2.1.9). The frequency and severity of such declines are increasing as human populations grow and climate change makes the occurrence of drought more likely. This water stress is particularly challenging in nations with few alternate sources of water.

Inadequate treatment: We have water treatment technologies to provide safe, high-quality drinking water. Unfortunately, adequate water treatment

FIGURE 2.1.9
The Elephant Butte Reservoir in New Mexico in 1994 (left) and in 2014 (right) provides about half the water used by the city of El Paso, Texas. *Source: NASA.*

FIGURE 2.1.10
Oxfam teams test the water quality to determine how much chemical to put in to make it safe to drink. *By Oxfam East Africa. [CC BY 2.0], via Wikimedia Commons.*

is not available in many less developed countries, leading to widespread disease and death from waterborne microbes (Fig. 2.1.10). About 6000 people a day, most of them children, die from waterborne diseases (Fig. 2.1.11).
Alternatives: While desalination of ocean water can provide a reliable source of drinking water in areas with limited freshwater supplies, the financial cost and energy consumed by salt removal are high. Similarly, the challenge of bringing filtration systems to remote locations limits their use for rural populations. Other alternatives like gray-water reuse, rooftop cisterns, and water conservation in agriculture, homes, and industries can help, but there still is much work to be done to provide a safe, sustainable water supply for billions of the Earth's inhabitants.

WATER CASE STUDIES

This text presents seven water-related case studies. These focus on current environmental problems from around the globe and range from location-specific [nutrient enrichment in the Gulf of Mexico (GOM)] to global [ocean acidification (OA)], and from comparatively straightforward (water diversion in South

FIGURE 2.1.11
The Congo River also serves as an important source of water for drinking, cooking, and washing. Conditions like this are perfect for the transmission of cholera and other waterborne diseases. *Oxfam East Africa [CC BY 2.0], via Wikimedia Commons.*

Florida and the Everglades) to complex (multiple stressors in the Mediterranean Sea). In each of these case studies, we present a variety of exercises that incorporate scientific data so that you can appreciate the role science plays in addressing environmental issues, including the complexities and connections important to consider when tackling issues like the protection of global water quality.

Before beginning these case studies, you may need to review some of the basics of water and aquatic ecosystems. On your Resource Page, we've listed some general websites that will help you review some of the basics of water resources. If, for instance, you're unfamiliar with the structure and function of aquatic food webs, you should review those sites identified for this area. In addition to the general resources provided, each case study will include specific resources chosen to help you understand the concepts associated with that issue.

Chapter 2.2

The Everglades: Changing Hydrology

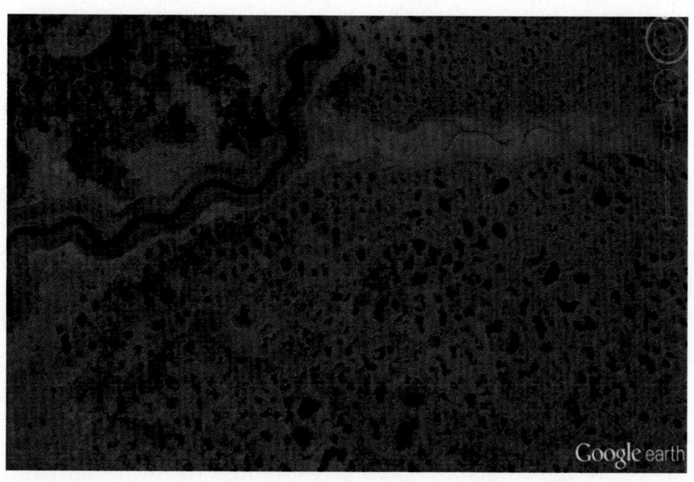

FIGURE 2.2.1
Satellite imagery of the complex hydrology of the Florida Everglades. *Source: Google Earth Images.*

OVERVIEW

This case study will consider the challenges faced by one of the world's truly unique ecosystems: the Florida Everglades (Fig. 2.2.1). These extensive subtropical wetlands are unlike any other in the world. Yet human activities, primarily water management programs and extensive agricultural and urban development nearby, have diverted freshwater flow away from the Everglades, changing both the quantity of water and the quality of the ecosystem. Currently, one of the largest restoration projects in the world is attempting to increase water flow to and improve conditions in this unique ecosystem.

As we consider the condition of today's Everglades, we'll focus on three aspects: (1) historic changes in the hydrology of the Everglades and the implications of these changes for the ecosystem, (2) the use of indicators to tell us how the Everglades are faring, and (3) efforts to increase freshwater flow to the Everglades. Before you begin this case study, go to your Resource Page and review some of the websites to become more familiar with the Everglades ecosystem.

BACKGROUND

The Florida Everglades represent a unique ecosystem, described in 1947 as a "river of grass" (Fig. 2.2.2) by writer Marjory Stoneman Douglas. The Everglades were created by the flow of fresh water from Lakes Kissimmee and Okeechobee southward toward Florida Bay and the Gulf of Mexico (GOM). Pronounced wet and dry seasons in the Everglades create a unique physical environment that supports highly specialized plant and animal communities.

While the Everglades cover an area of about 11,655 km^2, the entire drainage basin consists of about 23,310 km^2 (Fig. 2.2.3). Drainage from the Kissimmee River in Central Florida passes through tributaries and wetlands into Lake Okeechobee, and the lake occasionally overflows into the Everglades.

The Everglades contain a complex mosaic of wetland plant communities including extensive areas of sawgrass (McPherson and Halley, 1997). While the ecosystem has always experienced seasonal changes in water level and flow, it has adapted to these fluctuations. As we'll discover shortly, human activity has dramatically altered the hydrology of the Everglades, threatening many of the species that live there.

This unique landscape, recognized as a World Heritage Site and an International Biosphere Reserve, contains a variety of habitats, including mangrove swamps, pinelands, freshwater marshes, coastal estuaries, tropical hardwood hammocks, and tree Islands. Supporting the northernmost tropical environment in the world, the Everglades are home to a rich flora and fauna. Seven species, including the Florida black bear and the Florida panther, the most endangered mammal in the eastern US (Fig. 2.2.4), are unique to the Everglades.

FIGURE 2.2.2
The Everglades: a river of grass. *Source: USGS.*

THE PROBLEMS

The Everglades face a number of challenges. Because the hydrology of South Florida has been altered in past decades by various water management schemes, changing freshwater flows have resulted in a variety of environmental impacts, including altered water quality (Conrads and Benedict, 2013). Simply put, too much water during wet periods and too little during dry weather have stressed the ecosystem.

The Causes

Over the past century, the Everglades have become one of the most highly regulated and controlled watersheds in the world, with more than 2253 km of

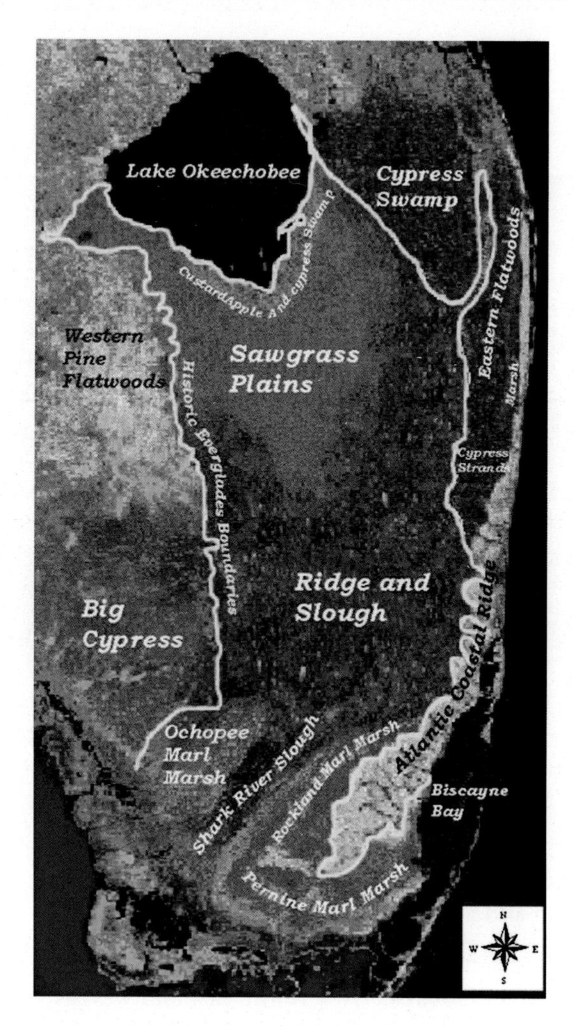

FIGURE 2.2.3

The Everglades watershed covers more than 23,000 km^2 across Central and South Florida. *Source: USGS.*

FIGURE 2.2.4
The Florida panther. There are only between 100 and 160 left, all in South Florida. *Source: US Fish and Wildlife.*

FIGURE 2.2.5
Various water management projects in the Everglades have transformed the structure and function of the landscape. Here, structure S-12A controls water releases into park areas. *Photo by: Fred Ward, (NARA record: 1109993), via Wikimedia Commons.*

dikes, canals, and levees (Fig. 2.2.5). Water diversions have been implemented to both reduce the risks of flooding and provide water to help meet Florida's substantial urban and agricultural needs.

Management actions have included diversions of water directly from Lake Okeechobee to the urban areas lying to the east and west as well as diversion of surface waters across the Everglades to agricultural lands to the south and east of the lake (Fig. 2.2.6).

FIGURE 2.2.6

Diversion of water to support agricultural and urban areas has altered the hydrology of the Everglades. *Source: USGS.* For a higher resolution version visit: http://sofia.usgs.gov/publications/posters/challenge/challenge_poster12x18.jpg

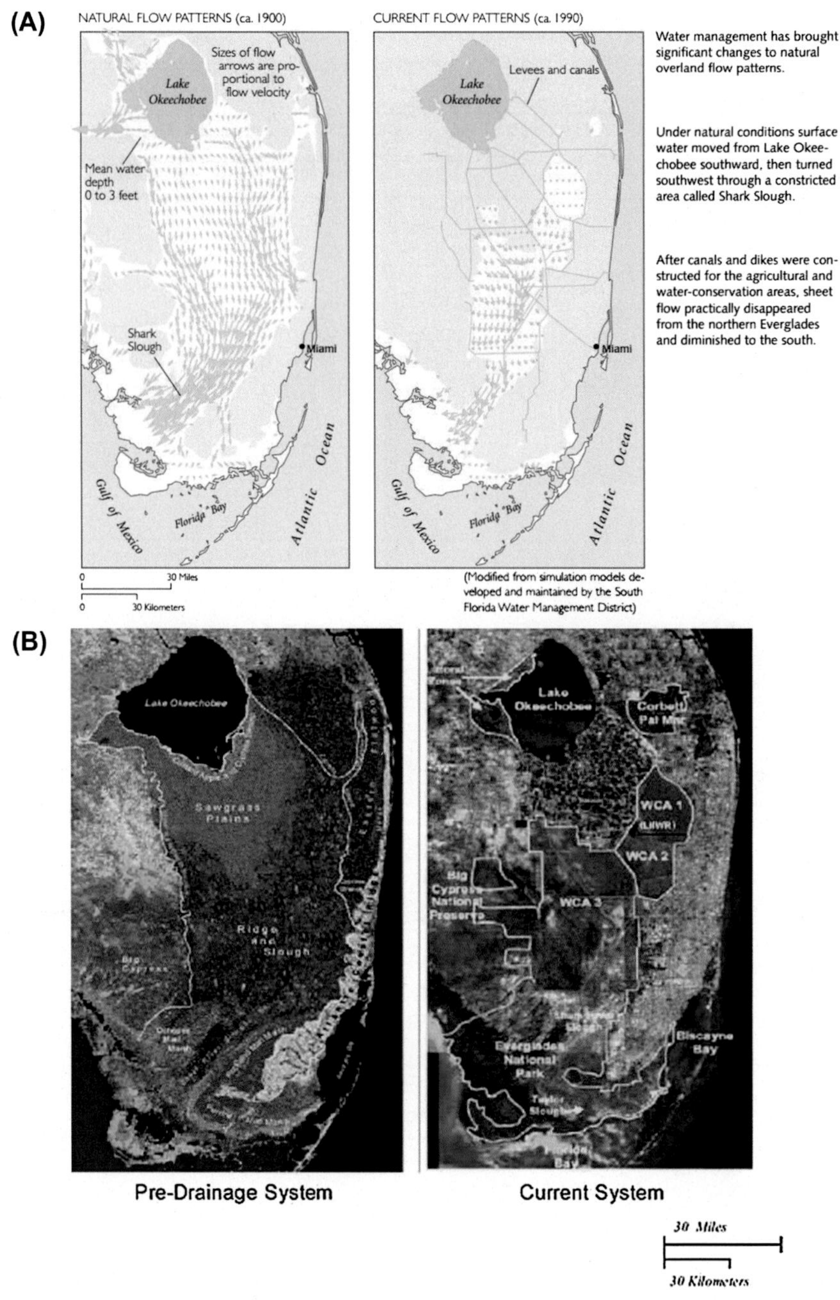

FIGURE 2.2.7

Schematic showing the effect of water management on flow into the Everglades including (A) pre- and postmanagement flow patterns, and (B) landscape characteristics. *Source: USGS.*

After canals and dikes were constructed, surface water flow was reduced to near zero in the northern Everglades, and was greatly diminished in the south (Fig. 2.2.7, top). This resulted in significant changes in the landscape, vegetation, and ecosystem structure and function (Fig. 2.2.7, bottom).

Environmental Sciences in Action

An important activity that can help demonstrate the hydrology–water quality–ecosystem connection is monitoring. Unfortunately, because of the time and expense involved in long-term monitoring, such data sets are relatively rare. But because of the intense use of, and interest in, the Everglades, long-term historical records are available to help scientists better understand changes in water quality and quantity over time.

Because of the unique qualities of the Everglades and the wide variety of water management efforts that have affected water quantity and quality in the system, the US Geological Survey (USGS) and other agencies have established a network of hydrologic monitoring stations in the Everglades (Fig. 2.2.8). Data describing surface water levels and water quality have been collected monthly at a number of sites in the Everglades since 1984 to provide baseline conditions against which to measure future changes caused by human activities in the watershed.

FIGURE 2.2.8
Airboat access to the USGS monitoring stations. *Source: USGS.*

CRUNCHING THE NUMBERS

Exercise 1: A number of state and federal agencies have been tracking the quantity and quality of the water in the Everglades for decades. We'll take a look a data set collected by the USGS. Fig. 2.2.9 shows the location of Site P-35 in the heart of the Everglades. This site has served as a long-term water quality reference location for the eastern Everglades for over 50 years.

Long-term monitoring at P-35 shows an interesting pattern in water levels during this period (Fig. 2.2.10).

- Considering both trends and variability shown in Fig. 2.2.10, describe these trends and how they differ over time.

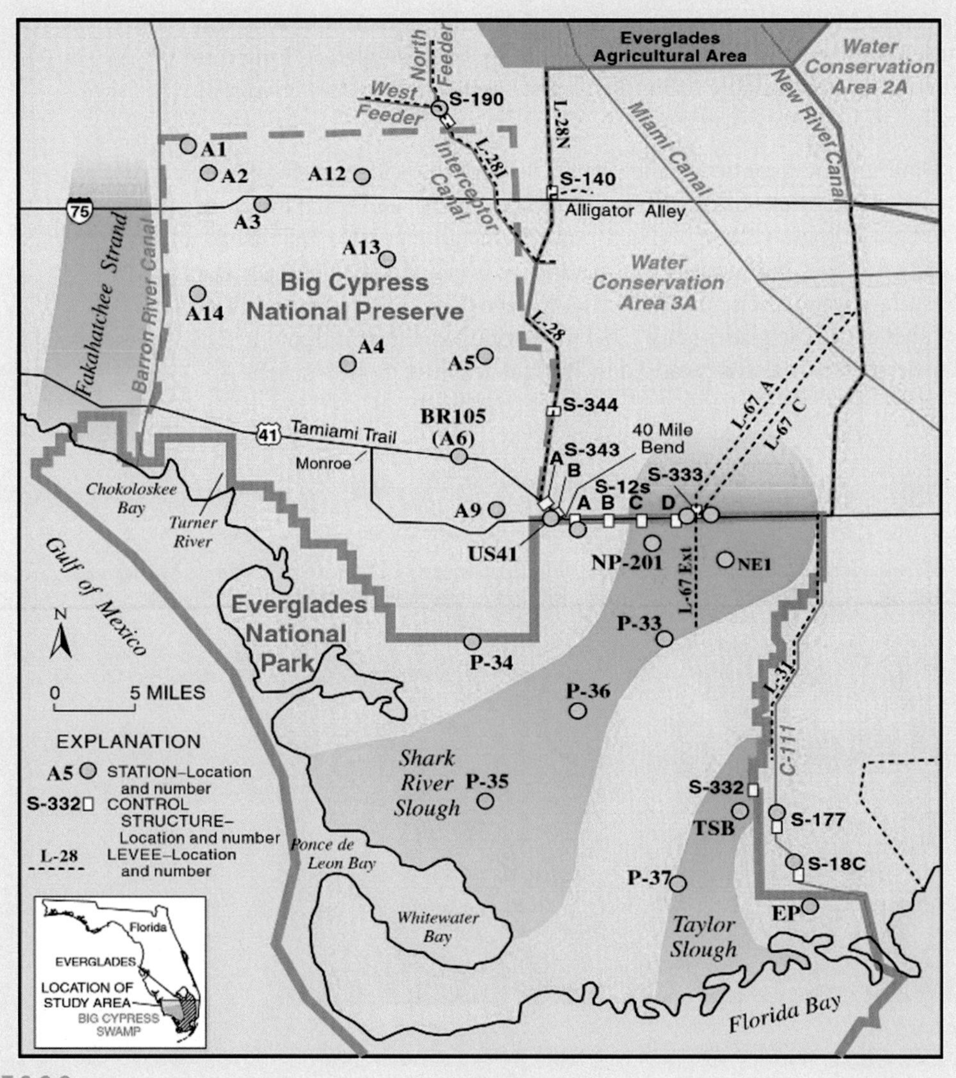

FIGURE 2.2.9
The USGS network of long-term monitoring sites across the Everglades (Miller et al., 2004).

CRUNCHING THE NUMBERS—Cont'd

- Using the **Everglades_Data.txt** data file downloadable from your Resource Page, calculate the slope of the trend line for each available decade.
  - Report this slope for each decade from 1960 to 2000.
  - How much variability is there in this relationship? How does this influence your interpretation of the trends?
- When did the increase in water level differ significantly from zero?
- If flow has decreased across much of the Everglades, why might it have increased so dramatically at P-35 in recent decades?
- Based on the timeline shown in Fig. 2.2.11, which changes do you think have had the greatest impact on water levels?

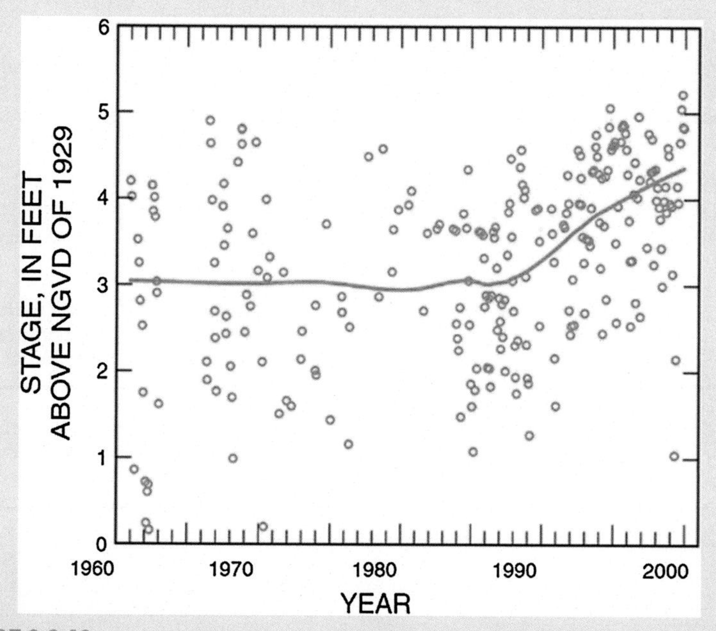

FIGURE 2.2.10

Historical National Geodetic Vertical Datum (NGVD) water levels at Taylor Slough (Miller et al., 2004).

Continued

CRUNCHING THE NUMBERS—Cont'd

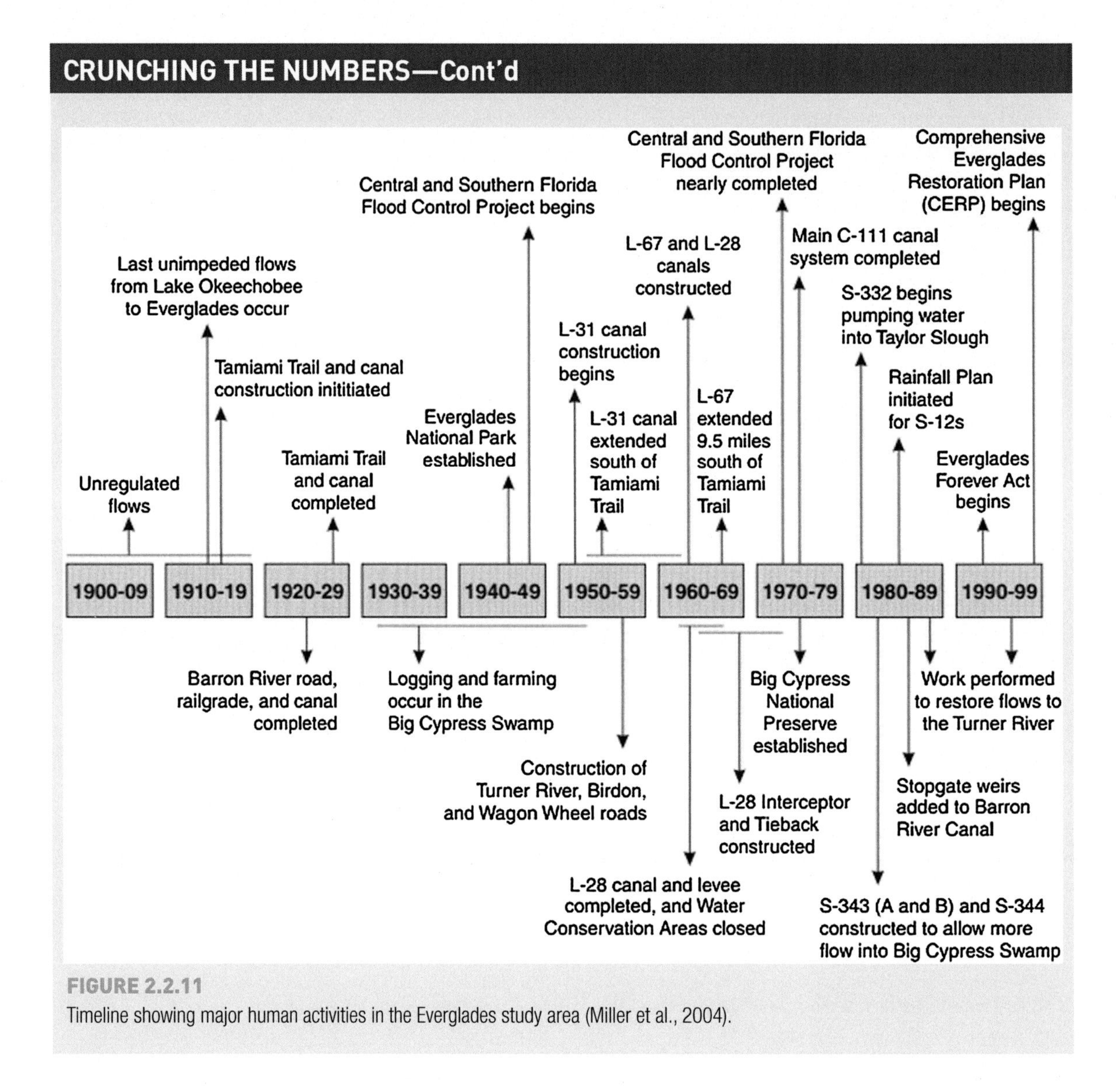

FIGURE 2.2.11
Timeline showing major human activities in the Everglades study area (Miller et al., 2004).

Because the hydrology of the Everglades wetlands is so critical to the functioning of the ecosystem, innovative approaches are being used to assess the condition of these habitats. Historically, water level measurements made by scientists in the field, such as those at the previously discussed Site P-35, have been used to track changes in wetlands over time. Now scientists are testing new remote sensing techniques to monitor wetland status throughout the Everglades.

EXPLORE IT

Exercise 2: Interferometric Synthetic Aperture Radar (InSAR)

Synthetic aperture radar (SAR) is an active remote sensing instrument that emits a pulse of long-wave radiation and then records the time it takes for the pulse echo to bounce back to the sensor from the Earth's surface. The differing times at which echoes return allow the distance (elevation) to different locations to be distinguished. To do this for locations on the ground very near to each other would require a very large antenna using a traditional radar configuration.

Because this sensor is mounted on a satellite with a continuous orbit and continuous emission of radar pulses, SAR radar can collect data while flying this distance and then process the data as if it came from one very long antenna. The distance over which the

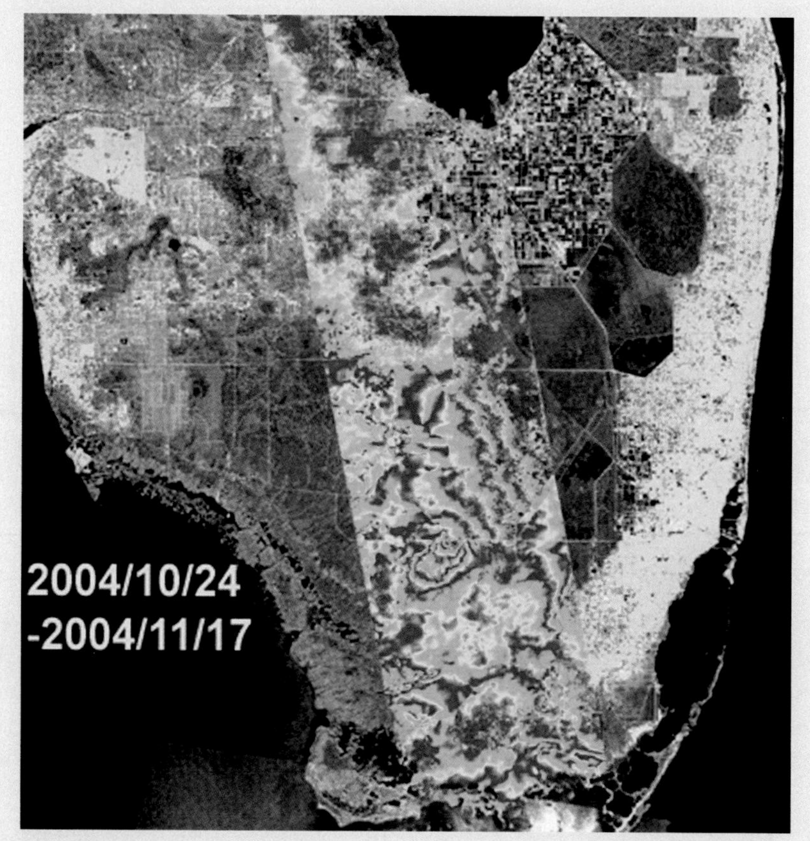

FIGURE 2.2.12
RADARSAT-1 interferogram of Everglades wetlands shows backscatter phase changes between the two RADARSAT-1 synthetic aperture radar (SAR) acquisitions. The observed phase changes measure centimeter-level changes in the wetland surface water level. Phase discontinuities are seen mainly along human-made structures (roads, levees), indicating uneven water level changes across the structures. *Source: Shimon Wdowinski, University of Miami.*

Continued

EXPLORE IT—Cont'd

signal is synthesized is known as the synthetic aperture. Using this interferometric technique, two or more SAR images are used to generate maps of surface deformation using differences in the phase of the waves returning to the satellite.

In the Everglades, InSAR technology is being used to provide high quality (5 cm accuracy and 1–2 cm precision), high-resolution maps of water levels in Everglades wetlands. The InSAR images not only track the extent of water in wetlands over wide areas of the Everglades, but can also yield direct observations of flow patterns (Fig. 2.2.12).

- The Radarsat-1 sensor used in this study is operated by the Canadian Space Agency. Visit their website (**http://www.asc-csa.gc.ca/eng/satellites/radarsat1/applications.asp**), and describe some other environmental applications for these types of data.
- What are the practical limitations when using these types of data for long-term monitoring?

SYSTEMS CONNECTIONS

One obvious direct effect that altered hydrology patterns have had on the Everglades is decreased habitat quality for various species that live in the ecosystem. Some other impacts may be less obvious. Mercury has been a concern in the Everglades for decades. Deposition of mercury contained in emissions from power plants and incinerators in South Florida and elsewhere has resulted in increased levels of this toxic element in food webs. Advisories regarding the consumption of fish from the Everglades due to mercury levels have long been in effect. Let's explore the connections between atmospheric inputs of mercury and ecological consequences in the Everglades.

Deposition rate isn't the only factor affecting mercury behavior in the Everglades. The problem is in part related to the changing hydrology of the system. It is true that the Everglades have always had dry and wet periods. Typically the dry season runs from November through May. Over the past 60 years, however, as we've seen, large portions of the Everglades watershed have been experiencing longer dry periods due to diversion of water to urban and agricultural areas.

It is well known that the conversion of inorganic mercury to highly toxic methylmercury is accelerated in wetland soils that dry out (do some exploring to find out why this occurs). The concern in the Everglades has been that the longer periods of wetland drying that have become more common in recent decades might lead to an increased conversion of inorganic mercury to methylmercury

FIGURE 2.2.13
White ibis. *Source: US Fish and Wildlife Service.*

and, subsequently, higher levels of the toxic form in fish and the birds which feed on them.

Herring et al. (2013) investigated the impact of hydrology and other factors on mercury levels in wading birds breeding in the Everglades. Of particular interest were the levels of mercury in adult white ibis (Fig. 2.2.13), a species that searches for its food in shallow wetland waters. A clear correlation can be seen between mercury levels in the red blood cells of ibis and the number of days that their feeding areas were dry during the previous dry season (Fig. 2.2.14). The authors concluded that the changing hydrology of the Everglades has increased the number of days that habitats are dry, favoring conditions that lead to enhanced production of methylmercury and a higher exposure of the ibis.

DRILLING DEEPER

Exercise 3: Using the link on your Resource Page, read the original article describing the Herring et al. (2013) study. Based on your reading, answer the following:

- The authors also measured mercury in ibis chicks and chicks of great egrets but found very different results. What did they find and what accounted for the differences?
- Setting up and conducting field investigations like this one are challenging. After reading through their methods section, identify and discuss several limitations of the approach used by Herring et al.

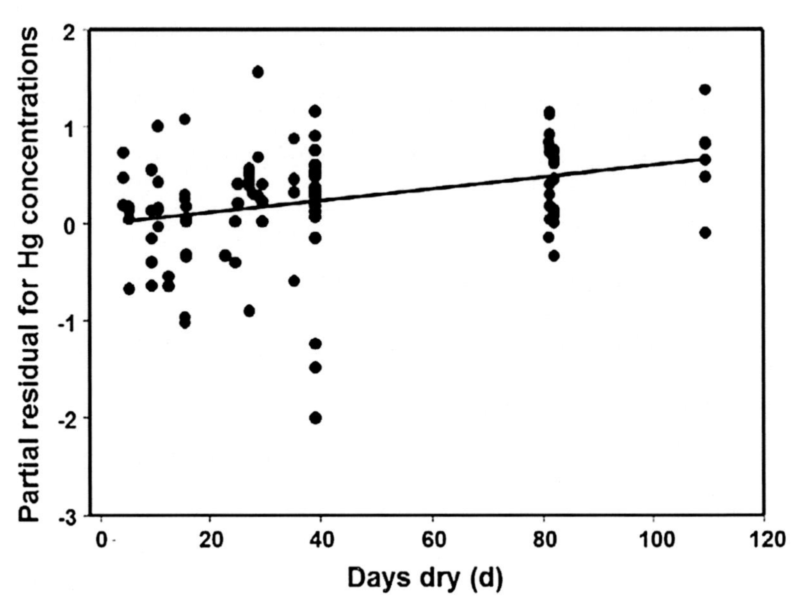

FIGURE 2.2.14
Mercury blood levels in adult white ibis versus days feeding areas were dry during preceding dry season (Herring et al., 2013).

ENVIRONMENTAL SCIENCE IN ACTION: MONITORING WITH INDICATOR SPECIES

In the environmental sciences, much of our time is spent trying to understand the effects of human activity on the environment. To help do this, we often measure variables like oxygen levels and nutrient concentrations that may be altered by the introduction of pollutants. The concept of indicator species, however, is a bit different. In this case, we use an organism known to be particularly sensitive to changes in environmental quality to tell us how a particular ecosystem is faring.

For example, in cold-water streams and rivers, we might use immature stoneflies as indicators of water quality. These invertebrates typically require cold water temperatures and high levels of oxygen to survive. While their absence doesn't necessarily indicate poor water quality (why not?), their presence tells us that the environment is likely in good shape.

There are several good examples of the use of indicator species in the Everglades. In Barnes Sound in Biscayne Bay, an estuarine ecosystem to the southeast of the Everglades, USGS scientists took core samples of sediment to look for the remains of various species of mollusks (Fig. 2.2.15). They then compared the density of mollusks known to be intolerant of salt in the water to other mollusks known to be tolerant.

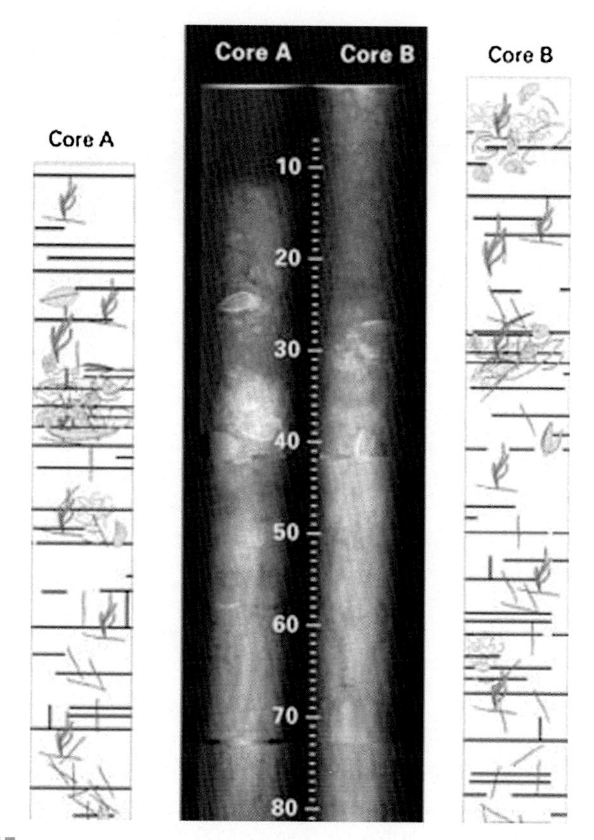

FIGURE 2.2.15

X-radiographs of USGS cores with schematic drawings of the occurrence of plant and shell remains.
Source: USGS.

Because deeper sections of the core contained sediments that were deposited long ago and upper sediments are from more recent times, scientists were able to determine how populations of the various mollusk populations had changed over time. As can be seen in Fig. 2.2.16, the density of freshwater mollusks steadily declined in more recent sediments, while species tolerant of low-salinity and fluctuating-salinity waters have increased in recent years. This provides strong evidence that the salinity of Barnes Sound has been changing over the past several hundred years, but with more dramatic shifts in recent decades.

There are two possible causes for the increasingly saline conditions noted in Biscayne Bay. One is likely related to our earlier discussion: water management efforts to the north have diverted freshwater flows to the south, meaning that there's less freshwater to dilute the saline bay waters. A second possibility, however, is that rising sea levels associated with climate change are contributing to the pattern.

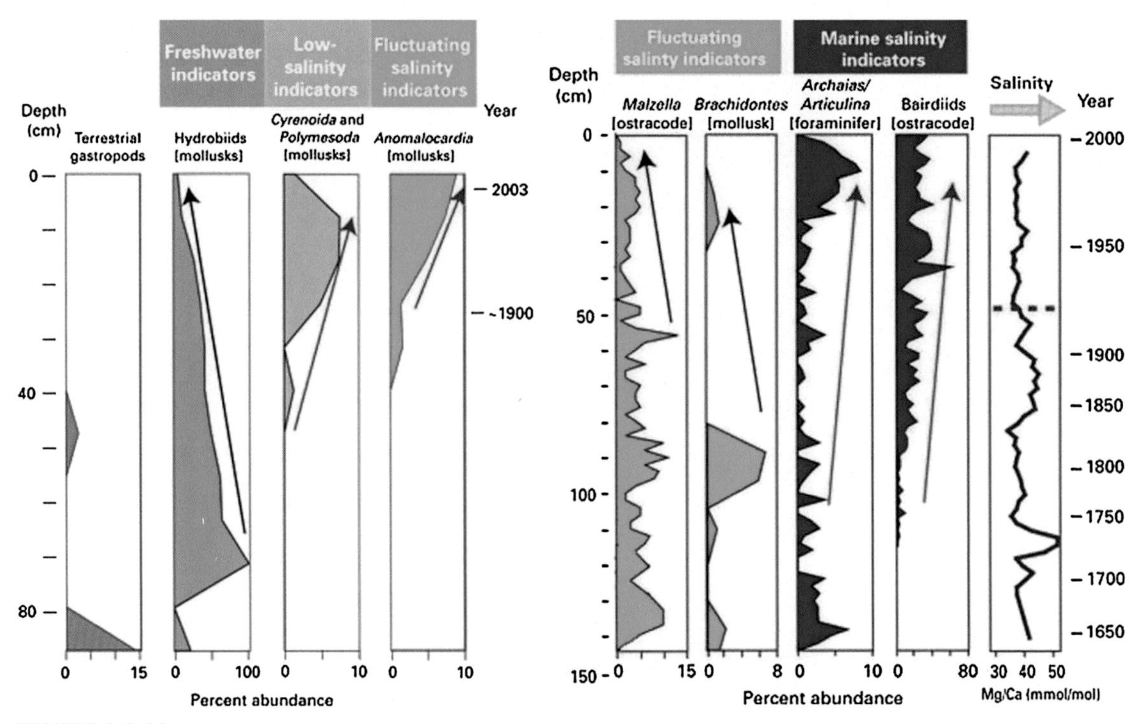

FIGURE 2.2.16

Percent abundances of indicator species in the Middle Key core are plotted against depth (in cm) on the left and approximate calendar year on the right. Red arrows track increased numbers of estuarine species. Black arrow shows decreasing influx of freshwater species. Combined, the indicators show a steady increase in salinity upwards in the core. *Source: (USGS, 2004).*

In fact, an assessment by the US National Park Service (NPS) (2013) indicated that sea level is rising about 13 cm per century off South Florida. This is already changing the patterns of vegetation, as salt-tolerant species move inland, replacing freshwater species unable to tolerate the increasing levels of salinity. Over the past 50 years, the coastal vegetation of red mangroves in this region has expanded its range inland (Fig. 2.2.17), displacing other freshwater species (Ross et al., 2000).

The NPS report provides similar analyses at other locations in South Florida, indicating that changes have also occurred outside the Everglades National Park. For example, northwest of the Everglades in the Ten Thousand Islands National Wildlife Refuge, mangroves have expanded inland into what were previously freshwater environments (Krauss et al., 2011). And in the southernmost Florida Keys, freshwater pine forests are shrinking and are being replaced by plants that can better tolerate saltwater environments.

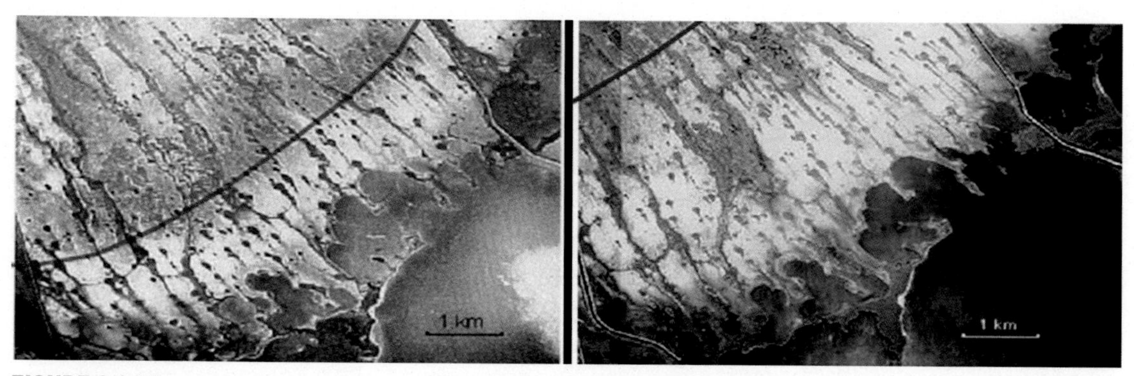

FIGURE 2.2.17
The red line in the photographs shows that the interior boundary of the saline glades has moved inland from 1940 (left) to 1991 (right). *Source: NPS.*

PUTTING SCIENCE TO WORK

Exercise 4: Understanding the complexity of the hydrology of the Everglades, think about how you might design a robust research project to address the following hypotheses:

How might you test the hypothesis that increasing salinity in a wetland over the past 50 years is related to the amount of fresh water entering that wetland?

How might you test the hypothesis that increasing salinity over the past 50 years is related to rising sea levels?

Would either of these experiments be able to prove that increased salinity is caused by decreased freshwater flow or increasing sea level? Why or why not?

THE BIG PICTURE

A much broader use of indicators is underway elsewhere in the Everglades. Led by Dr. Fred Sklar, the Everglades Division Director for the South Florida Water Management District, scientists have been tracking 11 ecological indicators annually to monitor the health of the Everglades ecosystem (Fig. 2.2.18). These measures focus on the status of key species, ranging from fish and macroinvertebrates to eastern oysters and juvenile pink shrimp, which are likely to respond to changing conditions in their habitats. To keep track of all of these different indicators and get an idea of the "big picture" condition of the ecosystem that they collectively tell, each indicator is given a red, yellow, or green light, depending on how well their population is doing. The number of red, yellow, and green indicators for each year can be summarized to compare and track the overall condition of the ecosystem over time.

Indicators at a Glance

This is a snapshot of the status of each indicator by geographic region (listed from north to south) for the last five years. Results shown here are consistent with an assessment done by the National Research Council (2012), reflecting the continued patterns of severely altered hydrology throughout the ecosystem. An exception is WY2011 in Lake Okeechobee where the Nearshore Zone Submersed Aquatic Vegetation exceeded the target level because of successive years where the Lake was near or below the lower end of the ecologically desired stage envelope with concomitant improved light penetration.

	Water Year 2008	Water Year 2009	Water Year 2010	Water Year 2011	Water Year 2012
Lake Okeechobee					
Invasive Exotic Plants Species					
Lake Okeechobee Nearshore Zone Submersed Aquatic Vegetation					
Northern Estuaries					
Invasive Exotic Plant Species					
Eastern Oysters					
Greater Everglades					
Crocodilians					
Fish and Macroinvertebrates (WCA-3 and ENP only)					
Invasive Exotic Plants					
Periphyton and Epiphyton					No species composition data
Wading Birds (White Ibis and Wood Stork)					
Southern Coastal System					
Crocodilians					
Southern Estuaries Algal Blooms**					
Florida Bay Submersed Aquatic Vegetation					
Invasive Exotic Plants					
Juvenile Pink Shrimp*	Data used as base	Data used as base	Data used as base		
Wading Birds (Roseate Spoonbill)					Prey community data not yet processed
Wading Birds (White Ibis and Wood Stork)					

FIGURE 2.2.18

The status of each indicator by geographic region (listed from north to south) for a 5-year period. Red means no progress, yellow is neutral, and green means improving status (Brandt et al., 2012).

DRILLING DEEPER

Exercise 5: The 2012 Indicators report includes historical data that summarize indicator status from 2008–2012. Based on this information presented in the summary table (Fig. 2.2.18):

- Which of the indicator species are most sensitive to perturbations in the ecosystem?
- How does the overall condition of the ecosystem change from year to year?
- What are some of the strengths and weaknesses of this approach to ecosystem assessment?

SOLUTIONS

Restoration of damaged ecosystems has become an established science. In another of your case studies, you'll read about efforts to restore two polluted rivers, the Thames in the UK and the Grand Calumet in northern Indiana (US). In the case of many rivers, it's fairly straightforward, involving removal or treatment of the sources of pollution. However, restoring flow to the Everglades is a much more involved process.

Probably the largest environmental restoration project ever undertaken, the Comprehensive Everglades Restoration Plan (CERP) is designed to restore some of the flow of fresh water into the Everglades. The plan, covering 16 counties in central and southern Florida, has more than 60 elements, and will take 30 years and more than $10 billion to complete.

The goal of CERP is "to capture fresh water that now flows unused to the ocean and the gulf and redirect it to areas that need it most. The majority of the water will be devoted to environmental restoration, reviving a dying ecosystem." To accomplish this goal, project personnel will capture and store water that can then be used to increase flows into the Everglades. Instead of being diverted to the ocean and gulf, water will be stored in new reservoirs and wetlands-based treatment areas. An additional 300 underground water storage wells will be used to hold water until it's needed. About 80% of the water captured will go to restore the Everglades. The specifics on several important features of CERP are shown in Fig. 2.2.19. You can also view the "Restoring the Everglades" video on your Resource Page to learn more.

The seven principal features of the Comprehensive Everglades Restoration Plan (CERP) will work to "get the water right" – to improve water quality, quantity, timing and distribution. Each CERP project includes one or more of these features:

◆ **Surface Water Storage Reservoirs**
181,300 acres of above and in-ground reservoirs are planned to store millions of gallons of water.

● **Aquifer Storage and Recovery**
More than 300 underground water storage wells are proposed to store up to 1.6 billion gallons of treated water a day in confined aquifers.

💧 **Stormwater Treatment Areas**
35,600 acres of manmade wetlands will be constructed to remove pollutants and other harmful contaminants from water before it is discharged to the Everglades.

■ **Wastewater Reuse**
Two advanced treatment plants are proposed to recycle more than 220 million gallons of wastewater a day, adding a new source of high quality water for the southern Everglades.

★ **Seepage Management**
Barriers are proposed to be built to stop the rapid underground seepage of water out of the Everglades, which today results in the loss of millions of gallons of water each year.

▲ **Removing Barriers to Sheetflow**
More than 240 miles of canals and levees may be removed to restore the historic overland sheetflow through the Everglades wetlands. Sections of Tamiami Trail will be elevated to handle increased water flows contributed by CERP project features.

✦ **Operational Changes**
Changes will be made in the regional water management system to benefit Lake Okeechobee, the Everglades and the coastal estuaries.

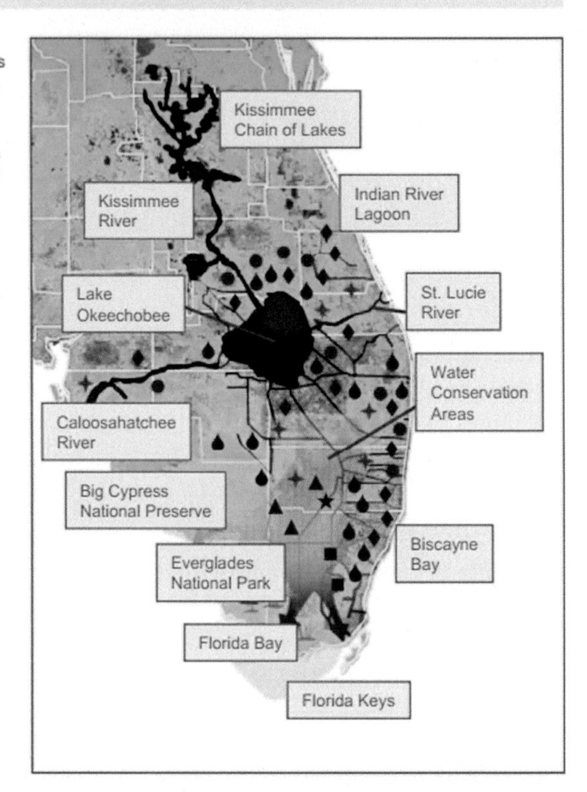

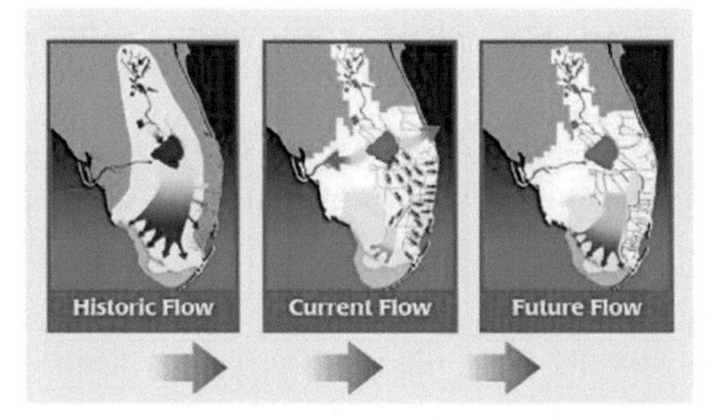

The Journey to Restore America's Everglades

A partnership of the U.S. Army Corps of Engineers, South Florida Water Management District and many other federal, state, local and tribal partners.

www.evergladesplan.org • 1-877-CERP-USA

FIGURE 2.2.19
A basic summary of the CERP plan. *Source:* www.evergladesrestoration.gov.

CRITICAL THINKING

Exercise 6: Go to the CERP website that's listed on your Resource Page at http://141.232.10.32/pm/ssr_2014/ssr_main_2014.aspx and review the basic steps proposed to help restore the flow of fresh water to the Everglades. As you ponder the future of the Everglades, consider the effects of climate change on this ecosystem.

- Describe three possible ways that climate change may impact the Everglades.
- Do these changes influence your opinion about the likelihood of restoration success?
- For some, the specter of climate change calls into question the wisdom of CERP. Form two groups to debate this topic: "Does the likelihood of climate change argue against continuing with the CERP?"

SOLUTIONS

Exercise 7: Download and read the most recent CERP System-Wide Ecological Status Report from your Resource Page. In this chapter, we've discussed some common ways to monitor and assess the Everglades ecosystem.

- List some of the specific indicators that were chosen to assess the progress and success of the CERP program.
- Are there other indicators you think should be included to get a more comprehensive assessment of the restoration effort?
- Based on this report, how would you describe the success of the CERP program so far?

CONSIDER THIS

Exercise 8: How can societies balance economic growth and quality of life with ecological preservation?

In the Everglades, how can land managers and decision makers successfully meet the water and agricultural needs of the surrounding population without further degrading this unique ecosystem?

REFERENCES

Brandt, L.A., Boyer, J., Browder, J., Cherkiss, M., Doren, R.F., Frederick, P., Gaiser, E., Gawlik, D., Geiger, S., Hart, K., Jeffery, B., Kelble, C., Layne, J., Lorenz, J., Madden, C., Mazzotti, F.J., Ortner, P., Parker, M., Roblee, M., Rodgers, L., Rodusky, A., Rudnick, D., Sharfstein, B., Trexler, J., Voiety, A., 2012. System-wide Ecological Indicators for Everglades Restoration. 2012 Report. Unpublished Technical Report, 90 pp.

Conrads, P.A., Benedict, S.T., 2013. Analysis of Changes in Water-Level Dynamics at Selected Sites in the Florida Everglades. U.S. Geological Survey Scientific Investigations Report 2012-5286, 36 p. http://pubs.USGS.gov/sir/2012/5286.

Herring, G., Eagles-Smith, C.A., Ackerman, J.T., Gawlik, D.E., Beerens, J.M., 2013. Landscape factors and hydrology influence mercury concentrations in wading birds breeding in the Florida Everglades, USA. Sci. Total Environ. 458–460, 637–646.

Krauss, K.W., From, A.S., Doyle, T.W., Doyle, T.J., Barry, M.J., 2011. Sea level rise and landscape change influence mangrove encroachment onto marsh in the Ten Thousand Islands region of Florida, USA. J. Coastal Conserv. 15 (4), 629–638.

McPherson, B.F., Halley, R., 1997. The South Florida Environment: A Region Under Stress. National Water-Quality Assessment Program. U.S. Geological Circular 1134.

Miller, R.L., McPherson, B.F., Sobczak, R., Clark, C., 2004. Water Quality in Big Cypress National Preserve and Everglades National Park—Trends and Spatial Characteristics of Selected Constituents US Geological Survey Water Resources Investigations Report 2003–4249.

National Park Service, US Dept. of Interior, 2013. Climate Change. www.nps.gov/ever/naturescience/upload/EVER-Climate-Change-05-2013.pdf.

Ross, M.S., Meeder, J.F., Sah, J., Ruiz, P.L., Telesnicki, G.J., 2000. The Southeast Saline Everglades revisited: a half-century of coastal vegetation change. J. Veg. Sci. 11, 101–112.

U.S. Geological Survey, 2004. Changing Salinity Patterns in Biscayne Bay, Florida. Fact Sheet 2004–3108. http://pubs.USGS.gov/fs/2004/3108/index.html.

Chapter 2.3

Mediterranean Sea: One System, Many Stressors

FIGURE 2.3.1
A satellite image showing the Mediterranean Sea. *Source: NASA.*

OVERVIEW

The Mediterranean Sea (Fig. 2.3.1) is an invaluable international marine resource in trouble. Home to an unusually diverse flora and fauna, the Mediterranean is impacted by a wide variety of stressors. In this case study, we'll focus on three very different threats to the Mediterranean ecosystem: a chemical pollutant (mercury), invasive alien species, and the physical impacts of the changing climate. We'll conclude by considering the possibilities and challenges faced by those trying to protect this resource. Before you start, review some of the Mediterranean Sea websites listed on your Resource Page.

BACKGROUND

Because the Mediterranean lies at a crossroads from both a geological and geographical perspective, it is a system both rich in resources, including both subtropical and temperate marine species, yet vulnerable to a dizzying array of stressors. Once the center of western civilization, these waters show ever-increasing impacts from human activity.

Considering all the challenges facing the Mediterranean and the implications for the diverse ecological and social systems it supports can be a daunting task. To personalize and contextualize this complex ecosystem, for this case study, we'll take the perspective of the Atlantic Bluefin tuna (*Thunnus thynnus*) that call the Mediterranean home (Fig. 2.3.2).

Bluefin tuna are a top predator and the only species of tuna that lives solely in the temperate waters of the Atlantic, including the Mediterranean. Bluefin tuna can live more than 25 years, reach lengths of up to 3 m and can weigh more than 400 kg. They are opportunistic predators, feeding on sardines, herring, mackerel, squid, and crustaceans.

Fast Facts About the Mediterranean

As a nearly landlocked sea, the Mediterranean has many unique features that shape its ecology:

- limited tidal fluctuations due to its narrow connection with the Atlantic Ocean;

FIGURE 2.3.2
Bluefin tuna. *Source: NOAA.*

- a high salinity due to evaporation rates that are greater than precipitation and river inputs;
- distinct circulation patterns driven by the pressure gradient between relatively cool, low-salinity water from the Atlantic that warms and becomes saltier as it travels east, and then sinks to travel back as a dense bottom current;
- a large surface-to-coastline ratio, with a surface area of 2.5 million km^2, and 46,000 km of coastline in 22 nations, excluding its more than 3000 Islands;
- two particularly deep basins, east and west, with an average depth of 1500 m, reaching greater than 4000 m in some locations;
- a relatively slow renewal rate of 80–100 years;
- inflows from the Atlantic Ocean and Black Sea and a connection to the Red Sea through the Suez Canal;
- a highly variable climate that changes from subtropical to temperate over a relatively short distance;
- about 17,000 species of marine life, including 20–30% that are endemic; and
- up to 18% of the world's known marine species are found in the Mediterranean, yet it contains only 0.82% of the volume and 0.32% of the surface area of the world's oceans.

As we work through this case study, you will learn how some of these features affect the sensitivity of the Mediterranean to environmental stressors.

EXPLORE IT

Exercise 1: Google Earth provides a unique platform to share and explore geo-referenced photos and information. Mission Blue is a global initiative to find and share such information to help build public support for a global network of marine protected areas called "hotspots." A collaborative effort involving more than 80 ocean conservation groups, companies, and scientific teams, Mission Blue has brought together images, videos, and information from across the world's oceans.

Let's explore the Mission Blue collection from the Mediterranean Sea in Google Earth to get an idea of the diversity of life in the Mediterranean. Visit the Mission Blue Global

Continued

EXPLORE IT—Cont'd

exploration website at http://mission-blue.org/google-earth/ linked on your Resource Page.

Use the site's search function to explore their content related to the Mediterranean.

- Explore the videos embedded within the Mission Blue Google Earth layer and report on at least three that demonstrate the biological diversity of the Mediterranean.
- From the videos you watched, what steps are being taken to protect deep-sea "hotspots?"

THE PROBLEMS

While the list of characteristics that make the Mediterranean unique is impressive, so is the array of stressors threatening this ecosystem. Signs that conditions in the Mediterranean are deteriorating are on the rise.

Consider that 82 million people live in cities on the Mediterranean coast with more than 100 million tourists visiting Mediterranean beaches each year. Even beyond the coastline, human activities can have significant impacts. Tributaries like the Po, Ebro, Nile, and Rhone discharge industrial and agricultural wastes into the Mediterranean. For example, data from 1976 to 2006 indicated that major point sources of pollutants like metal industries, oil refineries, tanneries, and chemical industries annually discharged an estimated 4 million tons (MT) of biochemical oxygen demand (BOD) and 85,000 tons of metals to the

FIGURE 2.3.3
Activist divers cut open the cages in which hundreds of young bluefin were being held. *By Danilo Cedrone (United Nations Food and Agriculture Organization) via Wikimedia Commons.*

Mediterranean. Oil is another important pollutant, with 1.96 MT discharged in 2003 alone (UNEP, 2006).

The Mediterranean faces many other challenges, including habitat loss, invasive species, sedimentation, and climate change. Evaluating the cumulative impact of the many stressors affecting the Mediterranean can seem overwhelming. To help in this assessment, let's take a look from the perspective of the bluefin tuna.

Unfortunately, overfishing has already removed many of the largest bluefin from the Mediterranean. As the global market for sushi and sashimi has increased in recent decades, demand and prices for bluefin tuna have soared. This has led to overfishing, causing populations to plummet. The fact that the species is late to mature and slow growing has made the problem worse.

More recent efforts to "farm" bluefin focus on capturing juveniles and fattening them in pens before harvest (Fig. 2.3.3). While farm-raised fish may sound appealing to many environmentally conscious consumers, removing individuals from the reproductive pool limits future bluefin populations and weakens the gene pool.

When it comes to the stressors that threaten bluefin, overfishing may have gotten the most attention. But there are many others to consider. We'll look at how mercury pollution, the invasion of alien species, and a changing climate also threaten the future of bluefin in the Mediterranean.

Mercury: Of the many types of pollutants entering the Mediterranean from point and non-point sources, few are more worrisome than mercury (Hg), a trace element that can be converted by microbes to highly toxic methylmercury (MeHg) in aquatic ecosystems. Sources of mercury to the Mediterranean include the usual suspects: discharges of industrial waste and sewage into tributaries and the atmospheric deposition of mercury released by fossil fuel– burning power plants and waste incinerators. An important natural source is runoff from the mining of cinnabar, the common ore of mercury. The Mediterranean watershed has the highest concentration of these mines in the world.

At levels encountered in the Mediterranean, mercury may not be acutely toxic to the tuna itself, but impacts on humans who consume fish containing mercury are well documented. Outbreaks of mercury poisoning have made it clear that adults, children, and developing fetuses are at risk from dietary exposure. Symptoms of mercury poisoning may include impaired neurological function; loss of peripheral vision; lack of coordination of movement; impaired speech, hearing, and walking; and muscle weakness.

Environmental Sciences in Action

We know that methylmercury can be concentrated by species at the top of marine food webs (Fig. 2.3.4). Let's consider two similar studies that evaluated how bluefin tuna in the Mediterranean accumulate mercury.

Storelli et al. (2005): In this study, scientists collected 73 bluefin tuna and 58 swordfish between June–August 2003 from the Ionian Sea (see Google Earth to locate this within the Mediterranean). Muscle and liver tissues were analyzed for a number of trace elements, including mercury. As you can see from Fig. 2.3.5 below, the mean Hg concentration in muscle tissues of the bluefin was 0.20 ug/g wet wt., while those in swordfish muscle averaged 0.07 ug/g wet wt.

An important finding in this study, consistent with many other similar studies in both fresh- and saltwater, was a positive correlation between Hg concentration and fish weight, with an r value of 0.54, $p<.0001$. However, in this study, even the highest Hg level measured in tuna muscle tissue (0.35 ug/g) was below levels deemed hazardous for consumption of Mediterranean fish.

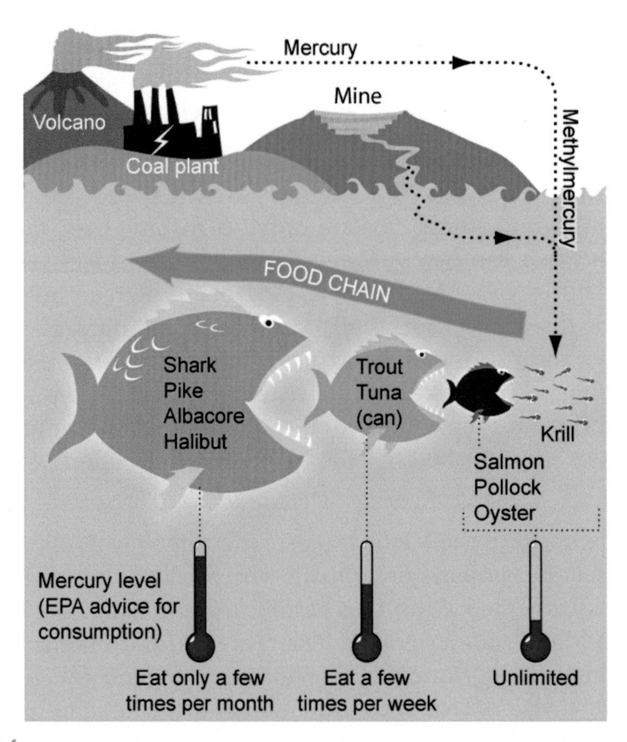

FIGURE 2.3.4
Mercury in the food chain. *By Bretwood Higman, Ground Truth Trekking. [CC BY 3.0], via Wikimedia Commons.*

CRUNCHING THE NUMBERS

Exercise 2: Understanding how Hg concentrations differ in organisms provides a valuable insight about how this element cascades through an ecosystem. It also helps set safety standards for human consumption of various food species.

Let's see how Hg concentrations compare between swordfish and bluefin tuna collected as a part of the Storelli et al. (2005) study. We know that Storelli et al. sampled 73 tuna and 58 swordfish.

- Using the mean and standard deviation for Hg in muscle tissue reported in Fig. 2.3.5, determine if there is a significant difference between the two species.
- Why might there be a difference in Hg concentrations between the two species?

Species	Tissue or organ	Hg
Swordfish	Muscle tissue	0.02–0.15
		0.07 ± 0.04
	Liver	0.10–0.37
		0.19 ± 0.09
Bluefin tuna	Muscle tissue	0.13–0.35
		0.20 ± 0.07
	Liver	0.27–0.60
		0.39 ± 0.10

FIGURE 2.3.5
Metal concentrations (range and means ± S.D. in micrograms per gram wet weight) in the muscle tissues and livers of swordfish and bluefin tuna (Storelli et al., 2005).

Renzi et al. (2014): In this study, 23 bluefin tuna were collected in the spring of 2012 off the coast of Sardinia (see Google Earth to locate this within the Mediterranean). Mercury and PCBs were measured in muscle tissue. As you can see in Fig. 2.3.6, Hg averaged 0.66 ug/g wet wt, with a range of 0.140–2.211.

As in the previous study, there was a strong positive correlation between size of the bluefin and Hg levels. As can be seen in Fig. 2.3.7, several of the larger specimens exceeded European guidelines for safe consumption.

		Mean	SD	Range
Fork length	cm	198.2	25.9	137.2 - 252.3
Body weight	kg	135.7	45.8	46.5 - 258.8
PCBs	mg/kg f.w.	0.732	0.354	0.155 - 1.403
Hg	mg/kg f.w.	0.660	0.585	0.140 - 2.211

FIGURE 2.3.6
Levels of mercury and polychlorinated biphenyls in bluefin tuna muscle (Renzi et al., 2014).

CRUNCHING THE NUMBERS

Exercise 3: The Storelli et al. and Renzi et al. studies occurred almost 10years apart. It would be useful to know if Hg contamination in bluefin is getting worse, better, or staying the same.

- To answer this, we first have to convert measurements of Hg to similar units between the two studies. Renzi et al.'s bluefin muscle Hg concentrations were in mg/kg, while Storelli et al.'s were in ug/g. Since ug/g is the equivalent of mg/kg, in this case, the numbers are directly comparable.
- Because the same locations were not measured in both studies, we cannot conduct a true dependent *t*-test for differences in Hg between the two time periods. But we can use the mean and standard deviation of both to calculate the 95% confidence intervals around each mean. Report these upper and lower confidence intervals for mean Hg concentration in mg/kg for both the Storelli et al. and Renzi et al. studies.
- Now you can determine if there is likely a significant difference between the two time periods. Examine the confidence intervals of the 2005 and 2014 studies.
  - Do they overlap?
  - What does this tell you about how Hg accumulated in tuna over this time period?

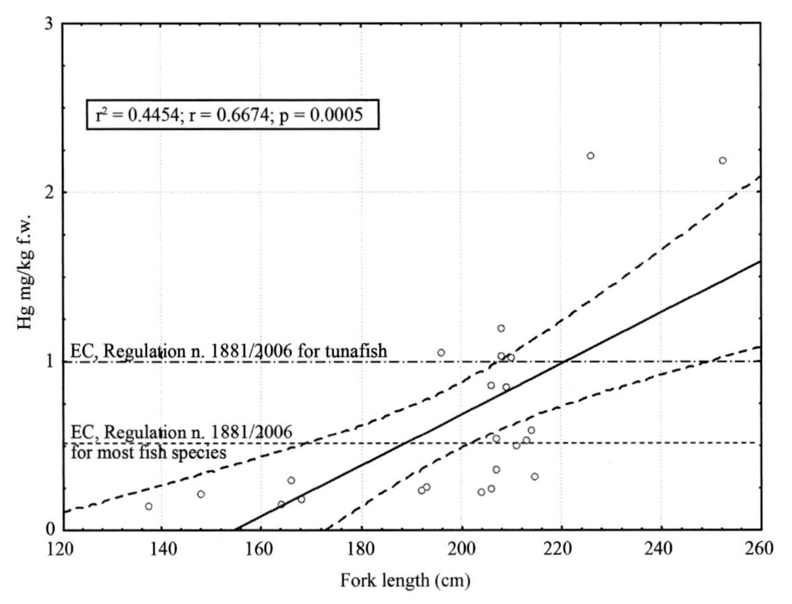

FIGURE 2.3.7

Relationship between fork length and mercury in bluefin tuna muscle, and a comparison to the maximum level set by the European Commission for fish species (Renzi et al., 2014).

Some good news? Renzi et al. noted that mercury levels in their study were somewhat lower than values from some of the older studies in the Mediterranean (excluding Storelli et al.). The bad news is that this may simply be an artifact resulting from removal of many of the older, larger bluefin (those with the highest Hg levels) by commercial fisheries.

A Take-Home Message: Sometimes environmental data are "all over the place." The challenge for the environmental scientist is to figure out whether very different data describing the same thing (in this case, Hg in Mediterranean bluefin tuna muscle) are due to statistical variability or some other factor that's causing the difference in the data.

DRILLING DEEPER

Exercise 4: Considering the complex cycle of Hg between land, air, water, and biota, drill deeper to uncover the most likely sources of Hg in the Mediterranean basin.

- Where might the bulk of Hg found in bluefin's tissues come from?
- Considering the information you found, what steps would you suggest to reduce Hg levels in bluefin tuna?

A Risk to Us or Bluefin? Bluefin are exposed to many different persistent pollutants. While some, like methylmercury, may not reach levels that harm the tuna themselves, they may still pose a threat to humans who consume them. Other pollutants, however, can harm bluefin tuna, even at very low levels.

CRITICAL THINKING

Exercise 5: The reality is that bluefin tuna are exposed to a veritable soup of pollutants in the Mediterranean.

- Do some digging into the scientific literature and identify one or two persistent pollutants (don't rule out the possibility of organic pollutants) that can affect key life properties like bluefin's growth rate or reproductive capacity.
- Where do these pollutants originate?
- What can be done to reduce their concentrations in the environment?
- How do you think the threat of chemical pollution compares to the other threats to this ecosystem?
- How might these chemical pollutants interact?

Invasive Species: Even if bluefin tuna aren't caught by fishing nets or exposed to mercury and other potentially harmful pollutants, there looms another increasingly serious challenge: "bioinvaders."

The Mediterranean Sea, now home to at least 900 introduced species (IUCN, 2012), is considered to be the most invaded marine environment on the globe. In most aquatic systems, alien invaders enter via shipping or are introduced by other human activities. While both these are important routes of entry into the

FIGURE 2.3.8
Bluespotted cornetfish (*Fistularia commersonii*), also known as the smooth flutemouth. *By Derek Keats [CC BY-SA 2.0], via Wikimedia Commons.*

Mediterranean, probably the greatest number of bioinvaders enter from the Red Sea via the human-made Suez Canal, which was completed in 1869. These Red Sea invaders are called Lessepsian migrants.

In some cases, these invasive species (e.g., the rabbitfish, *Siganus luridus*) have simply moved into an empty niche and become well established. Other invaders, however, are able to outcompete and eliminate native species. One of the most aggressive of the invaders is the bluespotted cornetfish (*Fistularia commersonii*) (Fig. 2.3.8), a top predator from the Red Sea. First captured in the Mediterranean in 2000 off Israel, the bluespotted cornetfish consumes almost exclusively small fish (Pinnegar et al., 2014).

When invasions first occur, it is often difficult to predict the long-term impacts of aliens like the bluespotted cornetfish. Research is being done to examine the response of native species to invaders across various trophic levels in the Mediterranean (Coll et al., 2010). Studies like this demonstrate how invaders from one trophic level can impact not only native species within their own trophic level, but also at higher trophic levels (Fig. 2.3.9). Based on these relationships, models that can predict broader ecological consequences over the long term can be constructed.

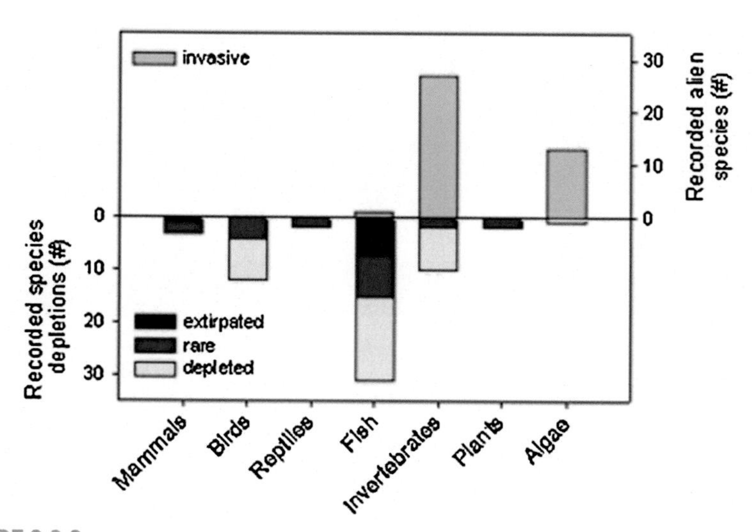

FIGURE 2.3.9

Impacts on populations are not always a direct impact on the same trophic level. Often, changes in one trophic level impact others where invasives are not as common (Coll et al., 2010).

Environmental Sciences in Action

J. Pinnegar of the UK's Centre for Environment, Fisheries and Aquaculture Science and his team of scientists (Pinnegar et al., 2014) developed a model to predict what bluespotted cornetfish invading the Bay of Calvi near Corsica in the Mediterranean Sea would eat. Initially, they looked at prey abundance in the bay and the characteristics of both this invasive predator and potential prey species. Their model predicted a substantial decline in plankton-eating fish, likely prey for the invaders.

Consider the implications of this research for the bluefin tuna. If the niches of the bluespotted cornetfish and the bluefin overlap, the availability of plankton-eating fish that are a staple in young bluefin's diet may decrease if cornetfish outcompete the tuna, possibly reducing growth rates of the young tuna.

Eventually, assuming that they survive and grow to adults, older and larger bluefin may experience a population increase as they feed upon the ever-growing populations of bluespotted cornetfish. In fact, Pinnegar et al.'s model predicts an increase in populations of top predators like the bluefin by later this century.

An interesting question is how a potentially negative impact on younger individuals within a population and potentially positive impacts on older individuals will influence the long-term population dynamics of the bluefin tuna.

SYSTEMS CONNECTION

Exercise 6: Using your knowledge of aquatic food webs and the place of bluefin tuna within the Mediterranean Sea food web, give an example or two of how the introduction of invasive species other than the bluespotted cornetfish might impact the bluefin.

DRILLING DEEPER

Exercise 7: As with our discussion about pollutants, there are many different bioinvaders in the Mediterranean Sea (and probably many more to come). However, the magnitude of invasive species is particularly high in the Mediterranean compared to the threat in other ocean waters.

- What physical or geographic characteristics might be responsible for the unusually high number of bioinvaders in the Mediterranean?
- Do your own research to find examples of invasive species that will:
  - threaten commercial fisheries in the Mediterranean Sea
  - reduce the use of Mediterranean beaches by tourists
  - likely make a positive contribution to the Mediterranean Sea ecosystem
- Among the many invasive species you read about, which do you consider to be the greatest threat to the Mediterranean? Why?

Climate change: By now it should be obvious that bluefin tuna are faced with many threats. The changing climate will add to that list. Models predict that annual temperatures in the Mediterranean region may increase from 2.2 to 5.1 °C by 2100 (UNEP, 2010). Already, water temperatures at all depths are increasing (Lejeusne et al., 2010). With these warmer temperatures, species will begin to shift their ranges, with warm-water species like the invasive species entering via the Suez Canal moving to the north and west (Fig. 2.3.10). Where will the cold-water species such as bluefin tuna go?

Already some changes in the behavior of bluefin have been detected; for example, they are staying for longer periods in the cooler western Mediterranean Sea. While bluefin have a fairly broad range of temperature tolerance, their ideal spawning temperature range is 22.5–25.5 °C (www.iucnredlist.org/details/21860/0).

Possible impacts of a changing climate on the Mediterranean go well beyond temperature increases and will likely affect far more than the bluefin tuna population. While there are many expected impacts, feedbacks, and interactions, some expected outcomes include:

- dramatic changes in ecosystem structure and function may occur. For example, unpalatable invasive aquatic plants favored by high water temperatures might replace valuable native species such as *Posidonia oceanica* (Fig. 2.3.11) and alter the structure of existing food webs;

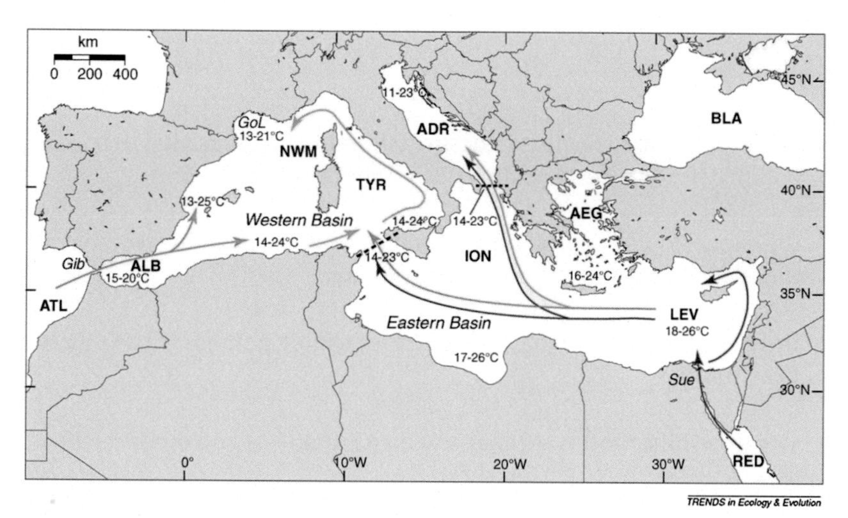

FIGURE 2.3.10
Geography of the Mediterranean Sea with the main routes of species range expansion. Bold capital abbreviations correspond to the main Mediterranean subregions. Abbreviations in italics correspond to some remarkable Mediterranean locations. Reported temperatures correspond to winter–summer mean sea surface temperatures. Arrows represent main routes of species range expansion according to their origin: Mediterranean natives (orange), Atlantic migrants (green), and Lessepsian migrants (red) (Lejeusne et al., 2010).

FIGURE 2.3.11
The threatened *Posidonia oceanica* meadows create critical habitat for hundreds of species, yet they are almost extinct in the eastern Mediterranean. *By Pino Bucca (*http://www.mondomarino.net*) [CC BY-SA 3.0], via Wikimedia Commons.*

- pathogens favoring warmer waters might threaten the health of marine species;
- sea level rise may disrupt or destroy shoreline habitats vital to species and may lead to greater salinity in estuaries;
- the expansion of marine phytoplankton species, including some toxic dinoflagellates, may degrade water quality;
- increased numbers of jellyfish may foul fishing gear and close tourist beaches; and
- Ocean acidification (OA) may threaten coral reefs and other calcifiers.

THE BIG PICTURE

This partial list of potential impacts of climate change on the Mediterranean Sea fits the pattern of many of our case studies: many different combinations of many different stressors threatening life in an impacted ecosystem. The existence of multiple stressors is very important to consider when trying to determine cause–effect relationships. While the impact of climate change on coastal ecology is linked, of course, to increased temperatures, it may also be accompanied by rising sea levels, increased salinity, increased numbers of warm-water invasive species and increased coastal erosion, to name a few.

Furthermore, while a strong correlation might be found between one event (e.g., *Posidonia* loss) and its cause (e.g., climate change), consideration of other factors might lead to the identification of additional causes (i.e., coastal development and increased turbidity and siltation).

CRITICAL THINKING

Exercise 8: To learn more about the potential effects of climate change on bluefin tuna, read the Muhling et al. article "Predicting the effects of climate change on Bluefin tuna (*T. thynnus*) spawning habitat in the Gulf of Mexico" found on your Resource Page.

- Summarize the likely implications of climate change for the bluefin tuna of the Mediterranean Sea.
- How might the negative effects on the bluefin tuna affect other members of the Mediterranean Sea food web?
- What steps might be taken to mitigate some of these impacts on the ecosystem?

The United Nations Environment Program (UNEP)'s Mediterranean Action Plan (UNEP, 2010) states that "It is important ... to carry out experimental studies to ascribe with some certainty a detected event to its putative causes. Comparisons among different areas of the Mediterranean might also lead to ... the identification of the real processes driving observed patterns. Because of the complexity of feedbacks and interactions between biotic and abiotic components of ecosystems, research is necessary to understand the impact of multiple stressors, many that likely act in synergy. Such understanding is necessary to ensure that any proposal of mitigation will prove effective."

PUTTING SCIENCE TO WORK

Exercise 9: Consider the research challenge posed by the UNEP program described above, and design one possible study to better understand the complex interactions of stressors at work in the Mediterranean.

Feel free to focus on a specific process or species, considering all of the interactions it has with its surrounding biotic and abiotic components of the Mediterranean Sea ecosystem.

In terms of the health and survival of bluefin tuna in the Mediterranean, the future is loaded with challenges: chemical, physical, and biological. However, there is no doubt that the Mediterranean Sea ecosystem will survive in some form. The broader question of interest to humans is what species will live there and how will the seemingly inevitable changes in food webs affect human users of the Mediterranean.

DRILLING DEEPER

Exercise 10: Many climate scientists believe that the future will see increased droughts in the Mediterranean Sea's watershed.

- Make a list of the ways that prolonged droughts might affect species like the bluefin tuna, being sure to explain the connections.

SOLUTIONS

Exercise 11: We've focused on three stressors that face Atlantic Bluefin tuna and the other species that live in the Mediterranean, but there are many others (e.g., oil pollution, sedimentation, habitat destruction, and even OA). Assume that the nations that lie in the Mediterranean watershed contribute money and hire you as Director of the Program to Protect the Mediterranean Sea. You're given an annual budget of $100 million euros.

1. Which of the current stressors that threaten the Mediterranean would you address first? Why? How would you address it?

Continued

SOLUTIONS—Cont'd

2. Which of the stressors facing the Mediterranean could be "fixed" most easily? Which would be most difficult? Explain.
3. What steps would you ask all the nations in the basin to take to help accomplish this task? Consider how likely nations would be to adopt each of these steps, along with the potential impact of the actions.
4. Are you optimistic that you could complete your task? Why or why not? What are the potential roadblocks to success? What might be the necessary catalyst for action?
5. How would you measure the success of the program? How long do you think it would take to see results?

CONSIDER THIS

Exercise 12: There are many seas around the world facing similar, if not more severe, environmental threats. How can you justify nonprofit and governmental organizations focusing their efforts to restore the Mediterranean in light of the worldwide need?

REFERENCES

Coll, M., Piroddi, C., Steenbeek, J., Kaschner, K., Lasram, F.B.R., Aguzzi, J., Ballesteros, E., Bianchi, C.N., Corbera, J., Dailianis, T., Danovaro, R., Estrada, M., Froglia, C., Galil, B.S., Gasol, J.M., Gertwagen, R., Gil, J., Guilhaumon, F., Kesner-Reyes, K., Kitsos, M.-S., Koukouras, A., Lampadariou, N., Laxamana, El, Lopez-Fe de la Cuadra, C.M., Lotze, H.K., Martin, D., Mouillot, D., Oro, D., Raicevich, S., Rius-Barile, J., Saiz-Salinas, J.I., Vicente, C.S., Somot, S., Templado, J., Turon, X., Vafidis, D., Villanueva, R., Voultsiadou, E., 2010. The biodiversity of the Mediterranean Sea: estimates, patterns, and threats. PLoS 5 (8), e11842.

IUCN, 2012. Marine Alien Invasive Species Strategy for the Medpan Network. Draft Strategy. www.medpannorth.org.

Lejeusne, C., Chevaldonne, P., Pergent-Martini, C., Boudouresque, C.F., Perez, T., 2010. Climate change effects on a miniature ocean: the highly diverse, highly impacted Mediterranean Sea. Trends Ecol. Evol. 25 (4), 250–260.

Pinnegar, J.K., Tomczak, M.T., Link, J.S., 2014. How to determine the likely indirect food-web consequences of a newly introduced non-native species: a worked example. Ecol. Model. 272, 379–387.

Renzi, M., Cau, A., Bianchi, N., Focardi, S.E., 2014. Levels of mercury and polychlorinated biphenyls in bluefin tuna from the western Mediterranean Sea: a food safety issue? J. Environ. Prot. 5 (2), 106–113. http://dx.doi.org/10.4236/jep.2014.52014.

Storelli, M.M., Giacominelli-Stuffler, R., Storelli, A., Marcotrigiano, G.O., 2005. Accumulation of mercury, cadmium, lead and arsenic in swordfish and bluefin tuna from the Mediterranean Sea: a comparative study. Marine Pollut. Bull 50 (9), 993–1018.

United Nations Environment Programme Mediterranean Action Plan, 2006. Pollution Data in the Med Sea Made Public-1976–2006: 30 Years of the Barcelona Convention. www.unepmap.org/index.php?module=news&action=detail&id=5.

United Nations Environment Programme Mediterranean Action Plan, 2010. Impact of Climate Change on Marine and Coastal Biodiversity in the Mediterranean Sea: Current State of Knowledge.

Chapter 2.4

Gulf of Mexico Dead Zone

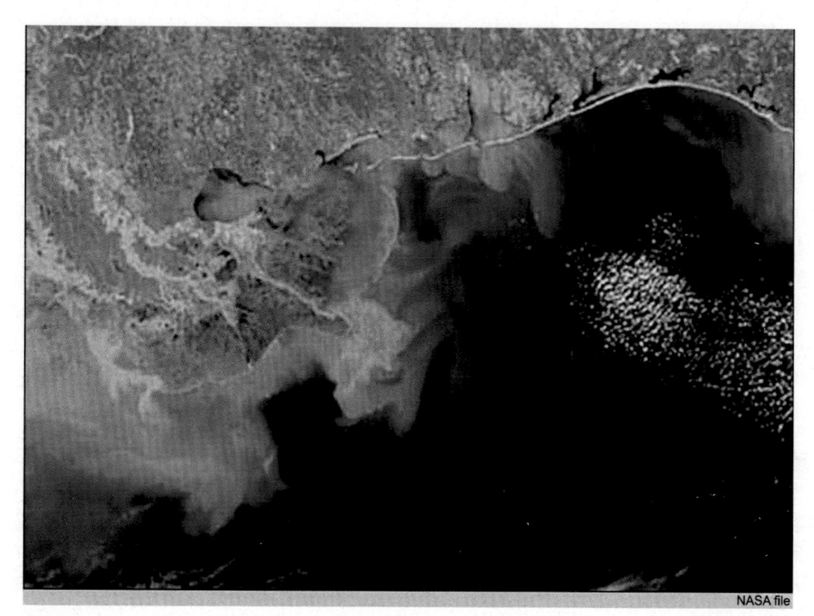

FIGURE 2.4.1
Satellite image of an algal bloom in the Gulf of Mexico. *Source: NASA.*

OVERVIEW

This case study examines a recurring problem that's plaguing the Gulf of Mexico (GOM): the annual formation of the Dead Zone, an area characterized by hypoxia, low DO levels in the water column that threaten Gulf marine life (Fig. 2.4.1). While there are many scientific aspects to consider, we'll focus on three of the most important: (1) the behavior of the key nutrient nitrogen in Mississippi River watershed soils, (2) the transport of nitrates through the Mississippi River to the Gulf, and (3) the impacts of elevated nitrate levels on the GOM ecosystem.

We'll conclude by looking at some options for reducing nitrate levels in the Mississippi. Along the way, we'll look in depth at research being done on various aspects of this recurring problem. Before you start, visit your Resource Page and follow the links for background on the Gulf.

BACKGROUND

The River and its Watershed: The Mississippi River watershed, draining more than 3.2 million km^2, or about 41% of the lower 48 US states (Strauss et al., 2011), is the third largest in the world. The watershed supports one of the globe's most productive farming regions, with about 58% of the basin in cropland.

Physical alterations of the river itself have also played a role in the formation of the Dead Zone. Over time, humans have modified the Mississippi for navigation. The river, 3700 km long, can be divided into two sections: the Upper Mississippi north of St. Louis, which is segmented into a series of navigational pools, and the open river to the south (Fig. 2.4.2). The upper river, featuring a lock and dam system created to facilitate navigation, has distinct habitats, including the main channel, side channels, impoundments, and backwaters. The role that these features play in Dead Zone formation will be discussed later.

THE PROBLEM

The GOM suffers from hypoxia (waters containing 2 mg/L DO or less) that occurs during warm months. The largest area of hypoxia to date occurred in the summer of 2002, when more than 21,755 km^2, roughly the size of Rhode Island and Connecticut combined, were affected (USGS, 2012). There are many studies investigating the impact of hypoxia on marine organisms. For example, O'Connor and Whitall (2007) found a negative correlation between brown shrimp catch and the severity and extent of the hypoxic zone, while Babcock and Kling (2008) estimated that up to 25% of brown shrimp habitat on the Louisiana shelf can be lost due to hypoxia, with serious implications for commercial fisheries. Craig and Crowder (2005) reported that Atlantic croaker, an important commercial fish species, moved to the edges of the hypoxic zone, likely to avoid the worst conditions. The impact of these episodes of hypoxia on the Gulf's food web can be devastating.

The Cause: A substantial percentage of all the fertilizers used in the US are applied in the Mississippi River watershed. Excessive amounts of nutrients, particularly nitrogen and phosphorus, move through the Mississippi into the GOM (Fig. 2.4.3). While scientists have recently suggested a role for phosphorus in Dead Zone formation, nitrogen has been pinpointed as the main trigger for the annual formation of this phenomenon.

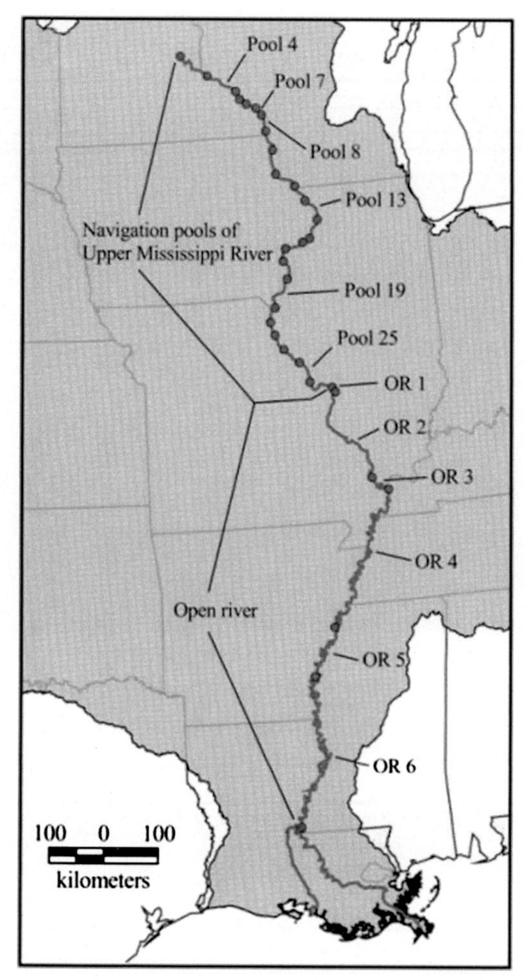

FIGURE 2.4.2
Morphology of the Mississippi River, including navigation pools (Strauss et al., 2011).

While there are many different sources of nitrogen in the Mississippi River watershed (Fig. 2.4.4), nonpoint sources have been identified as being particularly important, with 52% of the nitrogen coming from the use of fertilizers on corn and soybean crops alone (USGS, 2008). Note that nonpoint sources like crops and pasture dwarf point sources such as urban and population-related sources.

In addition to introducing large amounts of nitrogen through fertilizer use, humans have played another important role in enhancing nitrogen movement in the watershed. Large areas have been drained over the past century to make the land suitable for farming. Often, subsurface water from the saturated zone

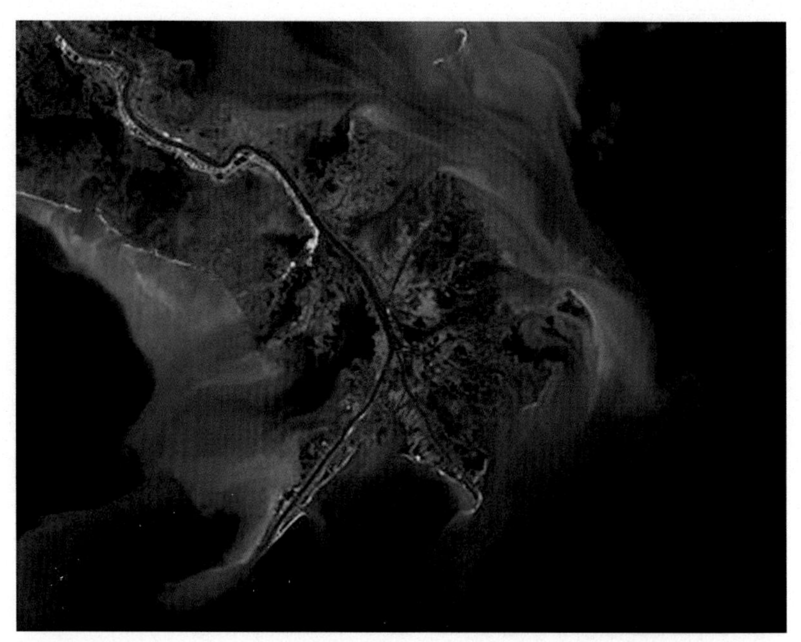

FIGURE 2.4.3
This 1999 NASA satellite image shows sediments deposited at the Mississippi River delta. These deposits include topsoil carried by farm runoff, which also transports fertilizers, which help deplete the surrounding water of oxygen. *Source: NASA.*

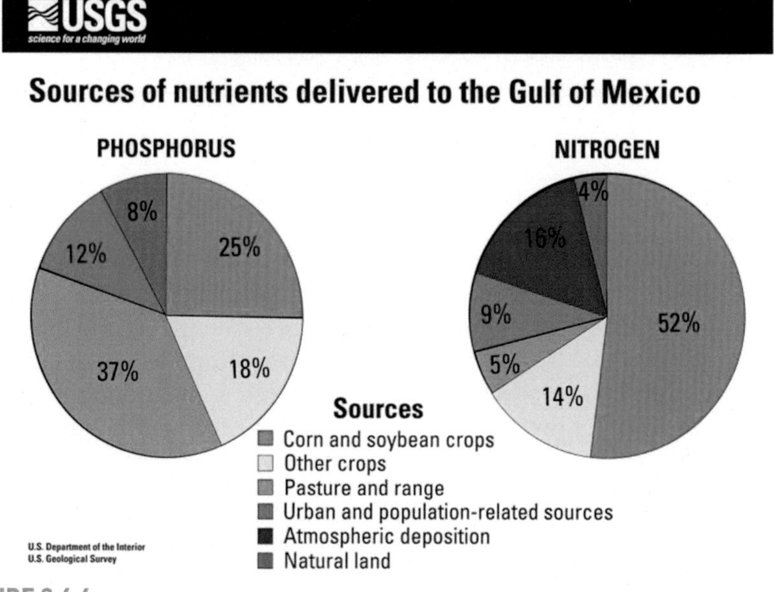

FIGURE 2.4.4
Estimated nitrogen contribution to the Gulf of Mexico from specific sources. *Source: USGS.*

of soils is drained into underground tiles, which may then discharge directly into adjacent streams, providing a direct route for any pollutants present.

DRILLING DEEPER

Exercise 1: Why are nonpoint sources like fertilizer applications so much harder to control than point sources like municipal wastewater discharges?

CRUNCHING THE NUMBERS

Exercise 2: Many scientists hypothesize that there is a strong connection between the amount of land area that supports active agriculture and surface water quality. Fig. 2.4.5 suggests that such a relationship exists at a large scale, but let's look at some data. We will test this relationship for several subwatersheds by comparing the percent of land in active agriculture to the nitrate concentrations measured at the subwatershed's outlet.

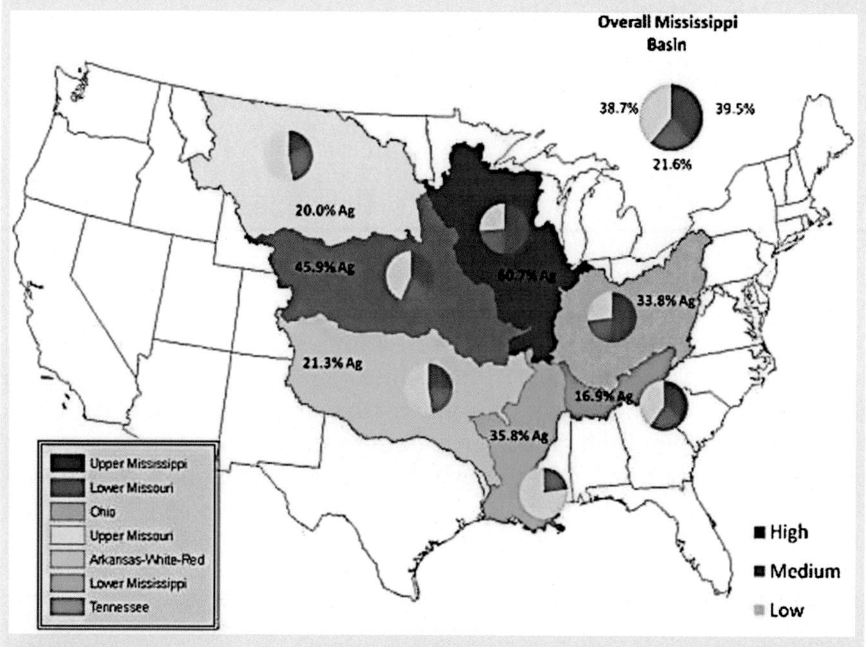

FIGURE 2.4.5
Percent of streams with high, medium, and low levels of nitrate and percent of land area in agricultural use. *Source: US EPA (2011).*

Continued

CRUNCHING THE NUMBERS—Cont'd

Using the data in the Ex2 tab of the **GOM_case_study_data.xlsx** file downloadable from your Resource Page, we will first create and examine a scatter plot of nitrate versus percent agriculture for the subwatersheds of the Mississippi.

- Do you notice any general pattern? How would you describe it? What does it suggest about the relationship between agriculture and nitrate in surface waters?
- Are there any unusual observations you would like to know more about (any extreme watersheds or data points that don't fit the overall pattern)?

Now we will quantify the strength and significance of this relationship.

- Run a Pearson's correlation test. Is there a significant relationship between agriculture and stream nitrate concentration? How strong is this relationship? Do you think it is ecologically meaningful? Explain.

IN-DEPTH: NITRATES AND THE GULF OF MEXICO: A SYSTEMS CONNECTION

In the following sections, we'll take an in-depth look at how nitrates move through the Mississippi River watershed and how this movement has affected the GOM food web.

The Chemistry of Nitrogen in Soils: How Does it Move so Easily from Field to River?

Sources: a variety of point and nonpoint sources release nitrogen in the Mississippi River watershed (Fig. 2.4.5). Data from recent decades show several interesting trends: (1) municipal and industrial point sources, while present in the watershed, are dwarfed by agricultural sources, particularly fertilizers; and (2) while inputs from most sources of nitrogen within the basin remained relatively constant from 1960 through the late 1990s, inputs from fertilizers during that time period rose dramatically. Rates of fertilizer application since the 1990s have remained fairly stable (Donner and Scavia, 2007) (Fig. 2.4.6).

CRUNCHING THE NUMBERS

Exercise 3: Research shows that inputs of nitrogen from fertilizer use in the Mississippi River watershed have remained fairly constant since the 1990s. Let's investigate whether or not surface waters have followed a similar trend, or if there are other factors that may be impacting nitrate concentrations in surface waters.

In Exercise 2, we used data in the **GOM_case_study_data.xlsx** file, downloadable from your Resource Page, to plot a time series of fertilizer application rates.

CRUNCHING THE NUMBERS—Cont'd

- Using the same data file, overlay a time series of surface water nitrate concentrations to compare the two trends.
- How closely does the time series for nitrate follow the time series for fertilizer application rates? Describe the patterns you see.
- Are there other factors that might be contributing to the variability you see in the time series for nitrate (for example, why does the concentration differ between 1985 and 1995, while fertilizer application rates remained constant)? Based on your knowledge, what might those factors be?

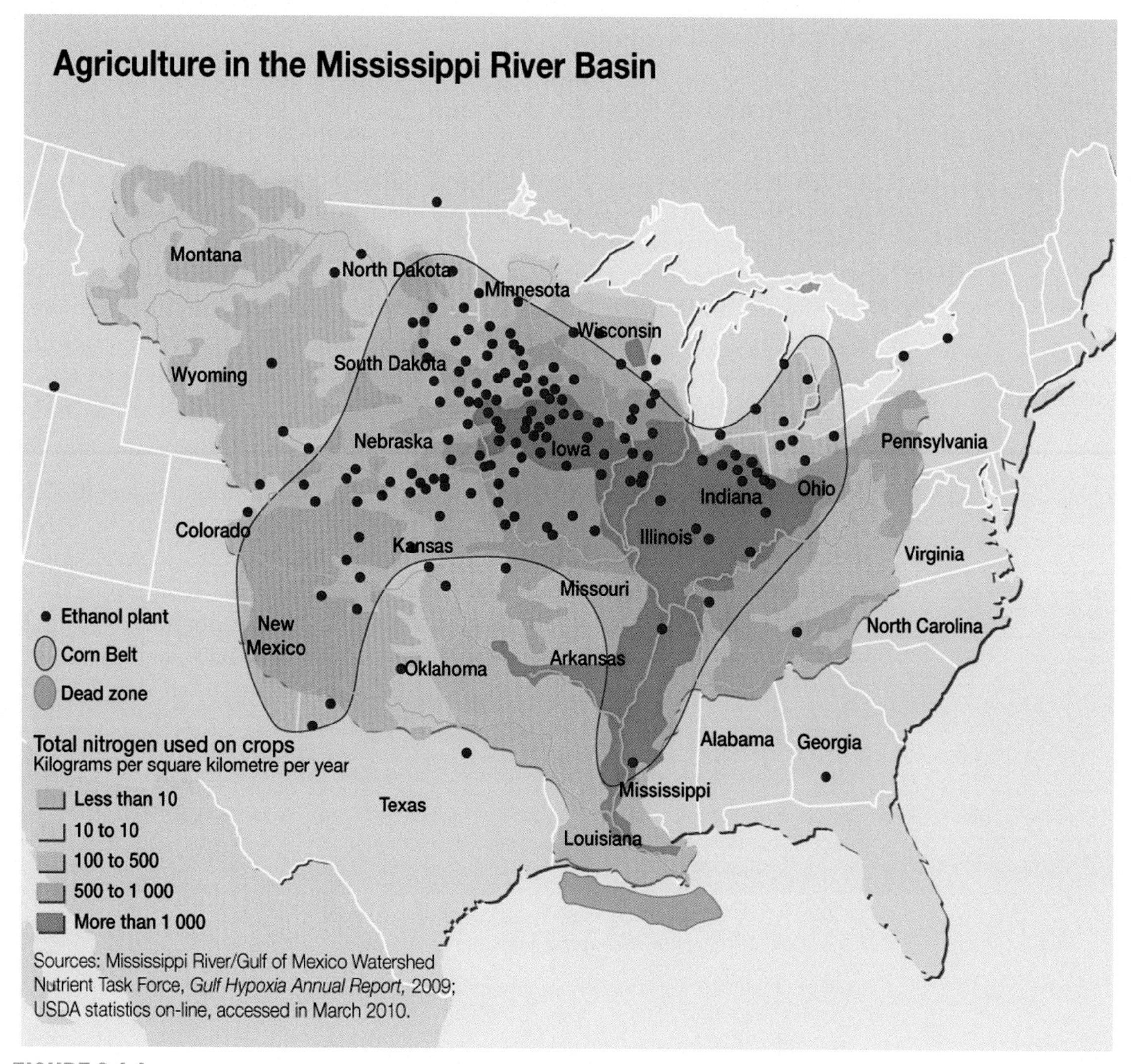

FIGURE 2.4.6

Nitrogen fertilizer application rates throughout the Mississippi Basin. *Source: Riccardo Pravettoni, UNEP/GRID-Arendal,* http://www.grida.no/graphicslib/detail/agriculture-in-the-mississippi-river-basin_45d0#.

Fate: From your review of biogeochemical cycles, you know that nitrogen is a very reactive element in the environment. Once atmospheric nitrogen is fixed by bacteria in soils, ammonia can be converted by *Nitrosomonas* and *Nitrobacter* bacteria into nitrate. Because most soil particles, organic matter, and nitrates are negatively charged, and like charges repel, nitrates tend to remain in solution where they can be easily transported by soil water into nearby ground and surface waters.

The hydrology of a particular watershed is an important factor in determining the movement of nitrogen. In those watersheds where water moves mostly over land, in very shallow subsurface flow, or through tile drains designed to remove water from fields, nitrates can move rapidly, particularly during wet weather, into adjacent streams.

Environmental Science in Action

Scientists have assessed the mobility of nitrates in different types of watersheds. Sudduth et al. (2013) examined data from 62 watersheds, including both undisturbed forested watersheds and those with substantial agricultural or urban development. They found a positive correlation between nitrate levels in soil water and those in streams draining those soils. The analysis also determined that undisturbed watersheds removed about half the soil nitrates, while a greater fraction of the nitrates reached adjacent streams in developed watersheds. Thus, the greater the degree of disturbance in a watershed, the more likely that nitrates in the soils will move to downstream areas.

With millions of tons of nitrogen inputs from fertilizer and the high solubility of nitrate in water, it should be no surprise that problems can result. While nitrates are readily available to support plant growth, they can also pose serious environmental threats. Not only can excessive nitrogen fuel algal blooms and subsequent drops in DO in receiving waters, nitrate contamination of groundwater supplies can lead to methemoglobinemia, or "blue baby syndrome."

EXPLORE IT

Exercise 4: Satellite imagery offers us the opportunity to examine land use patterns across a larger landscape. This provides a perspective often missed in field studies.

Let's explore the Mississippi and its subwatersheds in Google Earth to see how land use differs across the region and how that impacts surface water quality. First you will need to download and open the **USWatersheds.kml** file from your Resource Page.

Notice that this now adds a new layer to your Temporary Places in Google Earth called Major US Watersheds and opens up a shape file with watershed boundaries outlined in red with a labeled green icon at the center of each watershed.

EXPLORE IT—Cont'd

In the Places sidebar, click on the triangle to the left of the "Major US Watersheds" layer to expand the list of all watershed polygons included in the layer.

Scroll down the list and click on the Missouri Watershed. In the text box that opens, click "View the Missouri River Watershed." This will zoom Google Earth in to the level of the watershed and create a new layer in your Places sidebar.

Use the triangle next to the new "Missouri Watershed" layer to expand the list. Notice that you now have options to view the watershed characteristics or the watershed layers. Similarly, click on the triangle to expand the watershed layers.

- What information is available to you under the watershed layers for the Missouri River?

Turn off all of the Layer options except for Land Cover. Use the triangle to expand the list for Land Cover. Notice that you have two different dates available to you, 1992 and 2001, with descriptions of whether a location is covered by forest, crops, grassland, water, urban, etc. Turn on only the 1992 National Land Cover Database (NLCD) data and its corresponding legend.

- How would you describe the land cover for the Missouri River's watershed?

Toggle off and on the 2001 NLCD layer as well.

- How has land cover changed over this 9-year interval?

Click on the "Watershed Characteristics" layer. This opens a new window that quantifies the percent of land cover for each class.

- What is the dominant land cover type for the Missouri River watershed?
- What percent of the watershed is in cropland/pasture?

Following this same process, view the Red River watershed and open its Watershed Characteristics.

- How does the Red River watershed differ from the Missouri (in particular, consider how the percent of the major land cover types differs)?

Now we will see if there are any Geographic Information Systems (GIS) data layers that describe the geographic patterns in surface water quality. In Google Earth, use the Earth Gallery to search for "Mississippi Watershed."

Select the "Nitrogen Pollution in the Mississippi River Basin" layer and click on the "View in Google Earth" toggle. A new layer will be added to your Layers list. This data layer shows the proportion of streams in each subwatershed that falls into Low, Moderate, High, Very High, or Extremely High categories of Nitrogen Pollution Rank.

- How does the proportion of Extremely High watersheds in the Missouri River watershed compare to the proportion of Extremely High watersheds in the Red River watershed?

Turn off this data layer so that you can explore the satellite imagery base map for each watershed (e.g., Fig. 2.4.7). Zoom in to examine the types and patterns of crop/pasture activities in these two watersheds. Consider differences that might explain why nitrogen pollution is higher in the Missouri than in the Red River watershed.

- Considering that these two watersheds have roughly the same amount of land in cropland/pasture, why might these differences exist? Feel free to toggle these data layers on and off so that you can explore the satellite imagery to visually assess how these watersheds differ.

Continued

EXPLORE IT—Cont'd

FIGURE 2.4.7
A snapshot of land cover in the Missouri (left) and Red River watersheds (right). *Source: Google Earth.*

How Does Nitrogen Move Through the Mississippi River Into the Gulf of Mexico?

The Mississippi River is a very efficient conduit for nitrogen. The annual transport of nitrogen by the river increased at a rate of about 19,000 tons per year from 1955 to 1999. Most of the increase occurred between 1970 and 1983 (USGS, 2000). Because of the high solubility of nitrates in water, only 5 to 20% of the nitrogen entering the system is estimated to be retained by the river. In fact, Alexander et al. (2000) estimated that more than 90% of the nitrogen entering the Mississippi River makes it to the GOM.

One mechanism that results in some loss of nitrate from rivers is denitrification, during which nitrate in river water and sediments is converted by bacteria to atmospheric nitrogen (N_2) and nitrous oxide (N_2O), which readily move from the water into the atmosphere. Richardson et al. (2004) measured denitrification in the upper Mississippi River and found that it was highest during the spring and summer and that the greatest losses of nitrate came from backwater lakes and impounded areas created by the lock and dam system. The

CRUNCHING THE NUMBERS

Exercise 5: Denitrification can occur in many different habitat types. Below are data from Strauss et al. who measured rates of denitrification in various habitats along the Mississippi River. A digital version is available for download from your Resource Page as **GOM_case_study data.xls,** Ex5 tab. For mean denitrification rates, see Fig. 2.4.8.

CRUNCHING THE NUMBERS—Cont'd

	Denitrification (g N m^{-2} d^{-1})			
	Backwater	Impounded	Main channel	Side channel
Winter	0.143 (0.055)	0.251 (0.063)	0.109 (0.043)	0.085 (0.019)
Spring	0.325 (0.032)	0.366 (0.048)	0.074 (0.018)	0.142 (0.051)
Summer	0.398 (0.042)	0.457 (0.038)	0.133 (0.035)	0.189 (0.059)
Autumn	0.224 (0.023)	0.215 (0.022)	0.094 (0.027)	0.101 (0.018)

FIGURE 2.4.8

Mean denitrification rates for various water bodies along the Mississippi River (standard deviation in italics). *Source: (Strauss et al., 2011).*

- Using these data, plot seasonal denitrification rates for backwater and main channel habitats. What might explain the differences in seasonal denitrification rates between these two habitat types?
- Are there significant differences among seasons?
- Assuming that the averages given in the table accurately represent each 3-month period and that the total area of the side channels is 1.4 km^2, calculate the total kg of N released from the side channels annually.

total nitrate lost, however, via denitrification was estimated to be only 7% of the total entering the river.

Role of River Modifications: Humans have modified the Mississippi River. Large-scale engineering projects have included construction of levees, closing of dams, and sediment dredging. In the 1930s, a lock and dam system designed to facilitate navigation created open water impoundments, which inundated portions of the floodplain.

These modifications of the river's channel and the separation of the river from its floodplain, including wetlands, have decreased the river's ability to retain nitrogen at the very time that increased loadings of nitrate were occurring. By reducing the velocity of flowing water, natural floodplain features like wetlands can promote both physical removal of pollutants by settling and uptake by primary producers. Improved river management and restoration could increase habitat diversity, reconnect the river to portions of its historic floodplain, and thus increase the river's ability to retain more nitrogen (Strauss et al., 2011).

DRILLING DEEPER

Exercise 6: In addition to slowing the velocity of water and promoting biological uptake, what are some of the other ways that healthy floodplains can enhance pollutant removal?

The story for other pollutants is different: Unlike nitrates, which are soluble and move readily through rivers like the Mississippi, phosphorus and many toxic pollutants, like PCBs, are much less soluble and readily adsorb onto solids. They tend to accumulate in sediments behind dams, in pools, and in other areas where deposition is favored. To learn more, go to www.nywea.org/clearwaters/pre02fall/321070.html to discover what happens to PCBs in the Hudson River.

DRILLING DEEPER

Exercise 7: Given the different behaviors of nitrogen and PCBs in rivers, which is harder to manage? Why?

What Happens When Nitrates Enter the Gulf?

Once the nitrates are discharged into the Gulf, the cycle is completed. Just as nitrogen fertilizers enhance the growth of crops in the Upper Midwest, they perform the same function in the Gulf: they stimulate growth, but this time of algae.

DRILLING DEEPER

Exercise 8: Review the role of limiting factors in freshwater versus marine ecosystems. Why is it nitrogen and not phosphorus that limits primary production in the Gulf?

As algal populations in the upper layers of the Gulf's water column increase in response to elevated nitrate levels and these cells begin to die, organic matter levels in the water increase as the cells sink toward the bottom. Increased levels of organic matter in the lower half of the water column promote the growth of bacteria, which consume increasing amounts of oxygen. Additional organic matter is introduced as fecal pellets from zooplankton and other primary consumers feeding on the algae settle to the bottom (Fig. 2.4.9).

In the warmer months of the year, water holds less oxygen, so microbial consumption is soon followed by decreasing DO levels in Gulf waters. While there is substantial annual variability in the extent of midsummer bottom water hypoxia, the areal extent of hypoxia has exceeded 15,000 km^2 on a number of occasions.

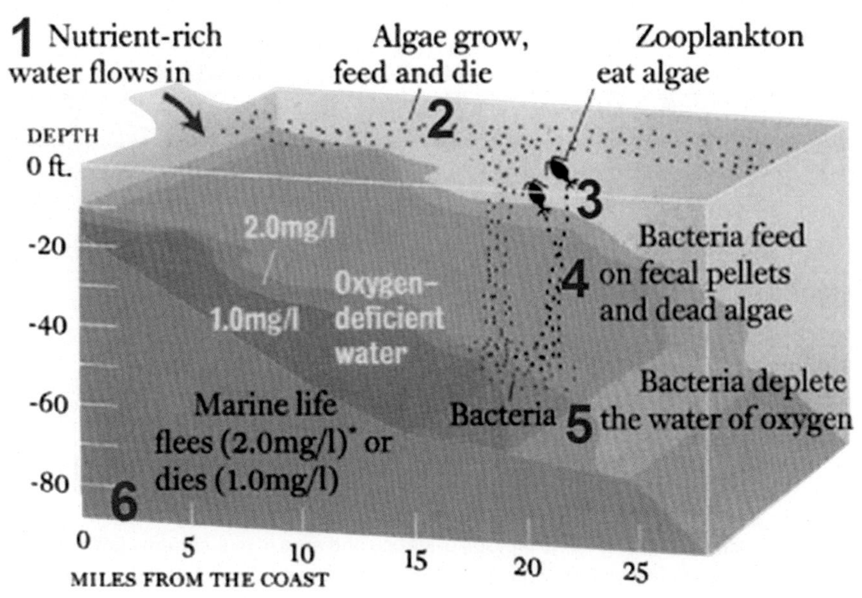

FIGURE 2.4.9

Chain of events leading to the occurrence of Gulf of Mexico hypoxia. *Source:* www.gulfhypoxia.net/.

SYSTEMS CONNECTIONS

Exercise 9: The two figures below display the variability of hypoxic events in GOM. Fig. 2.4.10 demonstrates year-to-year variability, while the Fig. 2.4.11 demonstrates the spatial extent of hypoxia in 2010. Examine these closely and use your knowledge of the environmental processes at work to determine the potential drivers of both temporal and spatial variability in hypoxic events.

- Why might 2001 have had such a low dissolved nitrate flux and limited hypoxic areas compared to other years?
- What characteristics of the GOM help determine where areas of high and low DO occur?

Continued

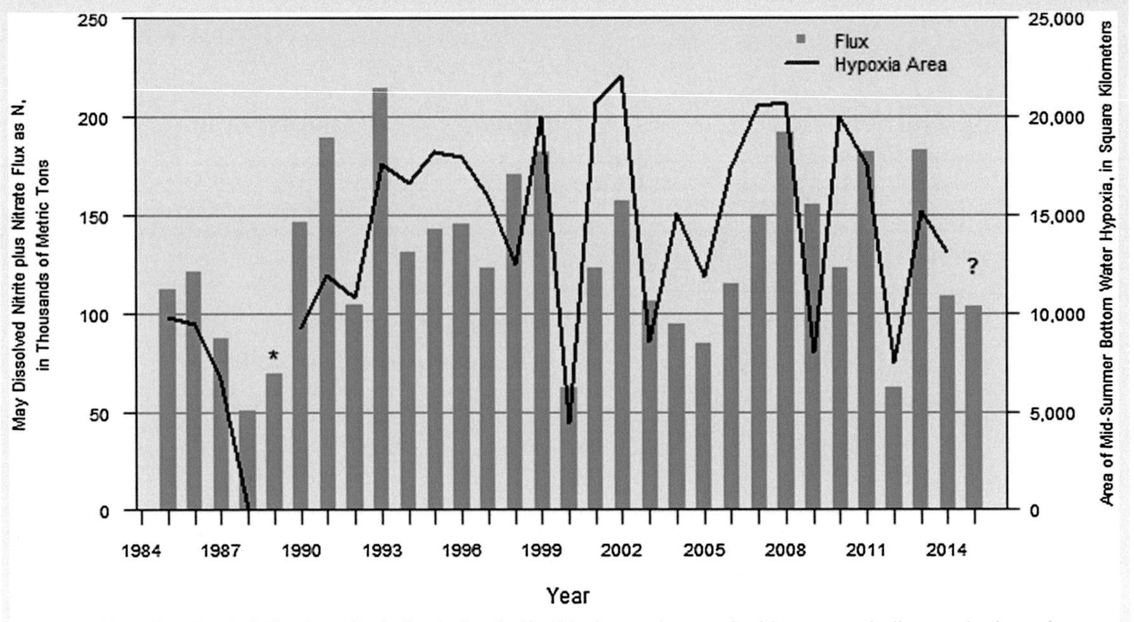

FIGURE 2.4.10

Area of midsummer bottom water hypoxia plus nitrate flux to the Gulf of Mexico, 1985–2014. *Source: USGS, 2014. Preliminary Mississippi–Atchafalaya River Basin Flux Estimate.*

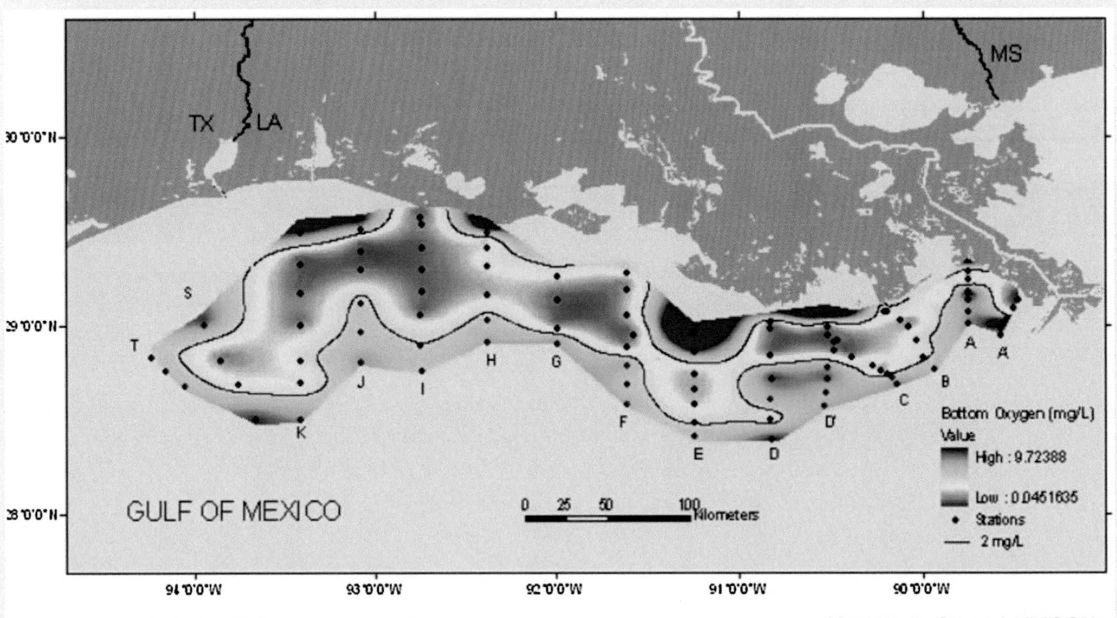

FIGURE 2.4.11

Map of ocean floor dissolved oxygen (2010). *Source: N. Rabalais, LUMCOM,* gulfhypoxia.net.

Most marine life avoids areas with 2 mg/L DO or less, while levels of 1 mg/L DO or less can be lethal. Some species, such as starfish and crabs, have limited mobility and may succumb to low DO levels in the Dead Zone. Other species, like jellyfish, are more tolerant of low DO levels and may increase their populations to nuisance levels as competitors die off.

Environmental Sciences in Action

Concerns about the Dead Zone go well beyond the immediate effects of low DO levels. Research scientists P. Thomas and M.S. Rahman, marine biologists at the University of Texas Marine Science Institute, studied a more subtle effect of Gulf hypoxia. A comparison of reproductive parameters in Atlantic croakers collected from the hypoxic zone and a control area in 2007 showed widespread reproductive disruption, including reduced gonadal growth and gamete production, in fish from the hypoxic zone. Male germ cells were detected in about 19% of croaker ovaries in specimens from the hypoxic zone (Thomas and Rahman, 2012).

Laboratory experiments conducted by the scientists yielded similar results, showing that as few as 10 weeks of exposure to low DO conditions could cause masculinization of female croakers. The scientists concluded that ovarian masculinization was a specific response to hypoxic conditions and that severe reproductive impairment might occur in marine fish populations exposed to seasonal hypoxia. The approach used in this research underscores an important point: verifying field observations with well-planned and executed laboratory experiments is critical when attempting to establish cause–effect relationships.

CRITICAL THINKING

Exercise 10: We've covered some of the effects of hypoxia on the GOM, but the Gulf suffers from other stressors as well. Do some research on the 2010 BP oil spill in the Gulf.

- How do the impacts of the oil spill compare to the impacts of the annual episodes of hypoxia?
- Which do you believe has been the more serious threat? Defend your choice.

THE BIG PICTURE

Probably no one has been a more visible symbol of the research effort on the GOM Dead Zone than Dr. Nancy Rabalais, Executive Director and Professor at the Louisiana University Marine Consortium. To see her research team in action in the Gulf, watch the video at the link provided on your Resource Page (Figs. 2.4.12 and 2.4.13).

FIGURE 2.4.12
Dr. Nancy Rabalais studying low levels of dissolved oxygen in the waters of the Gulf of Mexico.

> "There are more and more people trying to reduce their carbon footprint. That is the mantra of the day with regard to global warming. But there's also a nitrogen footprint that has a more immediate effect on our daily lives – in terms of water quality, food availability, sustainable agriculture, and daily subsistence. While our carbon footprint is making a not so immediate, but profound mark on the existence of human life on the Earth's ecosystem, the nitrogen footprint is there, lurking in the background, with an equal potential to disturb the balance of a global ecosystem." - Nancy Rabalais, Ph.D.
>
> *https://gustavus.edu/events/nobelconference/2009/rabalais.php*

FIGURE 2.4.13
A quote from Dr. Rabalais' talk at the Nobel Conference.

SOLUTIONS

The concept of cause–effect is important in environmental sciences. Being able to tie a particular effect (in this case, formation of the Gulf Dead Zone) to a specific cause (in this case, nitrogen) is critical when considering control strategies. Elsewhere in this book, you'll see how difficult it can be to establish cause–effect when considering the effects of air pollution on human health. In the present case, however, there seems to be little doubt that excessive nitrate levels contribute to hypoxia in the GOM. Unfortunately, establishing a cause–effect relationship doesn't necessarily lead to an easy solution, as is the case with the Dead Zone.

We're left with three possible approaches when trying to manage Dead Zone formation: (1) controlling nitrates at their source within the drainage basin, (2) removing nitrates from the river itself, or (3) minimizing impacts in the GOM.

There is no obvious way to attack the problem once nitrates reach the GOM. Also, while river restoration, including reconnecting the main channel to its floodplain and increasing the number of wetlands, may result in some increased retention of nitrates in the Mississippi, this seems unlikely to occur on a broad scale, given the probable costs of such a massive effort. While also a most daunting task, the real solution is to reduce the movement of nitrates into the system in the first place.

Just how challenging this effort will be is underscored by an announcement in early 2015 by a federal–state Hypoxia Task Force that it would not meet its 2015 goal of reducing the GOM Dead Zone from about 15,000 to 5000 km^2 and that they were extending their target date for the Dead Zone reduction to 2035. The 5-year average size of the Dead Zone has changed little since 1994 (WEF, 2015), making it clear that additional steps will need to be taken to reach the nutrient reduction goal of 20% by 2025.

David (2012) identified several important hurdles facing the reduction efforts, including the major role of tile drains in the Corn Belt, with, in some cases, 100% nitrogen export from tiled fields. Additional factors include warmer winters and wetter springs that lead to longer growing seasons and more nitrate runoff, as well as increased acreage devoted to growing corn for biofuels production in recent years.

Despite these challenges, over the past decade, numerous basin-wide voluntary conservation efforts have been started in the Upper Mississippi River watershed to reduce the movement of pollutants into the river. The following are some of the management options that are being used to reduce nitrate movement into the river:

- Preserving and enhancing buffer strips along streams and rivers, using grassed waterways, and planting field borders: These steps can trap some of the overland flow of nitrates.

- Removing tile drains: This reduces subsurface discharges to waterways in areas where drains have been installed.
- Conservation measures: Crop rotation, conservation tillage, and other soil and crop conservation practices can keep more of the nitrogen in the soil and out of waterways.
- Changing practices: Banning fall application of nitrogen fertilizers and manure spreading on frozen ground, and minimizing access of livestock to streams can reduce nitrogen inputs into waterways.
- Financial disincentives: Charging a "fertilizer tax" can encourage more judicious use of the chemicals.

Comparing some options: Wu and Tanaka (2005) used a model to compare the effectiveness of three conservation practices (conservation tillage, cropland retirement, and corn–soybean rotation) to the use of a fertilizer tax as ways to reduce nitrogen runoff in the Upper Mississippi River watershed. While paying farmers to use conservation tillage was the most cost-effective conservation practice, it reduced nitrogen output by only 37%. The fertilizer tax, while less popular politically, was found to be the most effective way to reduce nitrogen losses from the fields, resulting in up to a 70% reduction.

Finding ways for farmers to reduce their use of nitrogen fertilizer not only benefits the environment, but the farmers themselves. One estimate from the Environmental Working Group (2009) stated that "...more than $391 million worth of nitrogen fertilizer is flushed down the Mississippi River in the spring of each year."

Innovative Programs: Several large-scale efforts to improve water quality in the Upper Mississippi River watershed are underway.

1. The Iowa Initiative is targeting restoration of wetlands in "highly strategic locations" within subwatersheds to intercept tile drainage water. This approach focuses efforts on the often small portion of agricultural land in a watershed that is responsible for most of the problems. By focusing on these small areas, scientists hope to reduce nitrate losses by 40–70% (Environmental Working Group, 2009). For more information, visit the Iowa Initiative website linked from your Resource Page at http://www.ewg.org/news/testimony-official-correspondence/dead-zone-action-needed-ewg-remarks-hypoxia-task-force.

2. The USDA's Natural Resources Conservation Service's Mississippi River Basin Healthy Watersheds Initiative is a watershed-wide program involving 13 states. Federal scientists are working with farmers to employ conservation practices to reduce nutrient loading throughout the basin. For more information, see the Mississippi River Basin Healthy Watersheds Initiative website at http://www.nrcs.usda.gov/wps/portal/nrcs/detailfull/national/home/?cid=STELPRDB1048200.

Biofuels the Answer?

Recent research by Smith et al. (2013) suggests that conversion of row crop agriculture, such as corn, to perennial biofuel crops like maiden grass, switch grass, and mixed prairie may substantially reduce the leaching of nitrates from agricultural fields. In addition, plantings of both maiden grass and mixed prairie can reduce the release of nitrous oxide, a greenhouse gas. If the biofuel

SOLUTIONS

Exercise 11: In tackling environmental issues, environmental scientists often find themselves working with groups of people with different perspectives.

Assume that you're a US Environmental Protection Agency (EPA) scientist charged with reducing inputs of nitrogen fertilizers from farms in an Iowa county lying in the Mississippi River watershed. Form a group of four students representing the following four perspectives: a farmer who uses nitrate fertilizers; a member of River Watch, a citizen's group trying to protect the river; you, the representative from the local US EPA office; and an average citizen living in the county. Your group is charged with coming up with a solution designed to protect the river.

- What is your group's recommendation?
- How did you resolve conflicts and disagreements among group members?
- What sort of data did you need to help you make your decisions?

solution focused on cellulosic sources, rather than corn-based ethanol, we could help alleviate several environmental concerns at the same time.

In this case study, we investigated how a pollutant (nitrates) introduced into a system can have serious consequences thousands of kilometers away. The problem, of course, in this case is that the "source" is really thousands of individual nitrate-generating activities spread over a wide area. This represents a real challenge for regulators trying to develop effective control strategies.

SYSTEMS CONNECTIONS

Exercise 12: When considering the impact of any particular human activity, such as increasing the use of nitrogen-containing fertilizers to grow corn for ethanol in the Mississippi River watershed, always try to anticipate the "systems connections" and what distant impacts may occur. There are some interesting relationships between nitrogen fertilizer use, climate change, and the GOM Dead Zone formation.
Using your systems thinking, discuss one way in which the use of nitrogen fertilizers affects global climate change and two ways that climate change can affect the Dead Zone.

CONSIDER THIS

Exercise 13: Maintaining a supply of food is critical, considering the shrinkage of the agricultural land base and growth of the human population.

- What responsibility do we have to weigh the benefits of increased yields of crops like corn and rice versus the environmental costs of such increases?
- What factors must be included in this assessment?
- Who should make such decisions?

REFERENCES

Alexander, R.B., Smith, R.A., Schwarz, G.E., 2000. Effect of stream channel size on the delivery of nitrogen to the Gulf of Mexico. Nature 403, 758–761.

Babcock, B.A., Kling, C.L., 2008. Costs and Benefits of Fixing Gulf Hypoxia. Iowa Ag. Review 14 (4), Article 4.

Craig, J.K., Crowder, L.B., 2005. Hypoxia-induced habitat shifts and energetic consequences in Atlantic croaker and brown shrimp on the Gulf of Mexico shelf. Mar. Ecol. Prog. Ser. 294, 79–94.

David, M.B., 2012. Overview-Nutrient Fate and Transport. Presented at Building Science Assessments for State-Level Nutrient Reduction Strategies.

Donner, S.D., Scavia, D., 2007. How climate controls the flux of nitrogen by the Mississippi River and the development of hypoxia in the Gulf of Mexico. Limnol. Oceangr. 52 (2), 856–861.

Environmental Working Group, 2009. Dead Zone Action Needed: EWG Remarks to Hypoxia Task Force. 7 p.

Louisiana Marine Sciences Consortium, 2008, 2010. Hypoxia in the Northern Gulf of Mexico. www.gulfhypoxia.net/.

O'Connor, T., Whitall, D., 2007. Linking hypoxia to shrimp catch in the northern Gulf of Mexico. Marine Pollut. Bull. 54 (4), 460–463.

Richardson, W.B., Strauss, E.A., Bartsch, L.A., Monroe, E.M., Cavanaugh, J.C., Vingum, L., Soballe, D.M., 2004. Denitrification in the Upper Mississippi River: rates, controls, and contribution to nitrate flux. Can. J. Fish. Aquat. Sci. 61, 1102–1112.

Smith, C.M., David, M.B., Mitchell, C.A., Masters, M.D., Anderson-Teixeira, K.J., Bernacchi, C.J., DeLucia, E.H., 2013. Reduced nitrogen losses after conversion of row crop agriculture to perennial biofuel crops. J. Environ. Qual. 42 (1), 219–228.

Strauss, E.A., Richardson, W.B., Bartsch, L.A., Cavanaugh, J.C., 2011. Effect of habitat type on in-stream nitrogen loss in the Mississippi River. River Syst. 19 (3), 261–269.

Sudduth, E.B., Perakis, S.S., Bernhardt, E.S., 2013. Nitrate in watersheds-straight from soils to streams? J. Geophys. Res. Biogeosci. 118 (G1), 291–302.

Thomas, P., Rahman, M.S., 2012. Extensive reproductive disruption, ovarian masculinization and aromatase suppression in Atlantic croaker in the northern Gulf of Mexico hypoxic zone. Proc. Biol. Sci. 279 (1726), 28–38.

US Environmental Protection Agency, 2011. Nitrogen and Phosphorus Pollution in the Mississippi River Basin: Findings of the Wadeable Streams Assessment. EPA 841-F-11–004.

U.S. Geological Survey, 2000. Nitrogen in the Mississippi Basin-Estimating Sources and Predicting Flux to the Gulf of Mexico. USGS Fact Sheet 135-00, 6 p.

U.S. Geological Survey, 2008. Differences in Phosphorus and Nitrogen Delivery to the Gulf of Mexico from the Mississippi River Basin. NAWQA Program.

U.S. Geological Survey, 2012. Dead Zone: The Source of the Gulf of Mexico's Hypoxia. Science Features (Posted on June 21, 2012 by A. Massefski and K. Capelli).

Water Environment Federation, 2015. Stormwater Report: Hypoxia Task Force Extends Deadline, Plans New Strategies for Reducing Gulf of Mexico Dead Zone. http://stormwater.wef.org/2015/03/hypoxia-task-force-extends-deadline-plans-new-strategies-reducing-gulf-mexico-dead-zone/.

Wu, J., Tanaka, K., 2005. Reducing nitrogen runoff from the Upper Mississippi River Basin to control hypoxia in the Gulf of Mexico: easements or taxes? Mar. Resour. Econ. 20, 121–144.

Chapter 2.5

Restoration: A Tale of Two Rivers

FIGURE 2.5.1
The Thames River running through London. *By Konstantin Papushin [CC BY-SA 2.0, via Wikimedia Commons].*

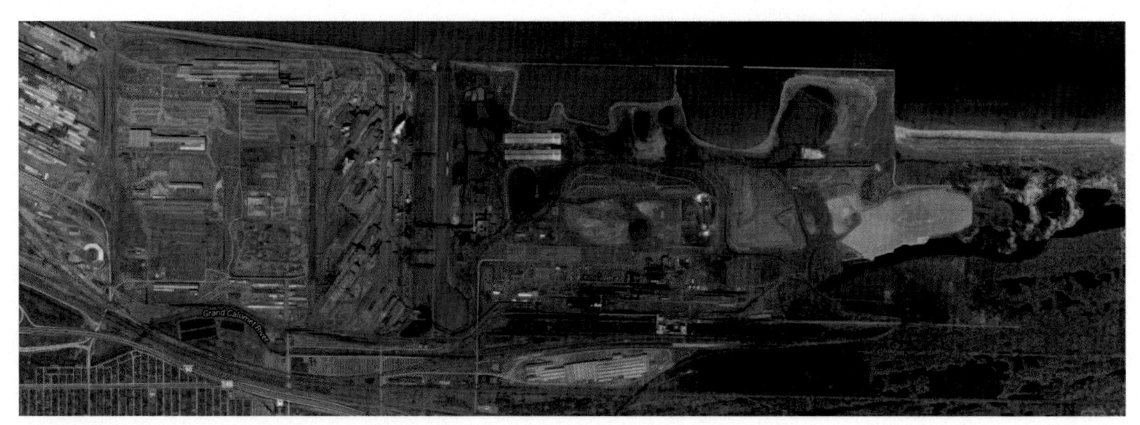

FIGURE 2.5.2
A Google Earth snapshot of the highly industrialized Grand Calumet River in Indiana.

OVERVIEW

While many of the world's rivers have been badly polluted, there have been a number of instances where such systems have been substantially cleaned up or "restored," often with sensitive species like salmon returning where once there were none. Unlike lakes, which tend to accumulate toxins and other pollutants, rivers, particularly those with substantial flow, may be able to cleanse themselves if pollution sources can be cleaned up.

We'll consider two very different examples of river restoration: (1) the River Thames (Fig. 2.5.1), which experienced one of the earliest major restoration efforts and has been held up globally as a success story; and (2) the Grand Calumet River (Fig. 2.5.2) and Indiana Harbor Ship Canal (GCR/IHSC) in northwestern Indiana (US). Although cleanup efforts here are well underway, the GCR/IHSC is a system still suffering from years of chemical pollution.

DRILLING DEEPER

Exercise 1: Before you get started, go to your Resource Page and review the basics of how pollutants move in rivers and lakes. Then view the videos on both the Thames and Grand Calumet Rivers. These videos provide an introduction to the important features of both rivers that influence the quality of their waters and the restoration steps to address their specific contaminants.

- List several ways in which the two rivers are similar.
- List several ways that the two rivers differ.

BACKGROUND

River Thames: At about 255 km from its source to the start of tidal influence near London, the Thames is the longest river located entirely within England. The Thames River basin is also the UK's most densely populated watershed. Because it flows directly through London, with its 7 million residents, the River Thames has always played an important part in the history and indeed the identity of England.

The headwaters arise from limestone seeps in the Cotswold Hills and reach their tidal limit at Teddington Lock in London. The upper reaches of the Thames are mostly rural, but as the river flows toward the North Sea, its watershed becomes increasingly urbanized. Important sources of pollutants include both direct discharges of industrial waste and sewage, and dozens of combined sewer overflows (CSOs) carrying both stormwater runoff and untreated sewage.

FIGURE 2.5.3
"The Silent Highwayman" (1858). Death rows on the Thames, claiming the lives of victims who have not paid to have the river cleaned up. *Source: By Punch Magazine [Public domain], via Wikimedia Commons.*

THE PROBLEMS

The most obvious problem in the Thames has historically been low DO levels caused by bacterial decomposition of the organic waste dumped into the river. As Cavenagh (2012) notes, the Thames has received waste for centuries.

As early as the 14th century, trash and human waste were dumped directly into the river. The discharge of untreated sewage from London's considerable population led to outbreaks of diseases like cholera (Fig. 2.5.3). In 1858, the river was so heavily polluted by sewage that all the fish, and consequently all the birds that fed on them, were killed.

The building of embankments and a gravity-fed sewer system improved conditions until World War II, when bombing caused major damage to the system. By the 1950s, the Thames was little more than an open sewer, containing no oxygen and emitting hydrogen sulfide with the smell of rotten eggs. Not until the 1960s were funds made available to repair the damage, resulting in gradual improvement in the water quality of the Thames.

By the 1970s, various species of marine life, including seahorses, began to reappear in the river, and salmon, a famously intolerant fish species but an historic resident in the river, were again seen spawning upstream. In 1975, juvenile salmon from hatcheries in Scotland were released into the river, and the presence of as many as 300 adults annually was documented by the 1990s. Now,

FIGURE 2.5.4
The Indiana Harbor and Ship Canal. *Source: US Army Corp of Engineers.*

the river attracts thousands of birds during the winter. Salmon, trout, and even seals have also been seen in the river.

THE GRAND CALUMET RIVER AND INDIANA HARBOR SHIP CANAL

Background

The GCR/IHSC is only 21 km long and has almost no natural watershed. This human-made river flows into Lake Michigan past one of the greatest concentrations of refineries and steel mills in the US (Fig. 2.5.4).

While the ship canal was completed to facilitate navigation between Lake Michigan and the industries along the GCR, it is no longer maintained for navigation (Martinez et al., 2010). While the GCR/IHSC system is not heavily used for recreational purposes, it is an important resting spot for wildfowl migrating south toward the unique habitats of the nearby Indiana Dunes National Seashore (Custer et al., 2000).

The Problems

Bordered by a wide array of chemical industries and steel mills, much of the GCR/IHSC has long been polluted by a broad range of substances, including heavy metals, petroleum aromatic hydrocarbons (PAHs), and PCBs, which have accumulated over time to great depths in the sediments of the GCR/IHSC.

Both the River Thames and the GCR/IHSC suffered from neglect and pollution for long periods of time, and efforts have been made to restore both. There are substantial differences in these two cases in both the nature of the problem and the cleanup approach used, so we'll consider each separately. At the end, we'll draw some comparisons and take a look ahead.

THE RIVER THAMES: A SUCCESS STORY

Environmental Sciences in Action

Although as recently as the late 1980s, hundreds of salmon were returning to the Thames each year, sharp declines began to occur, and populations nearly vanished by 2005. In an effort to learn more about why the salmon were disappearing, Griffiths et al. (2011) collected 16 salmon swimming up the Thames between 2005 and 2008 and performed genetic testing to determine their origin. Fig. 2.5.5 shows the weirs with fish passages along the Thames; most of the fish were captured at the Molesey and Sunbury weirs. Note the salmon stocking areas farther upstream.

Salmon captured in the Thames were not from hatchery stock, but came from populations living in other southern England rivers that drained into the Thames. The authors concluded that conditions in the Thames had deteriorated again to the point where salmon could not maintain reproducing populations in the river.

Cavenagh (2012) believes that the dozens of CSOs entering the river during storms are to blame. Carrying both untreated sewage and pollutants contained in urban runoff, these discharges introduce organic matter, which microbes break down, consuming DO in the process, and leading to unsuitable habitat conditions.

DRILLING DEEPER

Exercise 2: While data collected from the River Thames suggest that pollutants like suspended sediments, heavy metals, pesticides, and nutrients have continued to decline over the years, the recent downward trend in DO levels is troubling.

- Do some digging and identify three factors that likely have played a role in the oxygen decline (Hint: search for "London's Ecology—How Clean is the Thames?")

Concerns in the Thames go beyond oxygen stress. Research on an important commercial fish species, the European eel, suggests that persistent organic pollutants may also be affecting river life. For centuries, eels were harvested

SOLUTIONS

Exercise 3: CSOs are problems in many large cities around the globe. Study the approach that the Thames Water Company is using to deal with CSOs at their project website: http://www.thameswater.co.uk/about-us/10115.htm.

- What is their approach to eliminating the continuing CSO discharges into the River Thames?
- How are they accounting for changing climate conditions in their designs?
- What is the outlook for whether or not these measures will reduce or eliminate CSO's?
- Do a bit of web exploration to find information about other cities that have used similar techniques for controlling stormwater runoff (e.g., Chicago). What are the pros and cons of such approaches?

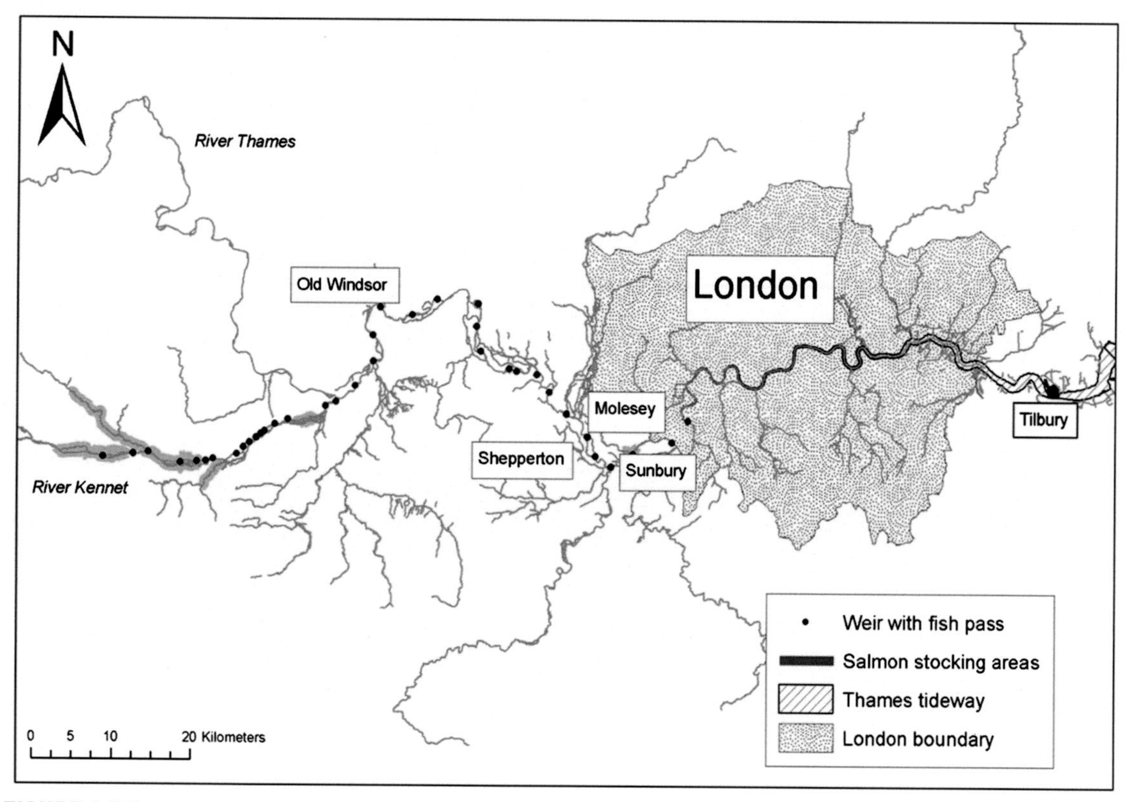

FIGURE 2.5.5

Locations of salmon sampling weirs and stocking areas along the River Thames. *Source: (Griffith et al., 2011).*

commercially along the Thames. Pollution eliminated them until the 1960s, when populations began to recover. Today, a small eel fishery still exists in the Thames.

CRUNCHING THE NUMBERS

Exercise 4: A team of scientists led by M. Jurgens (2015) collected eels from two locations, one tidal and one nontidal, in the London area of the Thames (Fig. 2.5.6) and analyzed them for persistent organic pollutants, including DDT (dichlorodiphenyltrichloroethane) and PCBs (polychlorinated biphenyls).

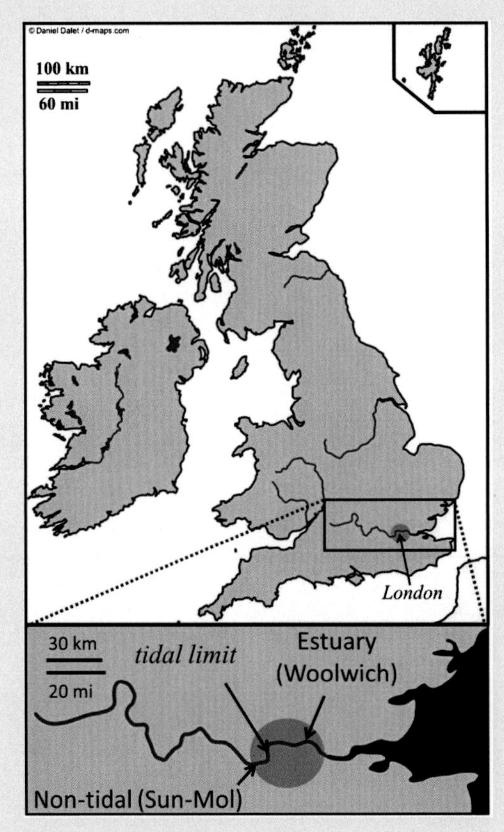

FIGURE 2.5.6
Sampling locations for eels. *Source: Jurgens et al. (2015).*

While only a limited number of eels were captured at the two sites, some interesting trends appeared in the data.

1. Examine the data in Fig. 2.5.7. For these 35 fish, which of the following parameters, weight, length, lipid content, or age, differed significantly between tidal and nontidal sampling sites? Can you think of any explanation for why these parameters would differ?

CRUNCHING THE NUMBERS—Cont'd

Parameter	Non-tidal	Tidal
Length (cm)	51 (27)	46 (31)
Weight (g)	228 (82)	186 (67)
Age (y)	12 (6)	9 (5)
Lipid Content (%)	10 (7.0)	16.5 (9.7)
PCBs ug/kg	63 (16.6)	113 (22.1)
Total DDT ug/kg	15.7 (2.4)	18.2 (3.5)

FIGURE 2.5.7
Mean (and standard deviation) of pollutant levels in 35 eels from two sites on the Thames. *Source: Modified from Jurgens et al., 2015).*

2. On your Resource Page, you will find a data file for download called River_restoration.csv. Using the data found in the Exercise 4 tab, identify which parameters are most closely correlated with PCB and DDT concentrations. What does this tell you about the mechanism of pollutant accumulation in eels?
3. PCBs and DDT were detected in every eel collected, despite the fact that both substances have long been banned in the UK. What might account for their presence in the eels?

SYSTEMS CONNECTIONS

Exercise 5: In addition to the threats described above, restored rivers may face an uncertain future in a changing climate. Johnson et al. (2009) used a hydrological model to predict how conditions in the River Thames might change as we move from the current climate to that predicted for 2080 under two scenarios: low and high CO_2 emissions (Fig. 2.5.8).

- Using the data in the Exercise 5 tab of the River_Restoration.csv file downloadable from your Resource Page, plot the predicted differences in soluble reactive phosphorus, nitrate, and NH_4-N between current and predicted 2080 climate conditions for each season.
- Which variables are most sensitive to a changing climate? Which seasons?
- What about future climate conditions might cause the changes you see?
- How much difference is there between the low and high emissions scenario predictions? What does this imply about our CO_2 reduction efforts?
- Visit the website listed for this article on your Resource Page; choose one group of organisms listed in Table 4 and describe how it will likely fare in 2080 in the River Thames.

Continued

SYSTEMS CONNECTIONS—Cont'd

River Thames (221 km main stem to tidal limit).

		Typical hydraulic residence time (d)		Mean SRP[b] (μg/L)[c]	Mean nitrate-N (μg/L)[c]	Mean NH_4 (μg/L)[c]	Typical mean Si (μg/L)[c]	Mean water temperature (°C)	Incoming shortwave radiation (W/m²)	Mean flow (m³/s)	Q95[a] flow (m³/s)	Q5 flow[d] (m³/s)
		Mean flow	Q95[a]									
Now	Winter[e]	7	31	324	9330	161	4440	6.2	46.4	113	32.3	269.0
	Spring	9	30	403	8301	182	2871	11	152.8	79	33.2	196.0
	Summer	17	41	835	6828	85	4262	17.5	209.2	43	24.2	81.9
	Autumn	20	47	831	7301	79	5324	12.5	90.4	37	21.4	149.0
2080 Low emission scenario (UKCIP)	Winter	10	47	440	13000	220	6000	7.5	46.3	83	21.5	263.0
	Spring	11	38	480	9900	220	3400	12.5	159.9	67	26.3	178.0
	Summer	24	59	1100	9300	120	5800	20.0	227.3	32	17.1	56.2
	Autumn	29	69	1200	10500	110	7700	14.5	96.1	26	14.6	67.6
2080 High emission scenario (UKCIP)	Winter	10	56	500	14000	250	6800	8.5	46.1	74	17.8	258.0
	Spring	13	43	540	11000	240	3800	14.0	168.2	60	23.5	170.0
	Summer	28	73	1400	11000	140	6900	22.0	244.3	26	13.8	46.8
	Autumn	34	82	1400	12500	140	9100	16.5	101.5	22	12.2	55.6

[a] Q95 Flow exceeded 95% of the time, for example might occur in dry summer conditions.
[b] SRP soluble reactive phosphorus.
[c] Concentration information from R. Thames at Howbery Park, Wallingford average of period 1999–2002. Concentrations for 2080 based on a pro-rata predicted change.
[d] Q5 Flow exceeded 5% of the time, for example very high flows in winter.
[e] Winter taken to mean Dec, Jan, and Feb.

FIGURE 2.5.8

Projected values for selected physical and chemical parameters in the River Thames under different climate scenarios. *Source: (Johnson et al., 2009).*

A VERY DIFFERENT STORY: THE GRAND CALUMET RIVER

Unlike the Thames, with its deep cultural and historic significance to the UK, the GCR/IHSC have generated comparatively little interest over time. The GCR is a victim of a double whammy: for years, it has been a dumping ground for industrial and municipal wastes, and it lacks a strong natural flow from upstream areas that might help cleanse the river if and when pollutant sources can be controlled. Today, 90% of the river's flow originates as municipal and industrial effluent, cooling and process water, and stormwater overflows. Although pollutant discharges have been reduced, a number of contaminants continue to impact the system.

Many of the problems seen in the GCR/IHSC stem from legacy pollutants found in the sediments at the bottom of the system. Contaminants present include PCBs, PAHs, and heavy metals such as mercury, cadmium, chromium, and lead. High fecal coliform bacteria levels, elevated BOD, suspended solids, and oil and grease create additional problems.

Nonpoint sources include 3.8 to 7.6 million cubic meters of contaminated sediment, industrial waste site runoff, hazardous waste sites, leaking underground storage tanks, atmospheric deposition, urban runoff, and contaminated

FIGURE 2.5.9
Levels of polychlorinated biphenyls measured in surface sediments along the Indiana Harbor Ship Canal (Martinez et al., 2010).

groundwater. Point sources include industrial and municipal wastewater discharges and CSOs. Because of its high degree of contamination, the GCR/IHSC system has been designated as one of 43 Areas of Concern (AOCs) in the Great Lakes in need of cleanup (http://www.epa.gov/greatlakes/aoc/grandcal/).

Environmental Sciences in Action

Two studies illustrate the degree of contamination of the GCR/IHSC. Martinez et al. (2010) measured PCBs in the surface sediments of the IHSC. Concentrations of PCBs in the 60 samples collected were very high, ranging from 0.053 to 35 ug/g dry wt., with a mean of 7.4 ug/g. Levels of PCBs in Lake Michigan sediments generally range from 0.01 to 0.02 ug/g dry wt. PCBs were fairly uniform in distribution, with several "hotspots" and lower values as the Canal neared its discharge point into Lake Michigan (Fig. 2.5.9).

A second study (Custer et al., 2000) looked at the accumulation of persistent pollutants by lesser scaup, a duck species known to overwinter on the waters of GCR/IHSC. Male scaup were trapped and analyzed for a wide variety of organic and inorganic substances. Levels of PCBs, a suspected carcinogen, in the carcasses of 88% of the scaup were high enough to pose a human health risk if the birds were to be consumed. In addition, analyses suggested that scaup had had a long-term exposure to petroleum aromatic hydrocarbons (PAHs), a group of substances that can cause a variety of adverse effects on both wildlife and humans if ingested.

DRILLING DEEPER

Exercise 6: Fig. 2.5.10 illustrates the elevated PCB levels in IHSC scaup compared to those from a clean reference location. Also note the number of analyses that exceeded guidelines for safe consumption of fowl.

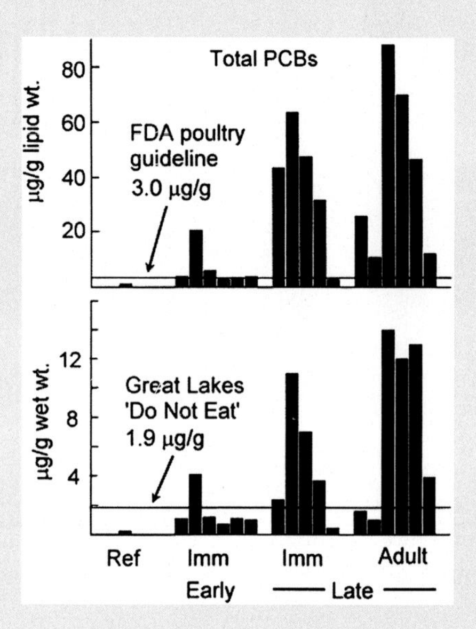

FIGURE 2.5.10

Levels of total polychlorinated biphenyls in scaup carcasses from Indiana Harbor Ship Canal compared to those from a reference site (Custer et al., 2000).

- What proportion of scaup exceeded the FDA poultry guidelines for PCBs based on lipid wt. (top graph)? What proportion exceeded the Great Lakes guidelines based on wet wt. (bottom graph)?
- Which metric (lipid wt. or wet wt.) do you believe is more conservative when considering human consumption of scaup?
- How do contaminant levels change as the scaup age from early immature to late immature to adult?
- What does this imply about the mechanism of contaminant accumulation in the scaup?
- What are the strengths and weaknesses of this approach to ecosystem assessment?

SOLUTIONS

Restoring river systems is a costly business. In the US, federal funds are often provided to support restoration of rivers like the Grand Calumet where there may have been damage to living resources (e.g., fish and wildlife). Determining if a river qualifies for these programs involves a federal evaluation called a Natural Resources Damage Assessment (NRDA). In the case of the GCR/IHSC, scientists from the US Fish and Wildlife Service conducted a series of tests and measurements to determine how badly damaged the living resources of the river were (MacDonald et al., 2002).

The NRDA of the condition of the fish community at sites throughout the system included laboratory measurements of the impacts of sediment exposure on fish, comparisons of concentrations of pollutants like PCBs in fish tissues to federal guidelines, and field assessments of how many fish lived in the river and whether or not individual fish showed signs of exposure to harmful chemicals (e.g., tumors). Additional evaluations were made of the invertebrate community.

The black dots in Fig. 2.5.11 show sites where their assessments led scientists to conclude that there had been significant damage to the living resources of the GCR/IHSC. The acronym SQG stands for sediment quality guideline, the concentration of a pollutant above which harmful effects can be expected. Note how the impact of pollutants extended from the GCR/IHSC out into Lake Michigan.

Additional assessments of conditions in the GCR/IHSC found other impairments, including loss of wildlife habitat and degradation of aesthetics. In fact, of the 43 AOCs in the Great Lakes, only the GCR/IHSC failed in all 14 of the impairment categories (USEPA, 2013). Based on these findings demonstrating substantial damage, in 2004, a court ordered eight responsible companies to pay $56.3 million in cleanup costs.

CURRENT RESTORATION EFFORTS: GRAND CALUMET RIVER LEGACY ACT CLEANUP

To address the problem of contaminated sediment in the Great Lakes, the Great Lakes Legacy Act of 2002 was passed by the US Congress and signed into law by then President George W. Bush on November 27, 2002. The Act authorized $270 million in funding over 5 years to help with the remediation of contaminated sediment in AOCs. For the Grand Calumet Project, the US EPA has been providing 65% of the project costs with funds from the Great Lakes Legacy Act, while the State of Indiana has been providing the remaining 35% using NRDA funds.

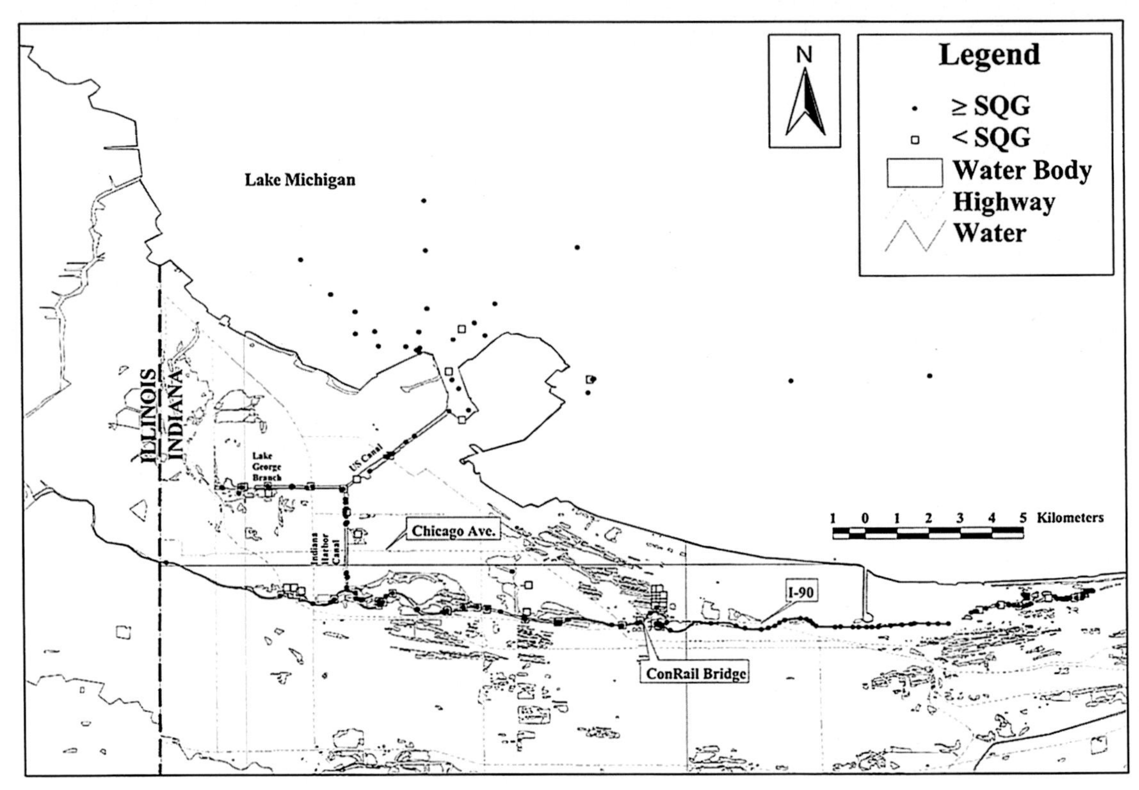

FIGURE 2.5.11

Extent of injury to fish and wildlife resources, based on comparisons of sediment chemistry data to bioaccumulation-based sediment quality guidelines. *Source: (MacDonald et al., 2002).*

FIGURE 2.5.12

Dredging and removing contaminated sediments is an effective, although costly, way to restore the Grand Calumet River. *Source: US EPA.*

Because many of the toxins in the GCR/IHSC are bound in river sediments, much of this funding has been used to pay for the removal of contaminated sediments. Dredging of these sediments is well underway, and sections of the GCR/IHSC are beginning the cleanup process (Fig. 2.5.12). It will, however, be a long and expensive effort.

To fully appreciate the restoration challenge posed by the GCR/IHSC, consider that there may be as many as 3.8 to 7.6 million cubic meters of contaminated sediments at the bottom of the GCR/IHSC. Once contaminants are removed, the river bottom must be capped with a layer of clay and sand to prevent the influx of any remaining toxins buried deep in the sediment. Wetlands and near-shore habitats can then be restored with native plants following the completion of the dredging (Fig. 2.5.13).

FIGURE 2.5.13

Restoration planting between 2012 and 2013 in the Roxana Marsh. *Source: US EPA.*

CRITICAL THINKING

Exercise 7: While this restoration work provides hope for a restored, functioning ecosystem in the GCR/IHSC, many question the value of taking on such a daunting cleanup challenge. For perspective on the financial costs of this work, for the West Branch portion of the IHSC alone, $81.1 million dollars were spent to successfully remove only 573,000 cubic meters of contaminated sediment (compare this to the estimated 3.8 to 7.6 million cubic meters present in the system) containing large quantities of PCBs, PAHs, heavy metals, and pesticides from the river. Consider that in addition to the financial costs of dredging, once contaminants are removed, they must be disposed of elsewhere.

1. Consider that you are presenting the case to allocate additional state funds for remediation of the GCR/IHSC waterway. How can you justify the cost of this restoration effort?

Continued

CRITICAL THINKING—Cont'd

2. Similarly, consider that you are presenting the case to allocate additional government funds for remediation in the Thames River. How would your justification differ? Which case do you believe is more likely to garner public support? Why?
3. In this case study, we have considered the impacts of sewage and oxygen depletion and the threat posed by PCBs and pesticides. Which of these pollutant types poses a greater challenge to those responsible for restoration efforts? Why?

CRITICAL THINKING

Exercise 8: Considering the expense of river restoration, those who would most benefit from its restoration and other noneconomic considerations, who should be responsible for funding restoration efforts?

How can these parties be convinced to provide the necessary funding?

REFERENCES

Cavenagh, S., 2012. The Thames: back from the dead? I, Science. The Science Magazine of Imperial College. Posted on September 2, 2012 www.iscience.mag.co.uk/features/thames-back-from-the-dead/.

Custer, T.W., Custer, C.M., Hines, R.K., Sparks, D.W., 2000. Trace elements, organochlorines, polycyclic aromatic hydrocarbons, dioxins, and furans in lesser scaup wintering on the Indiana Harbor. Canal. Environ. Pollut. 110, 469–482.

Griffiths, A.M., Ellis, J.S., Clifton-Dey, D., Machado-Schiaffino, G., Bright, D., Garcia-Vasquez, E., Stevens, J.R., 2011. Restoration versus recolonisation: the origin of Atlantic salmon (*Salmo salar* L.) currently in the River Thames. Biol. Conserv. 144, 2733–2738.

Johnson, A.C., Acreman, M.C., Dunbar, M.J., Feist, S.W., Giacomello, A.M., Gozlan, R.E., Hinsley, S.A., Ibbotson, A.T., Jarvie, H.P., Jones, J.I., Longshaw, M., Maberly, S.C., Marsh, T.J., Neal, C., Newman, J.R., Nunn, M.A., Pickup, R.W., Reynard, N.S., Sullivan, C.A., Sumpter, J.P., Williams, R.J., 2009. The British river of the future: how climate change and human activity might affect two contrasting river ecosystems in England. Sci. Total Environ. 407 (17), 4787–4798.

Jurgens, M.D., Chaemfa, C., Hughes, D., Johnson, A.C., Jones, K.C., 2015. PCB and organochlorine pesticide burden in eels in the lower Thames River (UK). Chemosphere 118, 103–111.

MacDonald, D.D., Ingersoll, C.G., Smorong, D.E., Lindskoog, R.A., Sparks, D.W., Smith, J.R., Simon, T.P., Hanacek, M.A., 2002. Assessment of injury to fish and wildlife resources in the Grand Calumet River and Indiana Harbor area of concern, USA. Arch. Environ. Contam. Toxicol. 43 (2), 130–140.

Martinez, A., Norstrom, K., Wang, K., Hornbuckle, K.C., 2010. Polychlorinated biphenyls in the surficial sediment of Indiana Harbor and Ship Canal, Lake Michigan. Environ. Int. 36, 849–854.

U.S. Environmental Protection Agency, 2013. Great Lakes Areas of Concern: Grand Calumet River Area of Concern. http://www.epa.gov/greatlakes/aoc/grandcal/.

Chapter 2.6

Ocean Acidification

FIGURE 2.6.1
The vibrant colors of living coral, one ecosystem at great risk from ocean acidification. *Source: NOAA.*

OVERVIEW

While many of us know about the likely effects of increasing carbon dioxide (CO_2) concentrations in the atmosphere on sea level rise and global surface water temperatures, fewer may be aware of the threat posed to the chemistry of our oceans. The connection between the Earth's oceans and its atmosphere is dynamic. As levels of CO_2 in the atmosphere increase, oceans absorb more of the gas, leading to ocean acidification (OA) (Fig. 2.6.1). As the pH of seawater decreases, there may be many impacts on both ecosystems and social systems, including effects on individual marine species and communities and economic and social impacts on human populations dependent on marine resources for their survival. Before you start this case study, go to one of the web sites listed on your Resource Page to review the chemistry of carbon in seawater.

BACKGROUND

As you recall from your review of pH and carbon chemistry, the adsorption of excess CO_2 from the atmosphere alters the balance of carbon species like

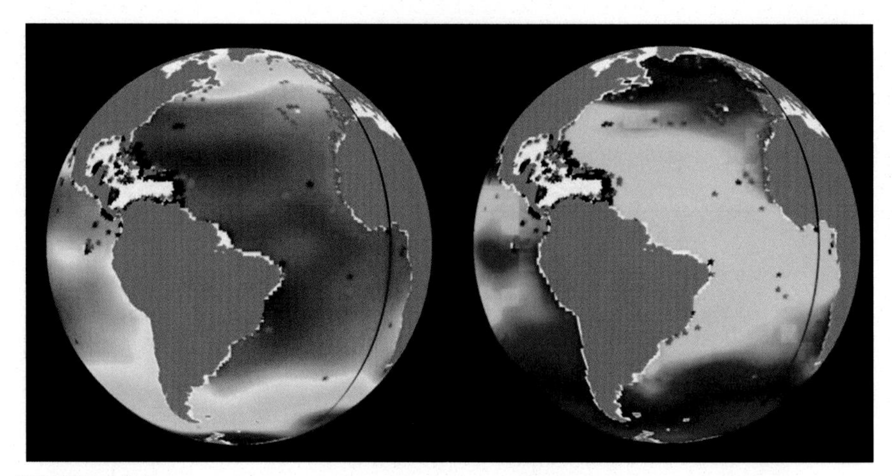

FIGURE 2.6.2

NOAA images of pH from modeled values for the 1700s (left) and 2100 (right). Warmer colors indicate higher pH, whereas cooler colors show increasingly acidic conditions. *Source: NOAA, Science on a Sphere.*

bicarbonate and carbonate, which play important roles in the structure and function of marine ecosystems. Previous National Oceanic and Atmospheric Administration (NOAA) studies estimate that the oceans have absorbed about one quarter of all the CO_2 given off by human activities since the Industrial Revolution at a rate of about one million metric tons per hour. The result is that oceans have become more acidic (Fig. 2.6.2).

CRUNCHING THE NUMBERS

Exercise 1: Let's practice some "back of the envelope" calculations to estimate how large a role the world's oceans play in mitigating our carbon emissions.

- Based on the above statistic that one million metric tons of CO_2 are removed from the atmosphere by the world's oceans each hour, how many pounds of CO_2 are absorbed each second?
- The most recent estimates are that approximately 2.4 million pounds of CO_2 are emitted globally each second. If, historically, the oceans have removed approximately 25% of our emissions, how many pounds of our current emissions are the oceans taking up each year? If about 19.64 pounds of carbon dioxide (CO_2) are produced by burning a gallon of gasoline that does not contain ethanol, how many gallons of gas is the uptake by oceans equivalent to?
- Considering that the average US household pumps 49 metric tons of carbon into the atmosphere each year and that the surface of the Earth's oceans totals approximately 360 million km^2, how many square meters of ocean are required to offset your household's emissions?

A Look at the Chemistry: The key issue in OA is the increasing acidity and falling pH of seawater. Note that OA doesn't mean that seawater will actually become acidic, dropping to below pH 7 (acidic by definition), just that values will become more acidic over time.

Historically, seawater has always been basic, with a pH averaging about 8.2. Since the beginning of the Industrial Revolution, the pH has dropped about 0.1 unit to 8.1 (Baker and Ridgwell, 2012). Scientists believe that, unless CO_2 emissions are substantially reduced, seawater pH may drop by another 0.3–0.4 units by 2100. Ridgwell and Schmidt (2010) noted that we'd probably have to go back 65 million years to experience changes in seawater chemistry as rapid as today's.

CRUNCHING THE NUMBERS

Exercise 2: Remember that pH is a logarithmic scale, so the drop of 0.1 unit represents what percentage increase in the acidity of seawater? Be sure to show your work!

As the pH of seawater decreases, the carbon chemistry begins to change. At a pH of 8.1, the predominant species of carbon in seawater are bicarbonate (91%) and carbonate (8%). We'll see shortly why carbonates are such an important part of this story.

The following reactions show what happens after excess CO_2 enters seawater (Fig. 2.6.3). CO_2 is dissolved and reacts with water to form carbonic acid (H_2CO_3). Carbonic acid then dissociates into hydrogen ions (H^+) and bicarbonate (HCO_3^-), a base. Seawater is naturally saturated with another base, carbonate ion (CO_3^{-2}), which neutralizes some of the H^+ released after the initial dissociation of carbonic acid, creating more bicarbonate. The net result of these reactions is that as more and more CO_2 enters seawater, levels of CO_2 and bicarbonate increase, while those of carbonate decrease. As more and more of the carbonate is used up, seawater becomes undersaturated with respect to the two carbonate minerals vital for shell building in marine organisms, aragonite and calcite.

1. $CO_2 + H_2O \leftarrow\rightarrow H_2CO_3$
2. $H_2CO_3 \leftarrow\rightarrow HCO_3^- + H^+$
3. $HCO_3^- \leftarrow\rightarrow CO_3^{2-} + H^+$

Because seawater is buffered, another reaction comes into play:

4. $CO_2 + CO_3^{2-} + H_2O \leftarrow\rightarrow 2\ HCO_3^-$

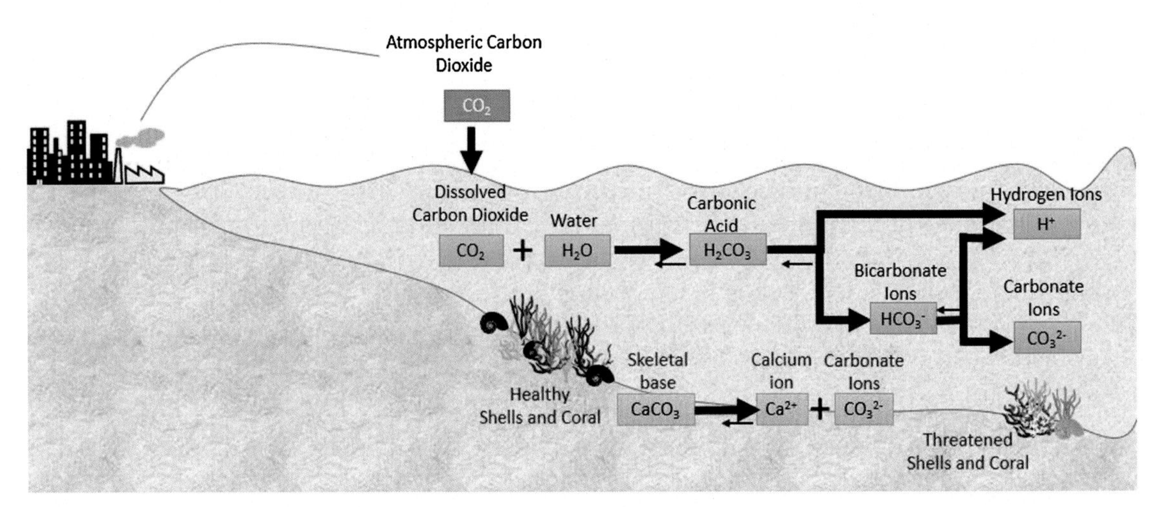

FIGURE 2.6.3
As carbon dioxide dissolves in seawater, carbonate ion neutralizes free hydrogen, causing carbonate levels to drop and acidity to increase.

In this manner, CO_2 is neutralized. The HCO_3^- produced in Reaction 4 dissociates as shown in Reaction 3, adding hydrogen ions and lowering the pH and carbonate content of the seawater (leading to undersaturation).

Why is this important? A reduction in carbonate levels in seawater has significant implications for marine ecosystems. Many marine species are calcifiers, as they rely on calcium carbonate in the seawater to form and maintain their shells and skeletons.

Calcium carbonate present in the marine environment will begin to dissolve as carbonate becomes undersaturated according to the following formula:

5. $CaCO_3 + CO_2 + H_2O \rightarrow Ca^{2+} + 2HCO_3^-$

An important term to learn in this case study is aragonite, one of the two crystalline forms of calcium carbonate (the other is calcite). Aragonite is the specific form produced by many marine organisms and is used to build skeletons and shells. Fig. 2.6.4 (USEPA, 2014) shows modeled changes in aragonite saturation in the world's oceans between 1880 and 2013.

The lower the concentration of aragonite, the more difficult it is for organisms to maintain their carbonate-based shells and skeletons. The yellows and oranges in warmer waters indicate decreasing levels of aragonite saturation and increased stress on carbonate-reliant species.

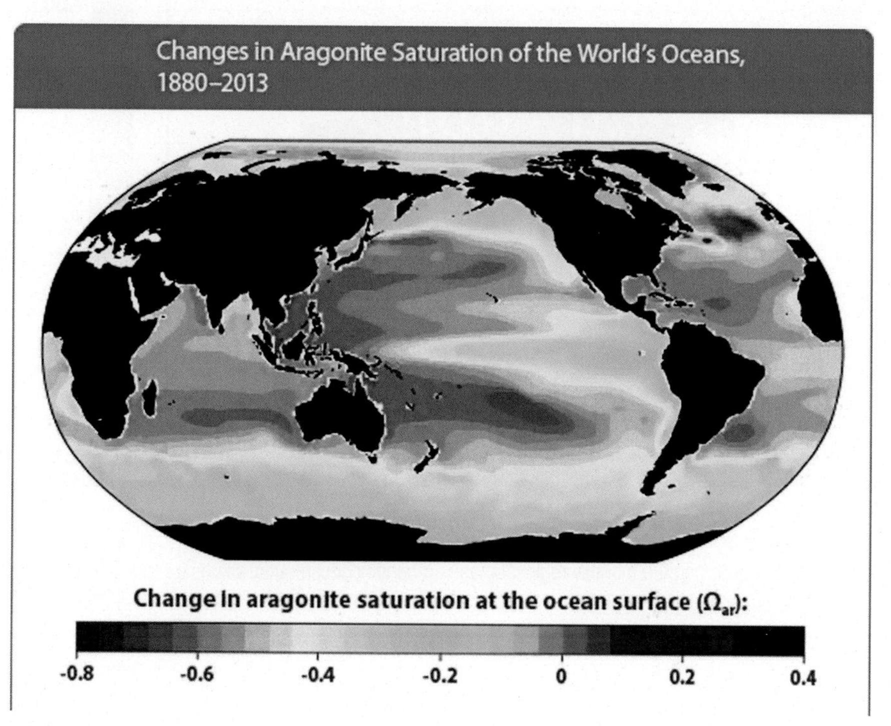

FIGURE 2.6.4

Changes in aragonite saturation of the oceans from 1880 to 2013. *Source: US EPA.*

DRILLING DEEPER

Exercise 3: Go to the NOAA Ocean Acidification interactive viewer at http://coralreefwatch.noaa.gov/satellite/oa/ (Gledhill et al., 2008). Here you can view an animation of a time series of modeled ocean acidity for the Caribbean region (e.g., Fig. 2.6.5)

- Why do they use aragonite saturation as an indicator of OA?
- What drives the seasonal variability within each year?
- Of what importance to Caribbean ecosystems might a gradual reduction in aragonite levels be?

Continued

DRILLING DEEPER—Cont'd

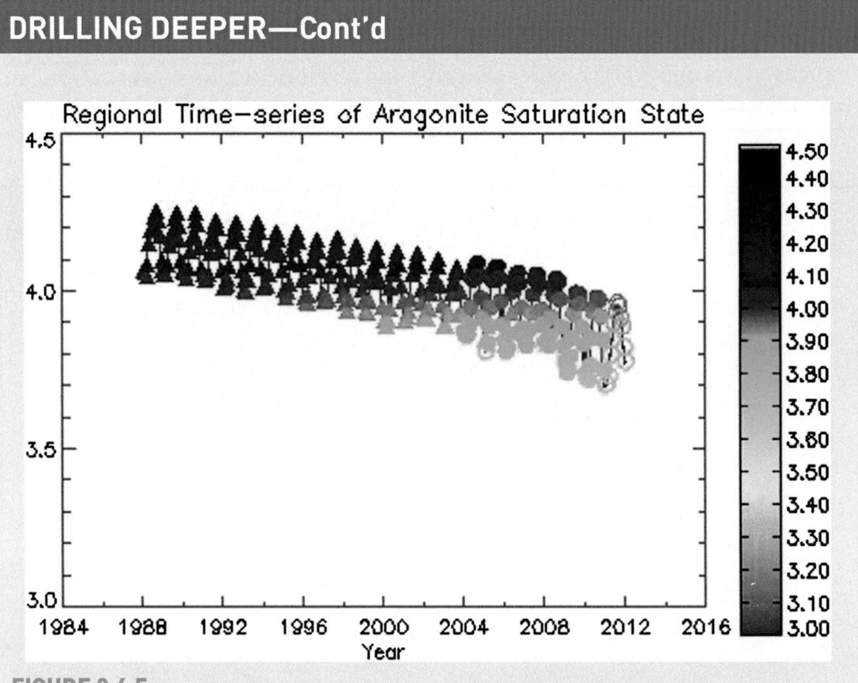

FIGURE 2.6.5

A time series depicting the monthly regional mean aragonite concentration values for the greater Caribbean region between 1984 and 2013 modeled by the NOAA Ocean Acidification Product Suite v. 2.0 shows the ongoing drop in aragonite saturation. *Source: NOAA.*

A loss of $CaCO_3$ means that not only will there be less for organisms like pteropods (sea butterflies) to use to build their shells and skeletons, but that the $CaCO_3$ already present in their bodies may begin to dissolve (Fig. 2.6.6).

Relatively little known to most, these "sea butterflies" are a key food item for marine species including krill, juvenile salmon, and some species of whales. They are also the major planktonic producers of aragonite in the oceans and are thus likely vulnerable to the effects of OA. In a laboratory experiment, Manno et al. (2012) exposed the pteropod *Limacina retroversa* to waters of pH 8.2, 8.0, 7.8, and 7.6, and shells were damaged at lower pH values. Scientists at NOAA's Pacific Marine Environmental Laboratory (PMEL) have done similar experiments showing complete dissolution of pteropod shells after 45 days when exposed to seawater pH levels expected to occur in 2100 (PMEL, 2014).

THE PROBLEM

Turley (2013) estimates that the Earth's carbon reservoirs (atmosphere, land, and ocean surfaces) hold fewer than 4000 PgC (a PgC is one billion tons of

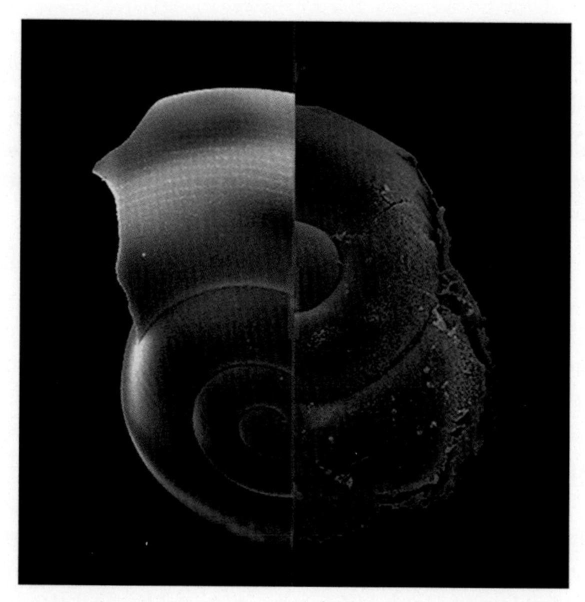

FIGURE 2.6.6
Photos show the impact of acidity on pteropods, or "sea butterflies," free-swimming sea snails about the size of a small pea. The left panel shows a shell collected from a live pteropod from a region in the Southern Ocean where acidity is not too high. The shell on the right is from a pteropod collected in a region where the water is more acidic (Photo credits: Bednaršek. *Source:* globalchange.gov).

carbon), while fossil fuel reserves are estimated to hold about 5000 PgC. Turley estimates that burning all 5000 PgC would result in a decrease in surface ocean pH of 0.77 from the preindustrial level of 8.2. A drop of this magnitude would have catastrophic implications for many marine organisms.

The Good News: Not all marine species will be losers as global CO_2 levels rise. In areas that are not nutrient limited, primary producers like marine algae and sea grasses may experience higher productivity with more CO_2 available. This increase in primary producers could cascade through higher trophic levels, allowing for increases in many fish and marine mammal populations that feed directly on such primary producers.

The Bad News: Among the species most directly affected by OA, in addition to pteropods, will be various shellfish and sea urchins. Let's look at some species already being affected by OA.

Oysters: An alarming development with OA has occurred over the past decade along the Pacific Coast of the US Northwest as huge numbers of oyster larvae, called oyster seed, have died off. Seed production in this region plunged by as much as 80% between 2005 and 2009. Larval oysters are the stage most vulnerable to OA, as corrosive acidic water inhibits shell formation. The Pacific Coast is particularly vulnerable to OA, as upwelling brings deep waters already low in aragonite levels to the surface.

Working with PMEL scientists and others, personnel at oyster hatcheries began to monitor the pH of seawater and use this water only when acidity levels were sufficiently low to protect the seed.

Additional corrective steps employed at hatcheries included adding sodium carbonate and eelgrass to help stabilize pH levels. For more information about how some hatcheries in the Pacific Northwest are working to maintain their livelihood, see the NOAA video "Oyster Farmers Facing Climate Change" linked on your Resource Page at https://vimeo.com/43828686.

Environmental Sciences in Action

A recent study of another carbonate-dependent species in Australian waters, the conch snail, shows just how subtle the effects of OA might be. Dr. Sue-Ann Watson from James Cook University in Australia and the ARC Centre of Excellence for Coral Reef Studies (Fig. 2.6.7) has been looking at how warming and acidifying ocean waters might affect the behavior of marine creatures like the Great Barrier Reef humpback conch snail.

An important defense mechanism that allows this snail to avoid a poisonous dart fired by a predator, another sea snail, is leaping (Fig. 2.6.8). It appears that this defense may be affected by OA. A laboratory experiment in which specimens were exposed to elevated CO_2 levels showed that snails were less likely to jump to avoid predators, and even when they did jump, it took nearly twice as long to do so (Watson et al., 2014). An important point underscored by this study is that changing a species' behavior may be just as lethal as killing the organism outright.

PUTTING SCIENCE TO WORK

Exercise 4: Go to the website listed on your Resource Page to read the Watson et al. (2014) article describing this experiment. Let's break down her study into its design components to better understand what we can and cannot infer from these results.

- What is the larger population that the sample in this study represents?
- Describe the experimental design that was used in her experiment.
- If you were to repeat the experiment, what might you do differently to better understand behavioral impacts on a larger population or under different environmental conditions?
- How would you design a study to test Dr. Watson's findings under natural conditions?

Coral Reefs: Coral reefs are unique marine communities that support a wide variety of life. Although they cover at most 1% of the ocean floor, perhaps as many as one-quarter of all marine species depend on reefs in one way or

FIGURE 2.6.7
Dr. Sue-Ann Watson, Research Fellow in Ocean Acidification and Adaptation at the ARC Centre, investigates how warming and acidifying ocean waters might affect the behavior of marine creatures.

FIGURE 2.6.8
A humpback conch snail (L) leaps backwards away from the predatory marbled cone. *Source: Dr. S Watson.*

another. Before reading further, go to coralreef.noaa.gov on your Resource Page to learn more about coral reef ecosystems.

Hard, or *scleractinian*, corals form a solid exoskeleton by secreting aragonite. Several studies have shown that a number of these coral species have increased difficulty forming hard skeletons as CO_2 levels increase and seawater pH drops. Models suggest that if OA continues, many reefs may be replaced by macroalgae and noncoral–dominated communities. Research has linked coral disease and decline to a suite of stressful environmental conditions, including increased ocean temperatures.

SYSTEMS CONNECTIONS

Exercise 5: The loss of coral reefs will have a number of negative impacts that extend well beyond the coral themselves.

- How might the loss of coral reefs affect ocean fisheries? Seaside economies? Coastal shoreline stability?
- Can you think of other impacts beyond these three?
- Coral reefs are affected by many other stressors besides OA. Identify and discuss three such stressors. How would you go about determining the cause of decline of a particular coral reef?

Adaptation: Recent research (McCulloch et al., 2012) suggests that some corals have the ability to adjust to decreasing pH values in seawater. By depositing calcium carbonate in their internal fluids, they appear to be able to regulate their internal pH levels. Even though some species may be able to survive in this manner, the authors note that "all coral species will likely have difficulty adapting to not only ocean acidification, but the combined effects of ocean acidification, change in ocean temperature, and the impact of human pollution."

And There's This

Several additional points about OA to remember (USEPA, 2014):

- It takes a long time for changes in carbon chemistry to diffuse down to deeper waters, so the full impact of OA won't be known for decades or longer.
- Ocean chemistry isn't uniform globally; for instance, cold water holds more CO_2 than warm water, so, over time, colder latitudes may be more affected by OA.
- Air and water pollution can influence OA in localized areas, including estuaries.

SYSTEMS CONNECTIONS

Exercise 6: While interactions with the atmosphere are closely linked to water chemistry, living organisms can also affect chemical parameters such as pH. For example, as algae and aquatic plants undergo photosynthesis, they remove CO_2 from the water, raising the pH. Conversely, during respiration, they release CO_2, lowering the pH.

- Given this activity by primary producers like algae, how might an increased release of nutrients from the landscape into an estuary like Chesapeake Bay also promote OA?

RISK ASSESSMENT

In your careers, some of you will be involved in estimating the risk posed by environmental actions. It is a complex process that includes a wide variety of elements that need to be considered and evaluated when determining risk. In the following, we'll examine and critique a risk assessment related to OA. Before you begin, review the basics of risk assessment in the Tools and Skills section.

J.T. Mathis and his colleagues at NOAA's PMEL have evaluated the risk that OA poses to Alaska, where vulnerable mollusks and other shellfish make important contributions to the state's vital commercial fisheries and traditional way of life. There will be direct effects of OA on key Alaskan species like the red king crab and the tanner crab. Of perhaps greater concern could be the loss of calcifying species like the pteropods, which we discussed earlier. Reductions in numbers of this important group, a key food item for juvenile finfish, could cascade through the ecosystem and lead to reduced numbers of commercial species like salmon.

In their study, Mathis et al. (2015) used chemical, biological, social, and economic data to estimate the potential risk OA will pose to Alaskans relying on marine resources. Three components of assessment went into their risk estimate (Fig. 2.6.9): (1) *hazard,* which in this case refers to OA and how it is expected to evolve over time in Alaskan waters; (2) *exposure,* which refers to OA impacts on those marine resources directly important to Alaskan communities; and (3) *vulnerability,* which combines the degree of human reliance on OA-sensitive species with adaptive capacity, which estimates a community's ability to adapt to any losses.

For each of the three components, Mathis and his team estimated the level of risk for different regions of Alaska. For instance, shellfish are an important part of subsistence diets in southeastern Alaska, while salmon play a key role in such diets throughout the state. Both shellfish and salmon are susceptible to the effects of OA. So in the category of exposure, areas of southeastern Alaska received high scores because residents there rely more heavily on species affected by OA than in other areas of the state.

For the hazard component, the team looked at which areas off the Alaskan coast were most vulnerable to OA and estimated how OA would impact fisheries in each area. To estimate vulnerability, the team looked at, among other things, the economic flexibility to shift to alternate food sources in different parts of the state.

Once Mathis and his team had developed a score for each area of Alaska for each of the three components of the risk analysis, they added the numbers to come up with an estimate of total risk.

According to the analysis of Mathis et al., residents in the southern rural regions of Alaska are at greatest risk because of a reliance on subsistence fishing, a more rapid rate of OA in waters adjacent to these areas, lower industrial diversity, greater economic dependence on fish harvests, lower incomes, and higher food prices (Fig. 2.6.10). Other areas of the state may have one or more elements of risk, but these are offset by such factors as higher income and more job opportunities.

CRITICAL THINKING

Exercise 7: Read through the Mathis et al. paper available on your Resource Page. Based on the article and your knowledge of the risk assessment process, answer the following questions:

- How much confidence can be placed in each of the factors that went into the determination of this index?
- What additional factors might they need to consider for a more thorough risk assessment?
- What other climate change–related impacts might add to the risk faced by certain Alaskans? Which Alaskans? Why?

Minimizing the Risks

The State of Washington in the US Pacific Northwest has recognized the particular vulnerability of its coastal waters to OA (State of Washington, 2012). Coastal upwelling of CO_2-rich, low-pH waters from the deep ocean, in addition to runoff of nutrients from the land and subsequent decay of organic matter in subsurface waters, are hastening the eventuality of OA in these waters.

A task force formed by the governor of Washington developed a series of 42 actions to address OA. Listed below are some of their specific recommendations:

1. Advocate for reductions in emissions of carbon dioxide;
2. Reduce nutrients and organic carbon in priority areas;
3. Request that the US EPA assess water quality criteria relevant to OA;
4. Increase public understanding of OA and its consequences; and
5. Consult with affected and interested Indian Tribes and Nations in Washington State.

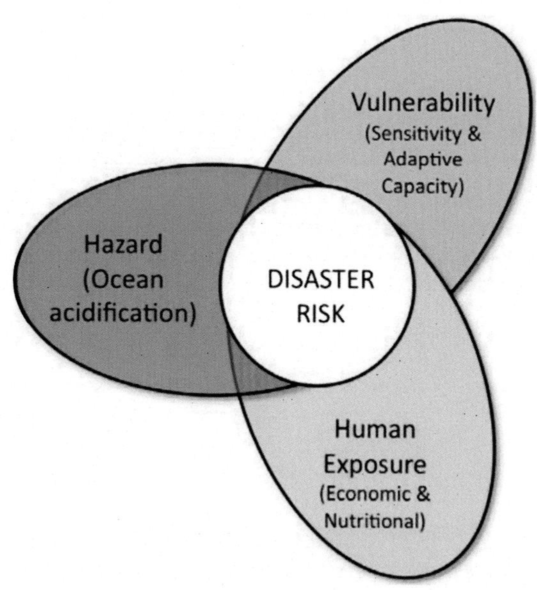

FIGURE 2.6.9
Components of the risk assessment of OA's impact on Alaskan communities. *Source: (Mathis et al., 2015).*

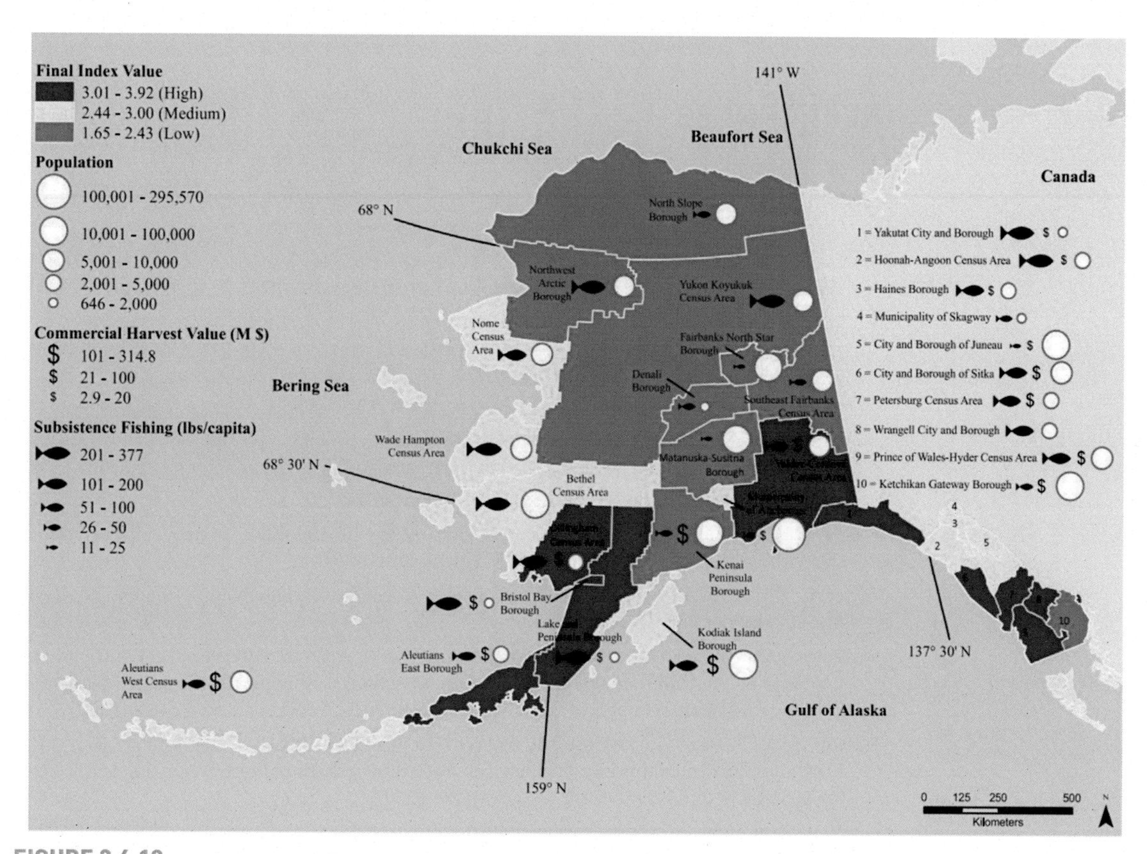

FIGURE 2.6.10
The Risk Index for Alaska developed by Mathis et al. Areas in *red* were assessed to be at greatest overall risk (Mathis et al., 2015).

SOLUTIONS

Exercise 8: You can find out more information about the entire list of recommended actions to minimize the risks of OA in Washington State in the full document (**Scientific Summary of Ocean Acidification in Washington State Marine Waters**) accessible through your Resource Page. After a more thorough review of their suggestions, break into small groups and answer the following questions:

- How effective do you think these approaches will be?
- If you were the governor, what additional steps would you recommend?
- Which of these recommendations would work well in Alaska? Which would not?
- Overall, how would you rank OA in terms of difficulty in finding a solution?

CONSIDER THIS

Exercise 9:

- Which nations stand to be the hardest hit by OA? Why?
- Is OA a good example of environmental justice? Why or why not?
- What is the single most important policy that could address OA?

REFERENCES

Baker, S., Ridgwell, A., 2012. Ocean Acidification. Nat. Educ. Knowledge 3 (10), 21.

Gledhill, D.K., Wanninkhof, R., Millero, F.J., Eakin, M., 2008. Ocean acidification of the Greater Caribbean Region 1996–2006. J Geophys Res.Oceans 113, C10031.

Manno, C., Morata, N., Primicerio, R., 2012. *Limacina retroversa's* response to combined effects of ocean acidification and sea water freshening. Estuar. Coast. Shelf Sci. 113, 163–171.

Mathis, J.T., Cooley, S.R., Lucey, N., Colt, S., Ekstrom, J., Hurst, T., Hauri, C., Evans, W., Cross, J.N., Feely, R.A., 2015. Ocean acidification risk assessment for Alaska's fishery sector. Prog. Oceanogr. 136, 71–91.

McCulloch, M., Falter, J., Trotter, J., Montagna, P., 2012. Coral resilience to ocean acidification and global warming through pH up-regulation. Nat. Clim. Change 2 (8), 623–627.

Pacific Marine Environmental Laboratory (PMEL), 2014. What Is Ocean Acidification? 6 p. www.pmel.noaa.gov/co2/story/What+is+Ocean+Acidification%3F.

Ridgwell, A., Schmidt, D.N., 2010. Past constraints on the vulnerability of marine calcifiers to massive carbon dioxide release. Nat. Geosci. 3, 196–200.

Turley, C., 2013. Ocean acidification. In: Managing Ocean Environments in a Changing Climate, first ed. Elsevier Press. 376 p. (Chapter 2).

Washington State Blue Ribbon Panel on Ocean Acidification, 2012. In: Adelsman, H., Whitely Binder, L. (Eds.), Ocean Acidification: From Knowledge to Action, Washington State's Strategic Response. Washington Department of Ecology, Olympia, Washington. Publication no. 12-01-015.

Watson, S.-A., Lefevre, S., McCormick, M.I., Domenici, P., Nilsson, G.E., Munday, P.L., 2014. Marine mollusc predator-escape behaviour altered by near-future carbon dioxide levels. Proc. Royal Soc. London B Biol. Sci. 281 (1774), 20132377.

U.S. Environmental Protection Agency, 2014. Climate Change Indicators in the United States, 2014, third ed. EPA 430-R-14-004. http://www3.epa.gov/climatechange/science/indicators/oceans/acidity.html.

Chapter 2.7

Groundwater: What Lies Beneath

FIGURE 2.7.1
Groundwater is the primary source of water for India. *Source:* globalwaterforum.org *Groundwater.*

OVERVIEW

It's hard to fix what you can't see. While much of the world's water supply comes from the ground (Fig. 2.7.1), it's a resource we've generally paid too little attention to in the past.

This case study will highlight several instances where groundwater quality and/or quantity have been impacted, look at new techniques used to track and manage this critical resource, consider some challenging policy implications related to its use, and discuss some of the difficulties faced in managing and protecting it.

BACKGROUND

Before tackling this case study, review one or more of the groundwater sites listed on your Resource Page. Several important points to remember about groundwater:

a. It moves but very slowly. Thus, soluble pollutants moving downward from the soil surface, if they reach the groundwater, may be carried in its flow. If there is a well lying in the path of this flow, contamination of the water supply may result.

b. The vulnerability of groundwater supplies to contamination depends on several factors: (1) *its proximity to the surface*: the deeper the groundwater, the longer it may take for pollutants to reach it; (2) *soil type*: materials like sand and gravel are porous and allow pollutants to move through more readily, while others, like clay, retard movement; and (3) *pollutant type*: some pollutants like phosphorus are more easily adsorbed and retained by soil particles than others like nitrogen, which may move more rapidly through soil into groundwater.

c. The availability of many groundwater supplies depends upon recharge/replenishment. The rate of recharge is influenced by many factors, including the amount of precipitation, whether or not recharge areas are covered by impervious surfaces like pavement, and the effect of climate change–related alterations in precipitation patterns.

d. In many parts of the world, groundwater is threatened by contamination and overuse, setting the stage for social conflicts over a much needed but dwindling resource. Fortunately, international agencies like the United Nations are working to help solve these problems.

KEY TERMS

Groundwater: subsurface water that fully saturates pores and cracks in soils and rocks (see Fig. 2.7.2 for a diagram of groundwater formation)
Aquifer: groundwater that can be tapped for human use
Unsaturated zone: upper layers of soil that contain both air and water
Saturated zone: soil containing only water and no air
Water table: top of zone that is saturated
Leaching: the movement of soluble substances through soils into surface and groundwater

FAST FACTS ABOUT GROUNDWATER (FAMIGLIETTI, 2014)

- Groundwater accounts for 33% of total water withdrawals globally
- More than 2 billion humans rely on groundwater
- About half of global irrigation water is derived from groundwater
- Groundwater is essential to meet human needs during droughts

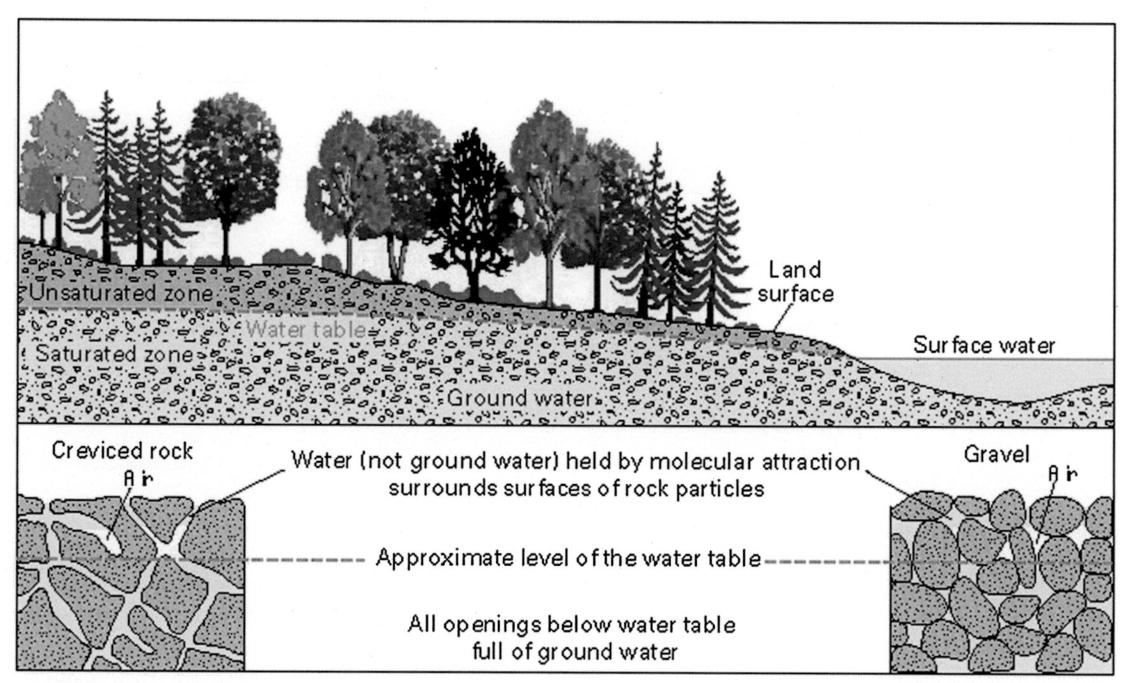

FIGURE 2.7.2
Diagram of groundwater formation. *Source: USGS.*

THE PROBLEM

Many of us around the globe use groundwater supplies for drinking water and irrigation. In fact, some regions experiencing rapid population growth, often in arid or semiarid regions, depend entirely on these reserves. However, with increased use and greater population density come new threats to what was once thought to be an infinite supply of clean water. Protecting the quality and quantity of groundwater supplies carries with it many challenges.

Later in the case study, we'll review several examples that illustrate some of the management options available and how, in spite of these tools, effective use of the resource remains an elusive goal. Many water experts believe that, to be successful, groundwater management must be done cooperatively at an international level.

Groundwater quality: contamination of groundwater differs from surface water pollution in several important ways:

1. Pollutants may enter surface waters directly (e.g., industrial waste discharges), while pollutants affecting groundwater often must travel some distance through soils. Fig. 2.7.3 shows an exception to this, as a plume of

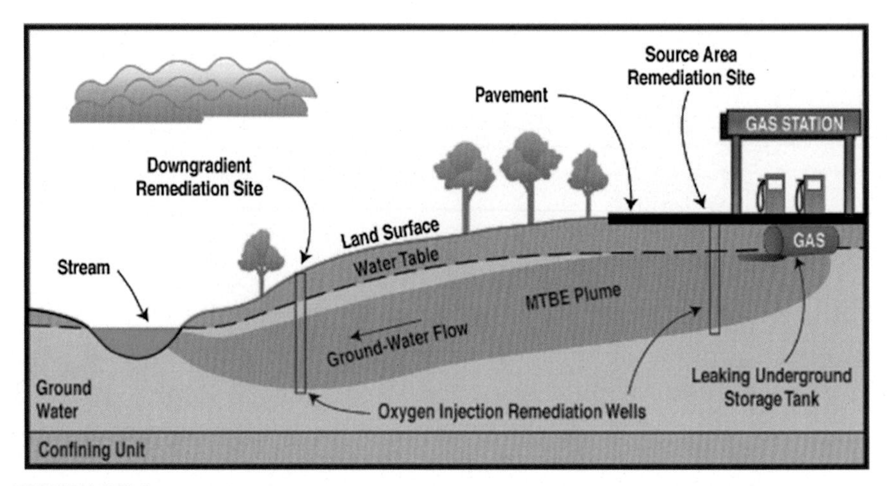

FIGURE 2.7.3
Diagram showing movement of the pollutant MTBE with groundwater flow. *Source: USGS.*

methyl tert-butyl ether (MTBE), a common organic groundwater pollutant, travels from a leaking underground storage tank with groundwater flow toward a stream. The figure also illustrates the use of remediation wells, one approach to cleaning up contaminated groundwater.

2. Pollutant behavior in the ground is different. Since there is usually little, if any, contact between the atmosphere and groundwater, pollutants like VOCs may remain intact for long periods of time once in groundwater without being oxidized as they would on the surface.

We'll look at two cases of groundwater pollution: nitrates in California's agricultural areas and arsenic in Bangladesh.

California: In a four-county area in the Tulare Lake Basin and in a portion of the Salinas Valley, 2.6 million residents rely on aquifers for their drinking water. This region also supports intensive agriculture (Fig. 2.7.4), and the heavy use of nitrogen fertilizers and the spreading of animal waste onto agricultural soils have led to widespread nitrate contamination of groundwater in the area (Helperin et al., 2001).

Remember that nitrates do not readily attach to soil particles and thus can move comparatively easily through soil systems with water. How much nitrate leaches into groundwater is influenced by how much water a soil can hold, how far the water table lies below the soil surface, how much nitrate is taken up by actively growing plants, and how much nitrate is already present in the soil. Regions like the Salinas Valley, where large amounts of nitrogen are applied to the soil, frequently experience nitrate contamination of groundwater.

FIGURE 2.7.4

The Salinas Valley region of California. Intense farming practices in this region result in an abundance of nitrates from fertilizers entering the groundwater. Note that both green and brown land covers represent different stages of agricultural plantings. *Image: Google Earth.*

CRUNCHING THE NUMBERS

Exercise 1: In 2015 in California, there were estimated to be 5.15 million cows raised for beef and dairy products (Fig. 2.7.5). Assume that each cow can produce 30 kg of manure a day.

Of this total organic waste, about 0.6 percent (0.18 kg) is composed of polluting nitrogen compounds (Helperin et al., 2001).

Based on typical soil texture and cation exchange capacity for the region, assume that 50% of these nitrogen compounds are converted to nitrate and transported through soils to groundwater.

Based on these values, answer the following questions:

- How many kg of waste (overall excrement) are produced by cows in California daily?
- Of this total waste stock, how many kg of nitrogen compounds are released into the soil system daily?
- Of this total supply of nitrogen waste, how much is leached into the groundwater each day?
- Knowing that fertilizer is typically sold in 50-pound bags, how many bags is this leaching equivalent to (in other words, in the absence of cows, how many bags of nitrogen fertilizer would you have to pour into the groundwater for the same pollution effect)?

Continued

CRUNCHING THE NUMBERS—Cont'd

FIGURE 2.7.5
Black cow at Freemont, CA. *By Mark J. Sebastian (IMG_23744) [CC BY-SA 2.0], via Wikimedia Commons.*

What's the Problem With Nitrates in Groundwater?

Ingestion of nitrates in drinking water poses a threat to infants and young children, whose digestive systems convert nitrates to nitrites, which can bind to hemoglobin and reduce the blood's oxygen-carrying ability. In severe cases, this can lead to potentially fatal "blue baby syndrome." Because of these concerns, California's drinking water standard for nitrates is 45 mg/L as NO_3^- or 10 mg/L as N.

EXPLORE IT

Exercise 2: Fig. 2.7.6 shows a GIS data layer of groundwater testing locations color-coded by whether they meet (blue, green, and yellow) or exceed (red) state water quality limits for nitrate. Note that dark red locations are twice the water quality limit.

There are several interesting questions that might be asked after examining this figure:

1. Considering all the sites tested, approximately what percentage of the public water supplies exceeded a safe level?
2. Approximately what percentage exceeded safe levels by at least a factor of two?
3. List three factors that might have contributed to the extreme variability in nitrate levels seen over short distances in the figure.
4. The study area includes some of the poorest communities in California; how might the prevalence of low quality water be linked to community income levels?
5. How does the nitrate contamination of the groundwater in agricultural areas of California compare to the ongoing lead poisoning issue in Flint, Michigan's drinking water supply? What are the similarities? The differences? Which problem do you think is more easily resolved? Why?

EXPLORE IT—Cont'd

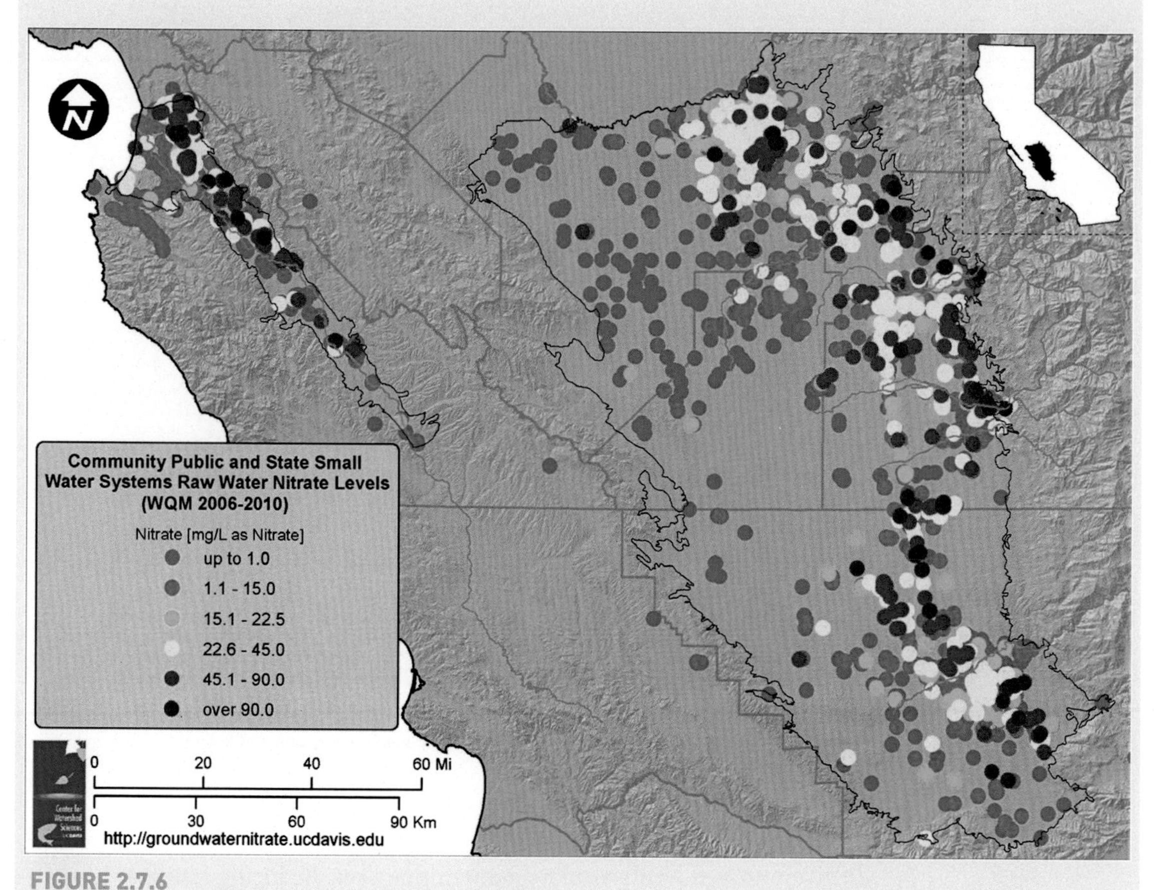

FIGURE 2.7.6
Nitrates (mg/L) in raw water from public water supplies in California's agricultural heartland (2006–2010). *Source: UC Davis.*

Solving the Problem: The extent of nitrate contamination in California's groundwater is substantial. A report from UC Davis' Center for Watershed Science (2010) estimated that 254,000 residents in the region are at risk of exposure to nitrate contamination, with most homes served by community or state public water systems.

Potential solutions are available but are challenging. Pumping up contaminated water and treating it to remove nitrogen is extremely expensive and is not a realistic approach in most settings. Instead, efforts have focused on reducing the amount of nitrate that enters the system.

DRILLING DEEPER

Exercise 3: Based on what you know about groundwater hydrology and nitrogen cycling, what are the primary pathways for nitrogen entry or removal from the system that must be evaluated when considering potential solutions to this problem?

FIGURE 2.7.7
Drip irrigation delivers water directly to roots to minimize water application and leaching of nutrients.
Source: USDA.

The State of California has been working with farmers to encourage more effective nitrogen fertilizer management practices. Reducing fertilizer loadings would be ideal, but California farmers deal with complex irrigation and fertilization requirements for "specialty" crops using diverse rotations.

Some of these "specialty" crops have a farm value of more than $4 billion per year. These high-value crops demand careful management of both water and nutrients to achieve high yield and consistently high quality (IPNI Guest Blog: Keeping Nitrate out of California Groundwater; see more at: http://www.cropnutrition.com/keeping-nitrate-out-of-california-groundwater#sthash.8o-C6Irc4.dpuf.) Altering fertilizer application rates could have substantial economic consequences for the region.

In addition to reducing the amount of fertilizer applied, farmers can alter the method and timing of fertilizer application. For many crops, yields have increased substantially in recent decades because of a switch to advanced techniques such as drip irrigation (Fig. 2.7.7). In a dry environment such as

California's, nitrogen management cannot be significantly improved without simultaneous improvement in water management.

Another alternative is "pump and fertilize," an approach in which farmers pump up nitrate-contaminated groundwater and apply it to their crops, essentially reusing the nitrates as a fertilizer. This can be risky, however, if salt levels in the groundwater (also a problem in many areas of the state) are too high for the crops to tolerate.

There are other technical approaches such as "blending" contaminated water with water from a clean source, but such approaches are often very costly, and small communities may not be able to afford to use these, even if a clean source is available. A fee charged to farmers who apply nitrogen fertilizers might also be used to offset some of the costs incurred by small communities faced with contaminated groundwater supplies.

A key step is education, as we know that once a groundwater supply is contaminated, there are usually no easy ways to improve the situation. Programs focused on helping farmers manage their land using a minimal amount of nitrogen can at least begin to reduce the extent of future contamination.

SOLUTIONS

Exercise 4: A basic understanding of the reactions of nitrogen in soils provides a solid foundation for making wise nutrient stewardship decisions. The International Plant Nutrition Institute has provided a series of fact sheets (http://www.ipni.net/nitrogennotes) that cover nitrogen fertilizer transformations in agricultural systems.

Read through their fact sheet entitled **Managing Nitrogen** and work in small groups to write a set of management suggestions designed to reduce the leaching of nitrate into groundwater in the region. Work in small groups to identify:

- Which of these are most likely to be successfully integrated into current farming practices? Why?
- Which of your suggested management techniques would have the greatest short-term impact on nitrate contamination levels? Why?
- Which would have the greatest long-term impact on contamination levels?

Bottom Line: The nitrates in California's wells likely entered the groundwater decades ago. More recently applied nitrogen is likely to continue to contaminate groundwater supplies for years to come. There are no easy answers to this problem. Solutions to long-term problems like this require careful planning and the cooperation of many entities, including regulators, scientists, and the public.

Bangladesh: A very different groundwater quality problem has occurred in Bangladesh. Concerned by the high incidence of illness among residents drinking contaminated water taken from shallow surface ponds, workers from international organizations began to drill tube wells to provide residents with what they assumed was a safer water supply. There are now about 10 million of these shallow wells accessed by hand pumps throughout the country (Fig. 2.7.8).

Unfortunately, the wells were sunk into shallow groundwater containing high levels of the carcinogenic trace element arsenic, which is soluble in anoxic waters (Columbia University, 2014). Starting as early as the 1970s, an estimated 51 million Bangladeshi were chronically exposed to arsenic at levels greater than the World Health Organization standard of 10 ug/L for safe consumption.

Chronic arsenic exposure is linked to a range of conditions, including cancers of the skin, bladder, kidney, and lung, as well as skin lesions (Fig. 2.7.9), arterial hypertension and cardiovascular disease, pulmonary disease, peripheral vascular disease, diabetes mellitus, and neuropathy (Smith et al., 2000).

While symptoms of arsenic poisoning were first seen in the mid-1980s, high levels of this odorless, tasteless toxin in Bangladesh's groundwater weren't discovered until 1993, years after people began drinking the polluted water. Flanagan et al. (2012) estimated that nearly 43,000 deaths occur annually in Bangladesh from cancer caused by chronic exposure to arsenic. This tragic case of good intentions gone awry is widely seen as one of the worst episodes of mass poisoning in history.

FIGURE 2.7.8
Tube well in a village of the province of Mazar in Afghanistan. *By Didiervberghe [Public domain], via Wikimedia Commons.*

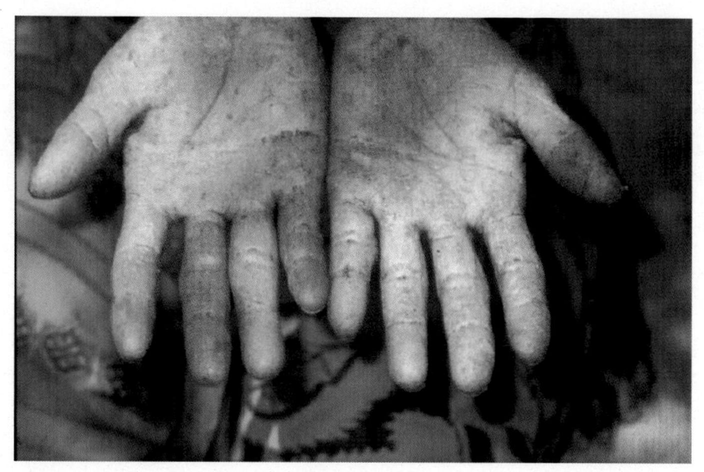

FIGURE 2.7.9
Arsenic-induced hyperkeratosis of the hands. *Source: CDC.*

CRUNCHING THE NUMBERS

Exercise 5: Fig. 2.7.10 shows the range of arsenic measured in groundwater supplies during two national surveys, one in 2000 and the second in 2009.

Arsenic concentration (µg/L)	2000 survey (n = 3 534) Proportion (%)	2000 survey (n = 3 534) Cumulative (%)	2009 survey (n = 14 442) Proportion (%)	2009 survey (n = 14 442) Cumulative (%)
0–10	57.9	57.9	68	68
10.1–50	17.1	75.1	18.7	86.6
50.1–100	8.9	84	7.2	93.8
100.1–150	4.2	88.2	1.4	95.2
150.1–200	2.9	91.1	1.4	96.6
200.1–250	2.1	93.2	1.1	97.8
250.1–300	1.8	94.9	0.4	98.2
300+	5.1	100	1.8	100

FIGURE 2.7.10
The range of arsenic measured in groundwater supplies in 2000 and 2009. BGS, British Geological Survey; DPHE, Department of Public Health Engineering; MICS, Multiple Indicator Cluster Survey. *Source: (Flanagan et al., 2012).*

Continued

CRUNCHING THE NUMBERS—Cont'd

Consider that 2000 and 2009 had means and standard deviations across all observations as follows. Was there a significant decrease in arsenic contamination between 2000 and 2009?

2000: $X = 4.52$, $s = 0.30$, $n = 3534$ and

2009: $X = 2.73$, $s = 0.21$, $n = 14,442$

Now let's test to see if there is an association between the number of wells that exceeded or did not exceed safe water quality standards (>10 ug/L) between the 2 years based on the count of wells in exceedance (Fig. 2.7.11).

1. Did your chance of drinking contaminated water differ between 2000 and 2009?
2. In which year were you more likely to have drunk contaminated water?

		Year	
		2000	2009
exceeds	no	2651	12507
	yes	884	1935

FIGURE 2.7.11
When data are reported as frequencies (counts of occurrences), a chi-square table can help identify any exceedance differences between the two surveys.

PUTTING SCIENCE TO WORK

Exercise 6: Fig. 2.7.12 shows the area of Bangladesh where exposure to contaminated groundwater has been the greatest.

Do some digging online to find a map of population density in Bangladesh.

1. How does the area of greatest exposure match up with population centers in Bangladesh?
2. What does the distribution of arsenic in wells tell you about the probable source of contamination?
3. Based on this distribution and the concerns for arsenic poisoning across Bangladesh, how would you design a systematic well sampling network to identify contaminated wells?

PUTTING SCIENCE TO WORK—Cont'd

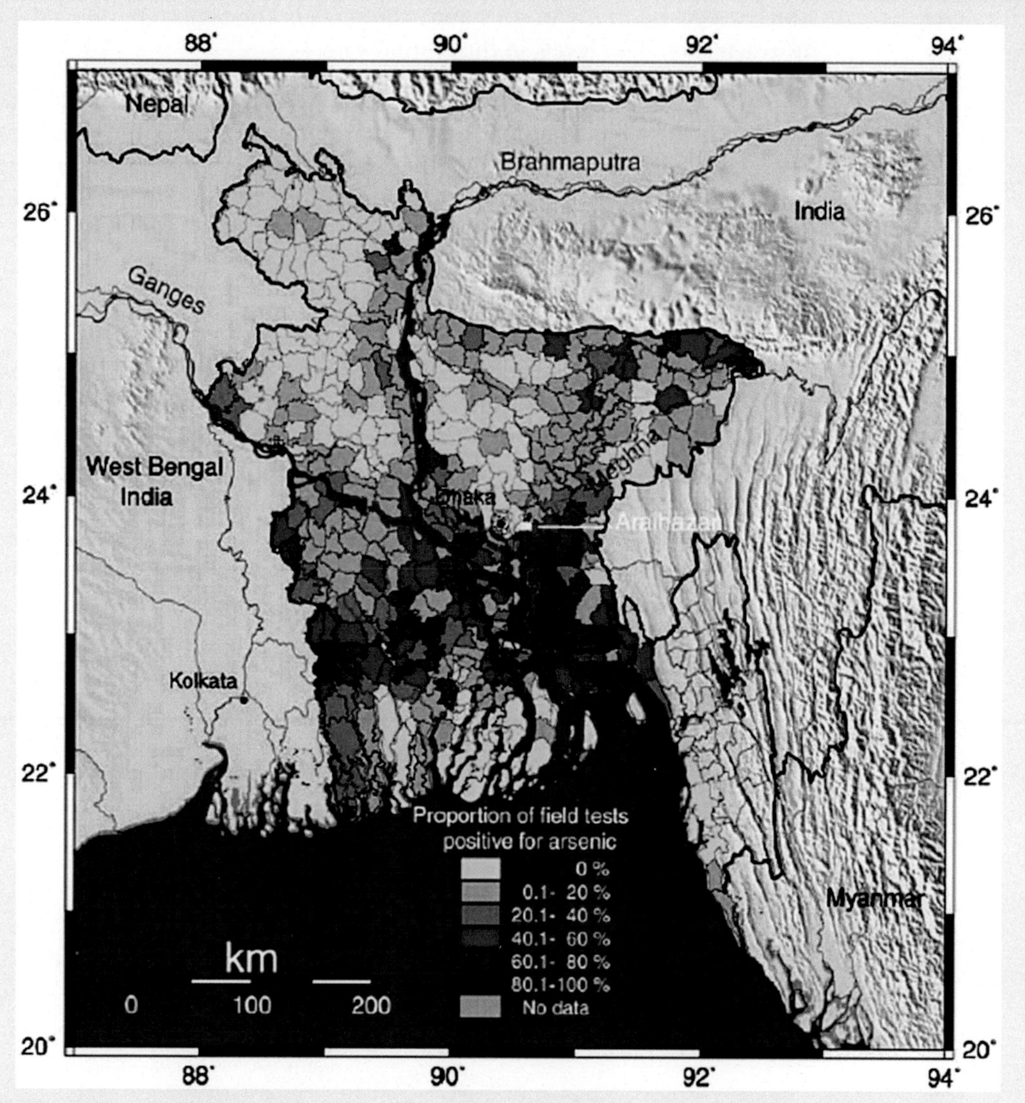

FIGURE 2.7.12

Percentage of wells that were yielding arsenic-contaminated water in Bangladesh. *Source: USGS International Programs.*

How did something this terrible happen? While the waters from the tube wells were initially tested for microbial quality, apparently no one measured arsenic, which occurs naturally in soils and sediments in the region. But how did arsenic reach such high levels in the groundwater?

Arsenic contamination of groundwater can often be considered a "natural process" (i.e., not directly added to a system by human activity). To understand how arsenic can naturally contaminate groundwater, one must consider the three major factors involved, namely, the source of the arsenic, its mobilization, and its transport into groundwater.

Environmental Sciences in Action

A team of scientists and engineers from MIT may have discovered the source of arsenic in Bangladesh's groundwater. Led by Professor Charles Harvey, the team from MIT began to study the thousands of ponds dug in Bangladesh to provide soil for flood protection (Neumann et al., 2010). Fig. 2.7.13 shows Rebecca Neumann, a member of the team, collecting a water sample for arsenic analysis. Harvey and his colleagues theorized that organic carbon from natural processes settled onto the bottom of the ponds and then seeped underground, where it was metabolized by microbes. The scientists believe that the oxidation of this carbon caused the release of arsenic from soils and sediments into the groundwater, resulting in

FIGURE 2.7.13

Rebecca Neumann hangs off the end of bamboo scaffolding built at the field site to draw water samples from the field. *Photo: Sarah Jane White, MIT.*

the high levels that have been found. This case clearly illustrates the need to look at the systems connections when trying to solve environmental problems.

Solving the Problem: Tragically, millions of Bangladeshi have been chronically exposed to arsenic in their drinking water. In too many cases, the damage has been done. Some steps are being taken to try to lessen the severity of the problem (Columbia University, 2014). Wells sunk more deeply into the ground tend

CONSIDER THIS

Exercise 7: Note that arsenic contamination of groundwater is not simply a problem for Bangladesh. In fact, some of the highest risk is in developed countries such as the US and Australia (Fig. 2.7.14).

FIGURE 2.7.14
A USGS map of arsenic concentrations in groundwater across the US based on samples from 1350 wells. In 2001 the US EPA lowered the maximum level of arsenic permitted in drinking water from 50 to 10 ug/L, making raw water from all orange and red locations on this map unfit for human consumption. *Source: USGS.*

Continued

CONSIDER THIS—Cont'd

Watch the following video from the Arsenic Mitigation and Research Foundation linked from your Resource Page at https://www.youtube.com/watch?v=OBHu8srmxsc.

1. Why is arsenic contamination a greater human health issue in Bangladesh than in other parts of the world?
2. How is this a social justice issue?
3. What steps could be taken to eliminate arsenic poisoning in poor regions?
4. How do developed nations like the US deal with arsenic in their drinking water supplies?

to have lower arsenic levels and thus may provide an alternative source in some areas. Also, treatment systems involving the use of plastic buckets, sand filters, and the addition of iron tablets can be used on individual wells to reduce arsenic levels, but, given the number of tube wells involved, the widespread application of such treatment schemes is a major undertaking. Experts believe that millions of Bangladeshi are still consuming arsenic-contaminated water.

GROUNDWATER QUANTITY: THE OGALLALA AQUIFER

FIGURE 2.7.15
Irrigation in the Ogallala Aquifer farming belt. *Source: NRCS.*

Groundwater supplies are not infinite. If a particular groundwater formation is heavily used and is being recharged only slowly from the surface, groundwater levels may begin to decline. If the water table drops enough, users may need to sink new wells to obtain their water. The Ogallala Aquifer in the US Great Plains represents a classic case of overuse of a vitally important groundwater supply.

The Ogallala Aquifer in the US Great Plains underlies eight US states and covers more than 440,298 km^2. It is a vital groundwater resource supporting both domestic use and agricultural production. The Ogallala supports nearly one-fifth of the wheat, corn, cotton, and cattle produced in the US, and accounts for 30% of all groundwater used for irrigation in the country (USDA-NRCS, 2015).

It also is heavily overused in some areas and is declining rapidly in areas like western Kansas and parts of Oklahoma and Texas. In some areas

(Fig. 2.7.16), the level of the aquifer has dropped 50 m or more. Once drained, it could take hundreds or thousands of years for this ancient aquifer to fully recharge with rainfall.

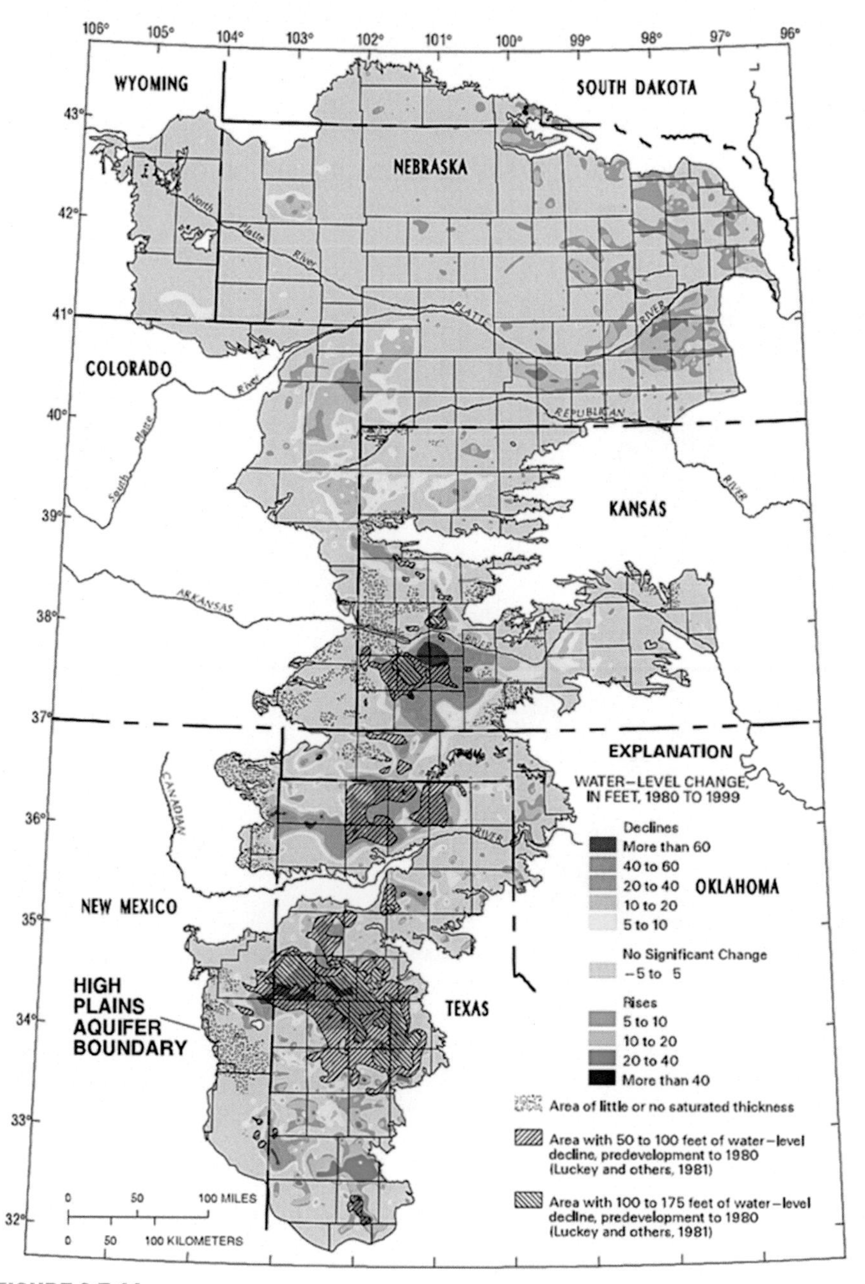

FIGURE 2.7.16

Changes in water levels of the Ogallala Aquifer between 1980 and 1995. *Source: USGS.*

EXPLORE IT

Exercise 8: How do scientists measure aquifer levels?

A joint mission between Germany and the US designed the Gravity Recovery and Climate Experiment (GRACE) to measure gravity from space. By studying gravitational anomalies caused by changes in locational mass over time, GRACE scientists can approximate changes in terrestrial water storage (see Fig. 2.7.17).

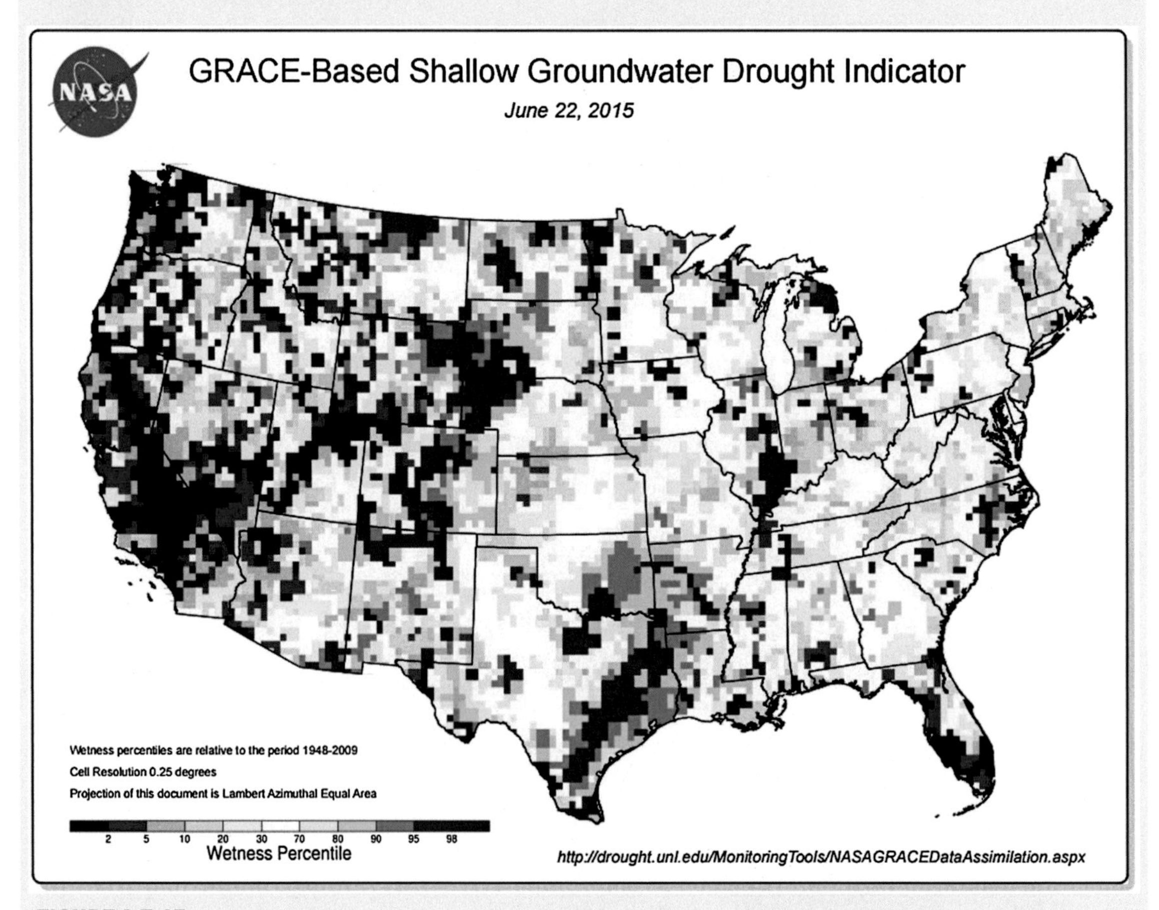

FIGURE 2.7.17

High-resolution groundwater indicator maps enable improved drought and flood risk monitoring across the continental US. These products integrate data from NASA's GRACE mission with other observations within a sophisticated land surface model in order to characterize wetness conditions. New maps are distributed weekly from http://www.drought.unl.edu/MonitoringTools.aspx. *Source: NASA.*

Scientists are using GRACE to monitor water levels in many of the world's most critical aquifers. Read the recent study of 37 aquifers around the world over the course of 10 years (Richey et al., 2015) linked on your Resource Page at **http://onlinelibrary.wiley.com/doi/10.1002/2015WR017349/pdf**.

- What is their assessment of the condition of these aquifers?
- How accurate are the GRACE assessments?
- Is it possible that their measurements could lead to incorrect conclusions over this 10-year study period? Why?

EXPLORE IT—Cont'd

In addition to technical articles, scientists at NASA's Goddard Space Flight Center generate groundwater and soil moisture drought indicators for the continental US each week.

Other visualizations are available online for many watersheds. Review this animation link **https://www.youtube.com/watch?v=zu4cBM4m5gU** found on your Resource Page to see the trends in aquifer height across Australia, California, the Middle East, and the Indian subcontinent.

- Based on this visual assessment, which area's aquifers have demonstrated the most significant loss?
- How do you assess loss (total area, proportion, severity of decline, etc.)? Justify your choice of measures to best capture the state of this important resource.

DRILLING DEEPER

Exercise 9: While many parts of the Ogallala Aquifer are experiencing significant declines, some areas are actually increasing.

- Based on your knowledge of groundwater hydrology, what might be causing levels to increase in a few areas?

Environmental Sciences in Action

Overuse and climate change are likely both contributing to the decreases noted in the level of the Ogallala Aquifer. Since the future is likely to bring both increased demand and the threat of diminished recharge because of climate change, scientists and land managers are working to slow the rate at which water is being withdrawn from the Ogallala.

Little (2009) described one high-tech approach to the problem. Scientists and engineers at a USDA research station in Texas have installed wireless infrared sensors on the arms of center-pivot irrigation systems in cotton fields (Fig. 2.7.18). The sensors measure leaf temperatures, and the plants' temperatures can "tell" the sensors when they need water. By applying water only when it is needed, less water needs to be withdrawn from the aquifer, slowing its decline.

Faced with the likelihood of further declines in levels of the Ogallala, some farmers are shifting production to dryland farming, growing crops that don't need irrigation. Others are growing crops like sunflowers, which require less water than most other crops.

Farmers are not alone in trying to solve this problem. Using a comprehensive set of conservation practices, the Ogallala Aquifer Initiative (OAI), coordinated

FIGURE 2.7.18
USDA-Agricultural Research Service agricultural engineers Susan O'Shaughnessy and Nolan Clark adjust the field of view for wireless infrared thermometers. *Source: USDA.*

by the US Department of Agriculture's Natural Resources Conservation Service (NRCS), is working to reduce aquifer water use and enhance the economic viability of croplands and rangelands served by the Ogallala. To achieve these goals, NRCS is working with landowners to develop and implement conservation practices. To encourage farmers to adopt such practices, conservation activities are carried out using NRCS's Environmental Quality Incentives Program with funding provided by state and local agencies.

SOLUTIONS

Exercise 10: Read more about the OAI at the link on your Resource Page: http://www.nrcs.usda.gov/wps/portal/nrcs/detailfull/national/programs/initiatives/?cid=stelprdb1048809. Consider that many farmers may be resistant to changing their agricultural practices. With their livelihoods dependent on large harvests of high value crops and many farms one failed crop away from foreclosure, it isn't easy to take a gamble on a new crop or agricultural system.

- How has the OAI worked to connect with landowners and ensure implementation of their suggested conservation techniques?
- How can organizations like the OAI assess their short-term success (considering that changes in aquifer levels occur over many years)?
- What other steps might the OAI consider in their efforts to restore groundwater levels in the aquifer?

SYSTEMS CONNECTIONS

Exercise 11: Climate change will likely have substantial impacts on groundwater systems around the world, not just the Ogallala. Read R.G. Taylor et al.'s review article "Ground water and climate change" available on your Resource Page (Nature Climate Change, November 25, 2012).

- Discuss three ways that climate change will likely affect groundwater systems and three ways that the climate is, in turn, affected by groundwater systems.

THE BIG PICTURE: WILL THERE BE ENOUGH TO GO AROUND?

Shrinking groundwater resources are a growing international concern. Goldenberg (2014) noted that, over the past decade, groundwater has been pumped out of the ground 70% more rapidly than in the 1990s. Overuse of groundwater in traditionally arid environments is increasing the risk of conflict. Even US national security experts have warned that overuse of groundwater and subsequent shortages can be a threat to the nation's security.

J. Famiglietti is a water cycle scientist with appointments at both NASA's Jet Propulsion Laboratory and the University of California at Irvine. He and his colleagues have used satellite data to assess groundwater levels around the globe. The team's analysis indicates that major aquifers in the world's arid and semiarid regions are seeing rapid rates of depletion.

Famiglietti and his team examined 37 of the globe's largest aquifers between 2003 and 2013 and found that eight, mostly in arid regions, were overstressed, with the Arabian Aquifer System, which serves more than 60 million people, ranked as being in the worst condition (Richey et al., 2015). What's most disturbing is that no one even knows how much water is stored in these aquifers and that estimates of "time to depletion" are likely just guesses.

Famiglietti suggests that ground and surface water systems must be managed as one entity, but this is complicated by the fact that water often crosses political boundaries. At the least, data on groundwater use and levels must be shared among countries (Famiglietti, 2014).

Because we currently lack the political infrastructure to effectively share groundwater across national boundaries, Famiglietti warns that climate change, population growth, and water overuse are threats to international peace and that a "full appreciation of the importance of groundwater to the global water supply and security is essential for managing this global crisis." Experts predict that by 2020, water problems will lead to instability in many countries across the globe. By 2040, they predict that water availability will no longer meet the demands of a growing population, with potential economic and human casualties.

Some Good News: Nations in the Near East and in North Africa, in an effort coordinated by the UN's Food and Agriculture Organization (FAO), have developed a partnership called the Regional Water Scarcity Initiative designed to boost agricultural productivity, improve food security, and use water resources sustainably.

CRITICAL THINKING

Exercise 12: A report from an FAO Regional Conference held in 2014 (FAO, 2014), linked on your Resource Page (http://www.fao.org/docrep/meeting/030/mj380e.pdf) contains a number of recommendations for improving water management throughout the Near East and North Africa. Go to your Resource Page, review the report, and answer the following questions.

- What is their definition of water scarcity? Do you believe this is an appropriate threshold to identify regions with water supplies at risk?
- Identify three specific steps that the FAO group identifies to better manage water in this region. Do they outline specific measures to quantify the impact of these steps?
- What are some of the political hurdles that must be overcome before regional water management can be successful?
- What are some additional steps that developed nations could take to help this initiative succeed?

REFERENCES

Columbia University, 2014. Health Effects and Geochemistry of Arsenic and Manganese. NIEHS Superfund Program Center for International Earth Science Information Network. http://superfund.ciesin.columbia.edu/.

Famiglietti, J.S., 2014. The global groundwater crisis. Nat. Clim. Change 4, 945–948.

Flanagan, S.V., Johnston, R.B., Zheng, Y., 2012. Arsenic in tube well water in Bangladesh: health and economic impacts and implications for arsenic mitigation. Bull. World Health Org. 90, 839–846. http://www.who.int/bulletin/volumes/90/11/11-101253/en/.

Goldenberg, S., February 8 2014. Why Global Water Shortages Pose Threat of Terror and War. The Guardian: The Observer. http://www.theguardian.com/environment/2014/feb/09/global-water-shortages-threat-terror-war.

Helperin, A.N., Beckman, D.S., Inwood, D., 2001. California's Contaminated Groundwater: Is the State Minding the Store? Natural Resource Defense Council. http://www.nrdc.org/water/pollution/ccg/ccg.pdf.

Little, J.B., March 1, 2009. The Ogallala Aquifer: Saving a Vital U.S. Water Source. Scientific American. http://www.scientificamerican.com/article/the-ogallala-aquifer/.

Neumann, R.B., Ashfaque, K.N., Badruzzaman, A.B.M., Ali, M.A., Shoemaker, J.K., Harvey, C.F., 2010. Anthropogenic influences on groundwater arsenic concentrations in Bangladesh. Nat. Geosci. 3 (1), 46–52.

Richey, A.S., Thomas, B.F., Lo, M.-H., Reager, J.T., Famiglietti, J.S., Voss, K., Swenson, S., Rodell, M., 2015. Quantifying renewable groundwater stress with GRACE. Water Res. Res. 51, 5217–5238. http://dx.doi.org/10.1002/2015WR017349.

Smith, A.H., Lingas, E.O., Rahman, M., 2000. Contamination of drinking-water by arsenic in Bangladesh: a public health emergency. Bull. World Health Org. 78, 1093–1103.

U.S. Department of Agriculture-Natural Resources Conservation Service, 2015. Ogallala Aquifer Initiative. http://www.nrcs.usda.gov/wps/portal/nrcs/detailfull/national/programs/initiatives/?cid=stelprdb1048809.

UC Davis Center for Watershed Sciences, 2010. Nitrate in California's Groundwater. 2 p. groundwaternitrate.ucdavis.edu.

Chapter 2.8

China's Three Gorges Dam: Costs Versus Benefits

FIGURE 2.8.1
The Three Gorges Dam. *Photo by Hugh via Flicker.*

OVERVIEW

Human activities often affect water resources, with impacts sometimes local and limited (e.g., if a farmer drains a wetland, there likely will be negative consequences, but, in some cases, they may be limited in scope) and other times, widespread and substantial. Some actions can even permanently alter the structure and function of major aquatic ecosystems.

There may be no better example of this later case than the construction and operation of large dams on major rivers, with both short- and long-term consequences possible above and below the dam. Sometimes, these dams are built on pristine rivers. In such cases, many of the more serious impacts result from

FIGURE 2.8.2
Two of the world's largest dams: Itaipu Dam (Brazil and Paraguay) (left) and the Terbela Dam (Pakistan) (right). *By Alicia Nijdam [CC BY 2.0] and Zakariahazro [CC BY-SA 3.0], via Wikimedia Commons.*

habitat alteration. In other cases, dams are built on rivers with large human populations in their watersheds. In these cases, pollutants entering the river from point and nonpoint sources within the watershed may play important roles.

There are many examples of dams spanning large rivers around the globe (Figs. 2.8.1 and 2.8.2). While we will focus on environmental impacts, which are generally negative, these massive structures may provide many important benefits, including flood control, improved navigation to facilitate the passage of large ships, and hydroelectric power generation.

One of the largest and most controversial major dam projects in recent history is China's Three Gorges Dam (TGD) (Fig. 2.8.1). The dam is about 2.3 km long and 185 m tall and is the world's largest hydropower station in terms of installed capacity. The Chinese government regards the project as an historic engineering, social, economic, and environmental success. However, the dam flooded archaeological and cultural sites, displaced about 1.3 million people, and has caused significant changes in the Yangtze River ecosystem.

In this case study, we'll look at how the TGD has altered environmental conditions in the Yangtze River. We'll focus on two important aspects: (1) how the TGD has affected fish populations in the river, and (2) how TGD operation has altered pollutant behavior in the ecosystem. We'll consider impacts both above and below the TGD.

BACKGROUND

Before you begin this case study, go to one of the websites listed on your Resource Page to review (1) how large-scale dams are constructed and how they generate electricity, and (2) background on the Yangtze River and its drainage basin (Fig. 2.8.3).

FAST FACTS

The Yangtze and Its Basin

- At 8380 km, it is the third longest river in the world.
- The watershed of 1.8 million km^2 covers about 1/5 of China and is home to 400 million people.
- About 350 fish species live in the river, while the watershed is home to as many as 6000 plant species and 500 terrestrial vertebrates (Fu et al., 2010).
- The Three Gorges Reservoir (TGR) has inundated 55,742 km^2 of the basin.
- The watershed is 74% mountainous and 22% low hills, with 46% of the land forested and 39% in crops.

Water Quality in the Yangtze

- Many sources of pollution lie along the river or within the watershed; for example, Sichuan Province (population: about 81 million in 2013)

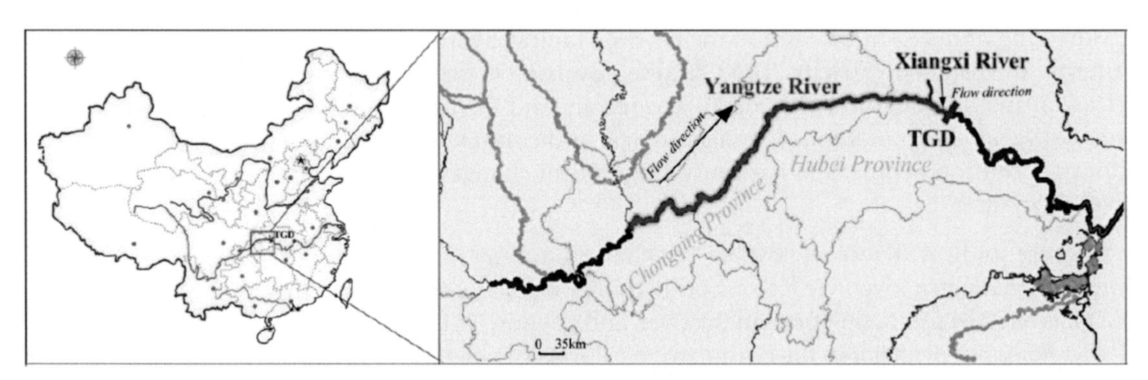

FIGURE 2.8.3
Location of the Three Gorges Dam and the Xiangxi River. *Source: (Lian et al., 2014).*

in the upper watershed generated 1204 MT of industrial waste and 1282 MT of sewage in 2010 (Xu et al., 2013)

- Industrial discharges, agricultural fertilizers, and sewage are important sources of pollution.

Three Gorges Dam and Reservoir

- The dam took 17 years to construct.
- Twenty-six turbines can generate 18,000 MW.
- The dam started storing water in 2003; it was completely filled in 2009.
- The dam can store 39.3 km^3 of water at full capacity.
- The surface water area of TGR is more than 1080 km^2.
- TGR stretches about 600 km behind the dam (Fig. 2.8.4).
- Many mountaintops have become Islands and up to 1.3 million people have been relocated.

A few key terms you should be familiar with:

- **lotic**: flowing water (e.g., a river)
- **lentic**: standing waters (e.g., a lake)
- **lacustrine**: lake-like
- **riverine**: river-like
- **dam**: the physical barrier across the river
- **reservoir**: the body of water held behind the dam

FIGURE 2.8.4
This image from April 15, 2009, is one of the first images that astronauts on the International Space Station were able to capture of the flooding behind the dam. *Source: NASA.*

EXPLORE IT

Exercise 1: Satellites have been capturing images of the Earth's surface for decades. Many of these sensors provide current and historical imagery that is crucial to monitoring and quantifying global change. Images of the TGD construction (Fig. 2.8.5) not only show dramatic changes in hydrology that accompanied impoundment of the Yangtze River, but also demonstrate changes in land use within the larger watershed.

On your Resource Page, you will find a Google Earth TGD.kmz file from the "UNEP: Atlas of our Changing Environment" that includes historical Landsat imagery from NASA from 1987 to 2004. Download and explore this imagery in Google Earth to better understand the extent of changes in hydrology and land use in the region upstream of the TGD.

- Using the Google Earth measurement tool, approximately how far upstream do you think the river impoundment extends?
- What is the primary change in land use in the upstream watershed? How extensive has this conversion been over the 17-year span of these images?

Turn off the TGD.kmz image layers and examine the most recent imagery along the Yangtze in Google Earth.

- Describe the types of development now common along the banks of Yangtze River.
- Where is agriculture primarily located? Use your photointerpretation skills to determine why agriculture is located in these areas.

EXPLORE IT—Cont'd

- You should be able to zoom in to see individual cargo ships. Choose several random stretches along the Yangtze River and approximate the average density of ships per kilometer along this major shipping route.

FIGURE 2.8.5
NASA earth observatory images of the TGD area before and after construction. *Source: NASA.*

THE PROBLEMS

Construction of large dams can have many impacts on the structure and function of riverine ecosystems, with cascading effects throughout the broader ecosystem. Environmentalists have warned that TGD is reducing downstream nutrient and sediment flow and impacting coastal fishing grounds and tidal wetlands.

The dam is also taking a toll on China's flora and fauna. Biodiversity has been threatened as the dam flooded some habitats and reduced water flow to others. Economic development has spurred deforestation and pollution in surrounding provinces in central China, endangering many sensitive plant species.

The submergence of hundreds of industrial centers upriver has led to an influx of silt, pollutants, and trash into the reservoir. Additional pollution of the TGR resulted from the growth of several major cities along the new river banks. Erosion of river banks is causing landslides and threatening one of the world's largest fisheries in the East China Sea. Critics have also argued that the reservoir has altered microclimates and that water retention may have exacerbated recent droughts.

The literature contains many articles about the effects of the TGD. Scientists have been assessing its impacts on aquatic biota ranging from microplankton to the largest fish species. We'll focus on the effects on fish populations, looking at two types of impacts, one above and one below TGD. We'll also look at water quality issues, with one focused on eutrophication of waters above the TGD and the other on the effects of TGD on downstream water quality.

FISH HABITAT IMPACTS

The filling of TGR occurred in stages: the first filling in 2003 raised the water level to 135 m, the second filling in 2006 increased the water level to 156 m, and the third and final filling raised the level to 175 m in 2009 (Fig. 2.8.6) (Gao et al., 2010). Altered water levels and flow rates can have serious impacts on species not able to adapt to the changing conditions. The nature of the impacts on individual fish species depends on where along the course of the river you look.

Upriver Impacts

A consequence of river impoundment is the transformation of lotic environments to lentic habitats, altering the conditions in which fish species have evolved to breed, feed, and find protection from predators. Increased predation on migratory fish is also linked to dams, largely because fish are concentrated and habitats become more favorable for predatory species. Another

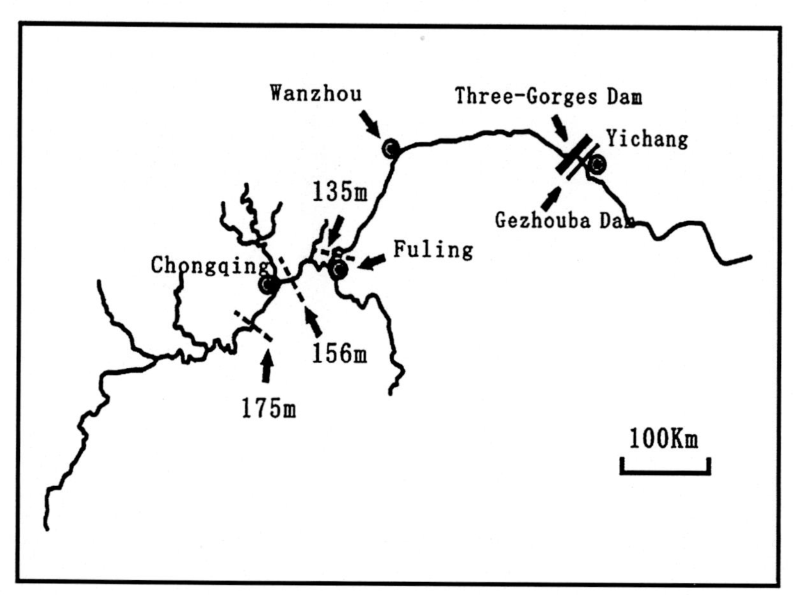

FIGURE 2.8.6
Upstream extension of Three Gorges Reservoir with each successive filling. *Source: (Gao et al., 2010).*

common effect of dam construction is the decline of anadromous species, fish that are hatched in freshwater, spend most of their life in the sea and return to spawn in freshwater. The dam prevents migration between feeding and breeding zones. The effect can become severe, leading to the extinction of species (United Nations FAO, 2001).

DRILLING DEEPER

Exercise 2: Gao et al. (2010) assessed changes in fish communities in the Yangtze River as the TGR formed by examining the catch from fishing boats at two sites: the Wanzhou reach, which became lacustrine with the first filling in 2003, and the Fuling reach, which was inundated by the second filling in 2006.

The team of scientists collected 12,952 fish from 57 different species in the Wanzhou reach (Fig. 2.8.7). The percentage of species favoring lacustrine conditions increased from 69% prior to filling to 94% after filling, a significant change in species composition.

In the Fuling reach, where 2451 fish of 22 species were collected, lacustrine species increased from less than 1% to 12% in abundance after inundation.

1. Why do some fish species favor riverine conditions, while others prefer more lake-like conditions?
2. The authors noted that fish species which favored riverine conditions moved upriver further into tributaries emptying into the Yangtze. Explain how this might have negative long-term consequences for the migrating species and the species already living there?

Continued

DRILLING DEEPER—Cont'd

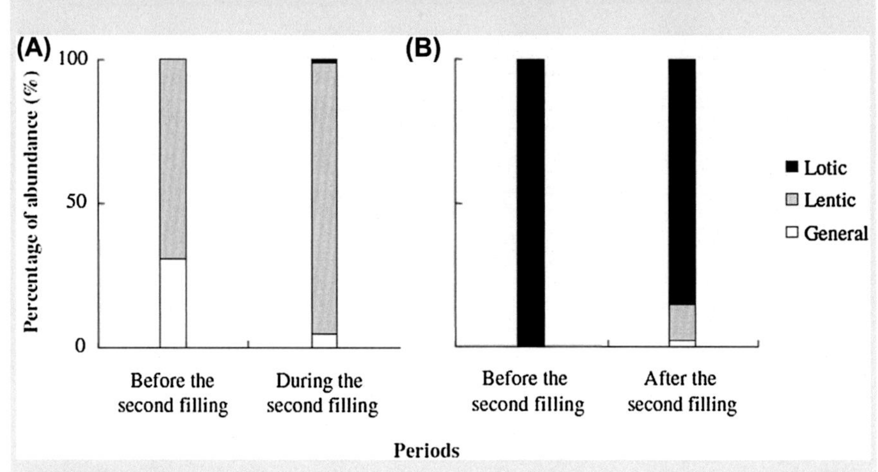

FIGURE 2.8.7
Abundance of ecological groups in the different stages in the Wanzhou reach (A) and the Fuling reach (B). *Source: (Gao et al., 2010).*

CRUNCHING THE NUMBERS

Exercise 3: Interestingly, species richness increased during and after the filling, from 34 to 49 species in the Wanzhou reach and 12 to 21 species in the Fuling reach. Fig. 2.8.8 shows species diversity and evenness indices calculated for the fish communities in the two reaches before and after the second filling.

Community index	Wanzhou reach		Fuling reach	
	Before filling (Oct., 2005)	During filling (Sept.–Oct., 2006)	Before filling (Sept., 2006)	After filling (Oct.–Nov., 2006)
Shannon–Wiener diversity index	1.04 (1.01–1.07)	1.57 (1.53–1.60)	1.38 (1.33–1.42)	1.81 (1.73–1.88)
Simpson's diversity index	0.53 (0.52–0.54)	0.68 (0.67–0.69)	0.70 (0.68–0.71)	0.75 (0.72–0.77)
Buzas & Gibson's evenness index	0.08	0.10	0.33	0.29

FIGURE 2.8.8
Various indices and 95% confidence intervals for fish communities in the different stages in the Wanzhou reach and the Fuling reach (Gao et al., 2010).

1. Are the differences you see in the Shannon-Wiener and Simpson diversity indices significant?
   a. Before and during filling at the Wanzhou reach?
   b. Before filling and after filling at the Fuling reach?
   c. Which site was impacted more? Explain and justify your response.
2. What might account for the increase in values for diversity indices at the two sites?
3. Are there significant differences in diversity between the two locations? Considering their locations, why might this be?

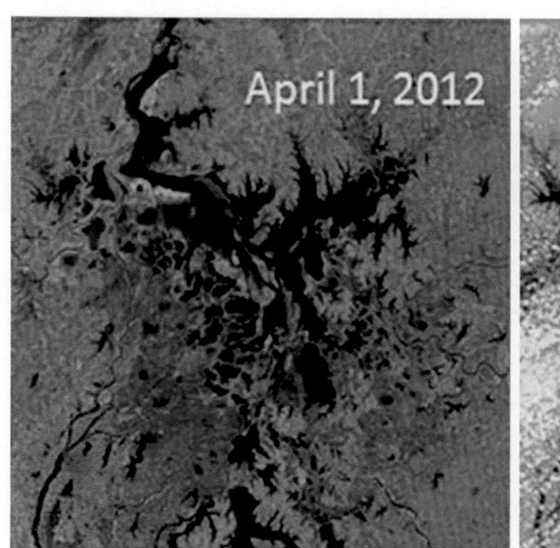

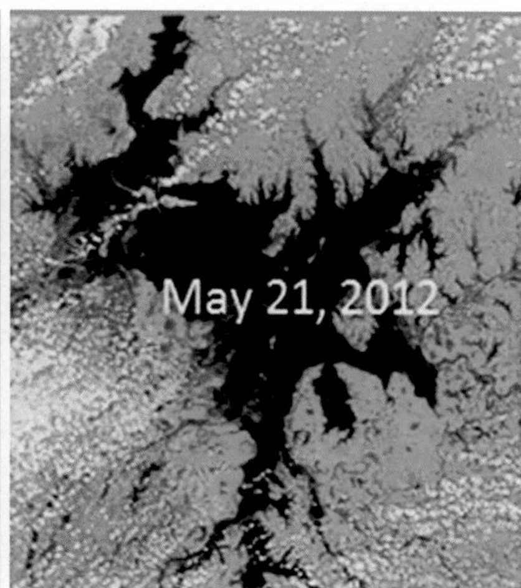

FIGURE 2.8.9
Water levels at Poyang Lake downstream of the Three Gorges Dam are controlled primarily by release of water from the dam. *Images courtesy of USGS.*

Downriver Impacts

The suppression of the flood regime below dams can deprive many downriver fish species of spawning grounds and valuable food supply. This modification of downriver flow can also reduce stimuli for migration, interfere with migration routes, and decrease survival of eggs and juveniles. Dams also affect the quality and temperature of water downriver. Release of anoxic water from the hypolimnion of the reservoir can cause fish mortality below the dam. Water that spills over the crest of the dam can become oversaturated with atmospheric gases (oxygen and nitrogen), reaching levels that can be lethal to fish (United Nations FAO, 2001).

Widely fluctuating water flows, particularly during times of flood, have reduced water levels in the lower reaches of the Yangtze, altering the habitat of many fish species there. These impacts can extend hundreds of km. China's largest freshwater lake, the Poyang, typically covers 3500 km^2, but, in April 2012, the surface area shrank to only 200 km^2 (Fig. 2.8.9). Caused in part by the worst drought in over half a century, this catastrophic decrease was also influenced by water retention above the TGD 500 km upstream.

Scientists at the Lake Science and Environment Laboratory at Nanking University found that the artificial regulation of TGR, which must be kept full to optimize electricity generation, significantly impacted the water level in the lower reaches of the Yangtze where Poyang Lake is located.

FIGURE 2.8.10
Algae on surface of Lake Taihu in the Yangtze delta. *Source: eutrophication&hypoxia Flicker.*

Also, with a decreasing flow of water into the lowest reaches of the river, saltwater from the East China Sea has moved further upriver, increasing salinity levels to the point where some freshwater species may have been eliminated (Hvistendahl, 2008). Scientists have even detected species of jellyfish normally seen only in the South China Sea in the lower reaches of the Yangtze.

The dam-related decrease in fish populations has repercussions all the way up the food chain to top predators, such as the Yangtze River Dolphin. Critically endangered during the time of dam construction, this historically sacred mammal was considered virtually extinct in 2007. A substantial commercial fishing industry was catching the dolphin's prey species, and the additional decrease in prey numbers after the start of TGD operation likely sealed the dolphins' fate.

The dam also interfered with the dolphin's technique for capturing prey, with rushing waters below the dam disrupting the nearly blind dolphins' use of sonar. This unique and fragile species sits on the edge of extinction, but many are still working to bring the population back to Chinese rivers. For more information on the Yangtze River dolphin, visit http://www.edgeofexistence.org/mammals/species_info.php?id=1.

WATER QUALITY IMPACTS

Upriver Issues

When water is retained behind dams, it undergoes many chemical and physical changes. The degree to which water quality is impacted is often related to the retention time of the reservoir. In general, the longer the retention time (e.g., where water is stored for many months or years), the greater will be the negative impacts on water quality.

Reservoirs trap both sediments and nutrients delivered from upstream sources. Increased nutrient concentrations can lead to increased populations of algae near the surface (Fig. 2.8.10). While algal blooms may provide food for some fish, they can also impart unpleasant tastes and odors that make the water unfit for either household or industrial use. As algae sink to the bottom, oxygen levels drop as microbes break down the organic matter, resulting in anaerobic conditions in some reservoirs.

The changing nature of the Yangtze River from riverine to lacustrine has presented an ideal opportunity for primary producers like algae to increase their populations. There have been frequent algal blooms in tributaries draining into the Yangtze River near TGD since 2003 when the first filling began (Lian et al., 2014).

Environmental Sciences in Action

To study the severity of these algal blooms, J. Lian and his research team focused on the Xiangxi River, a tributary only 31.3 km above the dam. The Xiangxi has experienced eutrophic conditions and algal blooms from March to October since 2003. In addition to odor and aesthetic concerns, some algal species in the blooms can produce toxic byproducts. A blue–green algae bloom in 2008 in the Xiangxi caused authorities to ban its use as a drinking water supply for local residents (China-Wire, 2008).

The cause of algal blooms in the Xiangxi and similar tributaries is twofold: first, there are plenty of nutrients to stimulate primary production. Sources in the Xiangxi basin include phosphorus mines, chemical processing plants, agriculture, and the discharge of human waste. Second, in addition to trapping water in the reservoir itself, the dam slows the flow of water through tributaries. This not only reduces the flushing of substances like phosphorus and nitrogen through the system, but it also creates lacustrine conditions, which promote high levels of algal growth.

CRITICAL THINKING

Exercise 4: Lian and his team modeled chlorophyll *a* concentrations in the Xiangxi under different TGD water release flow regimes (Fig. 2.8.11 upper). They concluded that by controlling the way the dam is operated, it should be possible to reduce retention time and increase dilution in the Xiangxi, thereby reducing algal blooms. Fig. 2.8.11 lower shows the spatial extent (coverage length) and duration of algal blooms under 11 different dam operating scenarios.

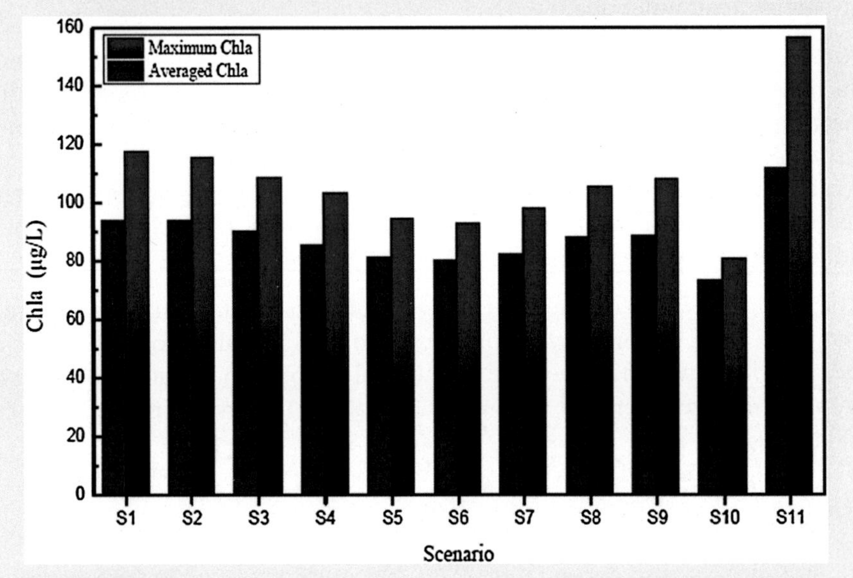

FIGURE 2.8.11

The average chlorophyl *a* concentration (upper figure) and coverage length and duration of algal blooms (lower figure) in the Xiangxi modeled for 11 different Three Gorges Dam flow operating scenarios (Lian et al., 2014).

1. Based on this figure, identify what you believe is the most effective TGD operating scenario.
2. Is this scenario a significant improvement over the others statistically? What information do you need from this figure to make that assessment?
3. In addition to controlling TGD operating conditions, what additional steps might help to alleviate these problematic algal blooms?
4. While pre-dam conditions on the Yangtze River promoted flushing and more rapid movement of pollutants, this doesn't mean that these pollutants didn't pose a threat to the environment. Choose a pollutant and discuss how this pollutant might have affected the Yangtze River ecosystem before TGD was built.

CRITICAL THINKING—Cont'd

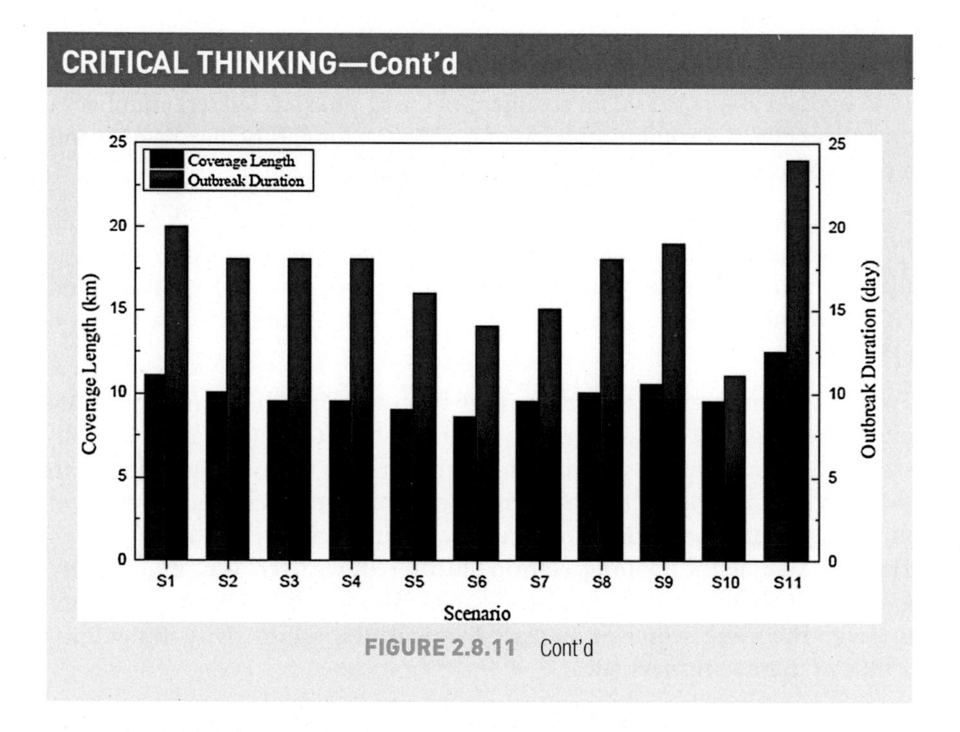

FIGURE 2.8.11 Cont'd

SYSTEMS CONNECTIONS

Exercise 5: Another interesting example of upriver impacts relates to the toxic pollutant mercury. Read the following article linked on your Resource Page: "Mercury contamination in aquatic ecosystems under a changing environment: Implications for the Three Gorges Reservoir" (Wang and Zhang, 2013).

- What are the "systems connections" that link mercury emissions from China's coal-fired plants to the Yangtze River?
- What additional information is needed to fully assess the severity of this problem?
- What solutions do the authors offer to mitigate the impact of mercury?

Downriver Issues

Water that has been retained behind large dams can impact water quality in a river far below the dam. Water released from the bottom of the dam is usually cooler in summer and warmer in winter than river water. Overspill water is usually warmer than river water year-round. These deviations from normal temperatures can impact DO levels and the life cycles of many aquatic organisms in downriver areas.

In addition to temperature impacts, the dynamics of sediment distribution below TGD have changed. In the first 3 years of its operation (2003–2006), TGD trapped about 151 MT of sediment per year. This has led to a number of downriver effects, including erosion of riverbed sediments below the dam and a recession of the delta at the river mouth (Yang et al., 2007).

Pollutant and nutrient concentrations have been affected as well. Sun et al. (2013) measured changes in dissolved inorganic nitrogen (DIN), which is the sum of nitrogen concentrations in NH_4^+ (ammonia), NO_2 (nitrite), and NO_3^- (nitrate), at three sites below TGD (Fig. 2.8.12) over a 20-year period (1990–2009).

Two interesting trends are obvious when examining the changes in DIN over time (Fig. 2.8.13) and in NH_4^+ and NO_3^- patterns (Fig. 2.8.14). Substantial increases in DIN occurred, particularly at the two sites closest to the dam site, around 1995. The researchers identified increased loadings of nonpoint source pollutants, particularly chemical fertilizers, as the cause. The increase at the Datong station, further downriver, was much more gradual, likely because the landscape above this sampling location did not undergo the same degree of increased agricultural productivity in the mid-1990s as the two upriver sites.

The second interesting trend, shown in Fig. 2.8.14, was the change in NH_4^+ and NO_3^- concentrations, particularly at the Yichang and Hankou stations, after 2003. Ammonia levels dropped dramatically after operation of the dam began, while NO_3^- concentration increased at the same time.

SYSTEMS CONNECTIONS

Exercise 6:

- What changes in the TGR led to the downriver changes in NH_4^+ and NO_3^-?
- How might these changes in the form of nitrogen be affecting the Yangtze River ecosystem below the TGD?

Alternate Scenarios: In the Tools and Skills section, we reviewed the basics of cost–benefit analysis (CBA). Reservoirs like TGD offer rich opportunities to investigate the challenges faced when using CBA in complex settings. Dams like TGD yield a wide range of benefits, including power generation, water supply, flood control, and transportation. They also carry with them a number of costs, financial, environmental, and social. In the case of TGD, environmental concerns also include increased risk of earthquakes,

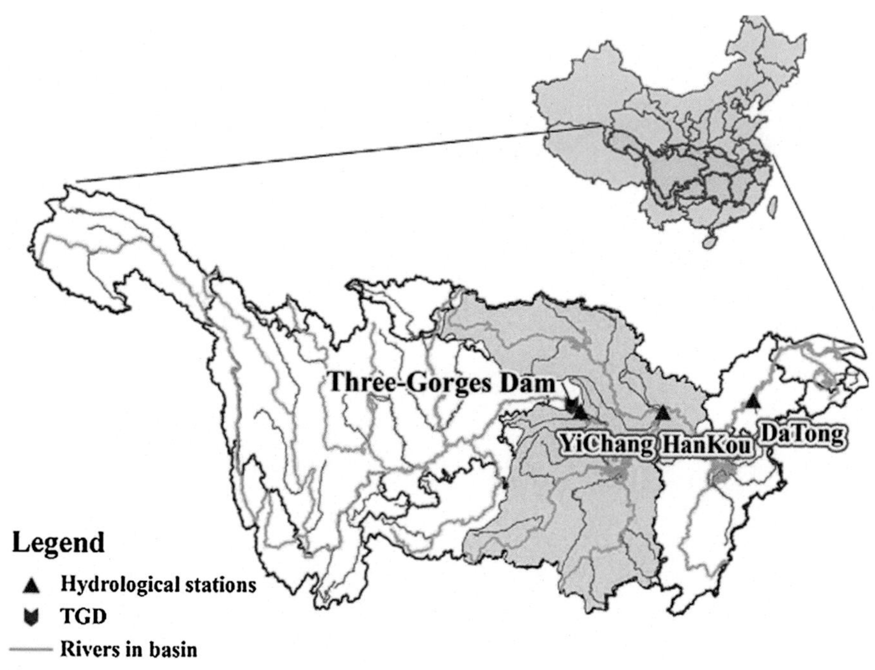

FIGURE 2.8.12
DIN sampling sites (*red triangles*) at three different distances from Three Gorges Dam. *Source: (Sun et al., 2013).*

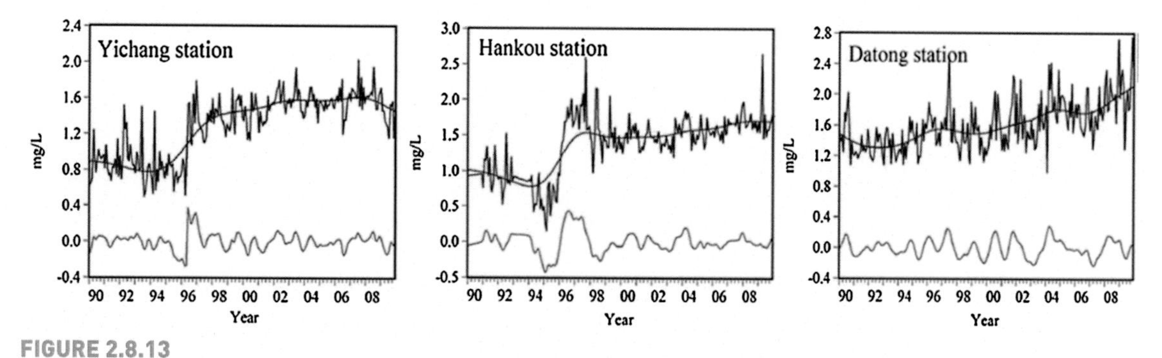

FIGURE 2.8.13
Dissolved inorganic nitrogen concentrations over time at the three downriver sampling sites. Note the dramatic increase in the final seasonal adjusted time series (blue) at the two closest sites when dam construction and impoundment began. Trend component (red) and cycle component (green) are also shown. *Source: (Sun et al., 2013).*

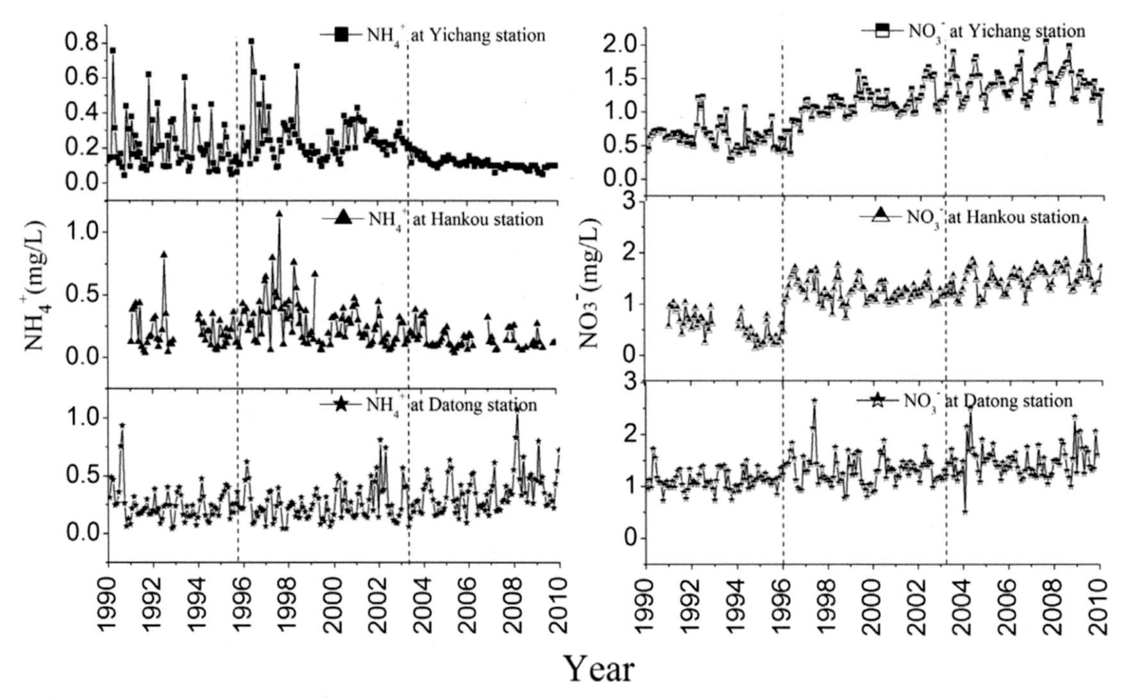

FIGURE 2.8.14

NH_4^+ and NO_3^- concentrations over time at the three downriver sampling sites. *Source: (Sun et al., 2013).*

landslides, and, as we've seen, changes in water quality, fisheries, and other impacts on the ecosystem. Social impacts include dislocation of residents, loss of productive lands, and increased incidences of disease. An additional benefit cited for TGD is that it has reduced emissions of CO_2 from coal-fired power plants.

PUTTING SCIENCE TO WORK

Exercise 7: Your challenge is to do a crude CBA of the TGD.

- List all the benefits and costs of TGD and give your best estimate of the monetary value for each.
- Sum the benefits and costs and draw a conclusion about whether the benefits of TGD have justified the costs.
- What aspects of this exercise did you find particularly challenging?
- What are your thoughts in general about the CBA process?

While quantifying impacts is a useful way to support decisions about whether or not to build a large dam, rarely are such economic/environmental choices as clear cut as the CBA might suggest. Let's dig a little deeper into the potential benefits and impacts of large dams in a setting similar to what occurs when planning and governmental bodies are making these decisions. Often, human stories and qualitative discussions can be more persuasive than the numbers we generated above.

SOLUTIONS

Exercise 8: Huge dams like TGD are one way to generate large quantities of electricity. But as we have seen in this case study, there are consequences to such large alterations to river ecosystems.

- Form teams and debate the pros and cons of a hypothetical dam construction project in your region.
- Do you believe such dams are a sustainable approach to meet human energy demands?
- What stakeholder groups must be engaged in this discussion?
- Do the benefits outweigh the costs? Can the calculations be debated by various stakeholders?

Often, if we are willing to "think outside of the box," we can see that many environmental decisions go beyond a simple "yes or no." Environmental scientists must always be ready to offer alternate scenarios to minimize ecological impacts of otherwise economically appealing projects. We presented one example of that with Lian et al.'s modeling of various water management scenarios for the TGD (Exercise 4). There are often many other plausible options that both meet human needs for resources and minimize ecological disturbance. We'll explore some of these options in our final set of case studies on green alternatives.

PUTTING SCIENCE TO WORK

Exercise 9: To avoid the impacts that large-scale projects like TGD have on the environment, one option is the use of microdams. These small-scale systems, while generating far less electricity, may provide enough for local needs and can likely avoid many of the environmental and sociological impacts that their larger cousins cause.

- Do some research in the literature and find an example of a microdam.
- Describe how it works, how much power it generates, and what, if any, impacts it has. Is a microdam a possible alternative to your hypothetical dam in Exercise 8?
- Investigate other possible "high-tech" solutions to help mitigate many of the environmental impacts of large dams.

CONSIDER THIS

Exercise 10: Many of the loudest voices against large dam construction are the same stakeholders who already benefit from a safe, dependable energy supply.

Is it fair to restrict the development of large dams for people who desperately need the energy they create?

REFERENCES

China-Wire, July 2008. Algae Infests River Near Three Gorges, twenty second ed.

Fu, B.-J., Wu, B.-F., Lu, Y.-H., Xu, Z.-H., Cao, J.-H., Niu, D., Yang, G.-S., Zhou, Y.-M., 2010. Three Gorges project: efforts and challenges for the environment. Prog. Phys. Geography 34 (6), 741–754.

Gao, X., Zeng, Y., Wang, J., Liu, H., 2010. Immediate impacts of the second impoundment on fish communities in the Three Gorges Reservoir. Environ. Biol. Fish 87 (2), 163–173.

Hvistendahl, M., March 25, 2008. China's Three Gorges Dam: an environmental catastrophe? Sci. Am.(On-line).

Lian, J., Yao, Y., Ma, C., Guo, Q., 2014. Reservoir operation rules for controlling algal blooms in a tributary to the impoundment of Three Gorges Dam. Water 6 (10), 3200–3223.

Sun, C.C., Shen, Z.Y., Xiong, M., Ma, F.B., Li, Y.Y., Chen, L., Liu, R.M., 2013. Trend of dissolved inorganic nitrogen at stations downstream from the Three-Gorges Dam of Yangtze River. Environ. Pollut. 180, 13–18.

United Nations Food and Agriculture Organization, 2001. Dams, fish and fisheries: opportunities, challenges and conflict resolution. In: Marmulla, G. (Ed.), FAO Fisheries Technical Paper 419. ISBN: 92-5-104694-8.

Wang, F., Zhang, J.Z., 2013. Mercury contamination in aquatic ecosystems under a changing environment: implications for the Three Gorges Reservoir. Chin. Sci. Bull. 58 (2), 141–149.

Xu, X., Tan, Y., Yang, G., 2013. Environmental impact assessments of the Three Gorges Project in China: Issues and interventions. Earth Sci. Rev. 124, 115–125.

Yang, S.L., Zhang, J., Xu, X.J., 2007. Influence of the Three Gorges Dam on downstream delivery of sediment and its environmental implications, Yangtze River. Geophys. Res. Lett. 34 (10), L10401.

CHAPTER 3

Air Quality and Atmospheric Science

Science and the Global Environment. http://dx.doi.org/10.1016/B978-0-12-801712-8.00003-2

Chapter 3.1

Introduction

FIGURE 3.1.1
Dense smog hangs low over Halde Hoheward, Germany. By Rainer Halama [CC BY 3.0], via Wikimedia Commons.

As critical as water is for human survival, arguably the quality of the air we breathe is even more important for several reasons: (1) we breathe air constantly, and if our air is polluted (Fig. 3.1.1), our exposure to pollutants may also be constant; (2) while some air pollutants like ammonia are immediately detectable if inhaled, others, like radon and carbon monoxide (CO), are not; and (3) if our drinking water is polluted, we might have an alternative like bottled water available. Bottled air isn't really an option.

Some interesting facts about our atmosphere and the air we breathe:

- Our atmosphere keeps us warm. If it weren't for the greenhouse effect, our climate would be like that on Mars; the problem is that increasing levels of greenhouse gases are destabilizing the climate.

- Ozone in our stratosphere protects us from the most damaging forms of UV radiation (UVR); we'll talk more about ozone depletion in one of our case studies. Ozone at ground level is considered a pollutant, however.

- Air is the ultimate recycler. You may have inhaled a molecule of air once breathed by George Washington.

- The atmosphere even helps make it possible for us to hear. Sound is simply waves of energy or vibrations transmitted as moving molecules of air.

THE BASICS OF OUR ATMOSPHERE

Structure: there are several layers of our atmosphere (Fig. 3.1.2), with temperature being a defining factor. We'll focus on only two: the troposphere and the stratosphere.

Troposphere: this layer of the atmosphere, closest to the Earth's surface, extends up to about 10 km. It's in this relatively thin layer of air that clouds form and our weather happens. It's also where most of our pollutants are found.

Stratosphere: Above the troposphere, this layer extends up to about 50 km. Temperatures gradually increase with altitude in this layer. The concern here is with the protective ozone layer that lies within the stratosphere. Remember that ozone also occurs in the troposphere, where it is considered an important air pollutant.

Composition: our atmosphere contains a mixture of gases (Fig. 3.1.3), but the two dominant ones are oxygen (20.95%) and nitrogen (78.08%). Traces of argon (0.93%), carbon dioxide (0.40%), and small amounts of other gases are also present. Our atmosphere also contains variable amounts of water vapor, around 1% at sea level and 0.4% averaged throughout the atmosphere.

Key Air Quality Concepts

Before beginning the air quality and atmospheric science case studies, visit one of the websites listed on your Resource Page for more information about the science of the atmosphere. To get you started, here are some key points to remember about air quality and the atmosphere:

1. There are important natural sources of air pollutants. For example, the amount of mercury released during a large forest fire can be substantial, and particulate matter (PM) given off during major volcanic eruptions can temporarily cool the Earth's atmosphere. We'll focus our attention on human sources of air pollution. After all, there's not that much we can do about most natural sources.

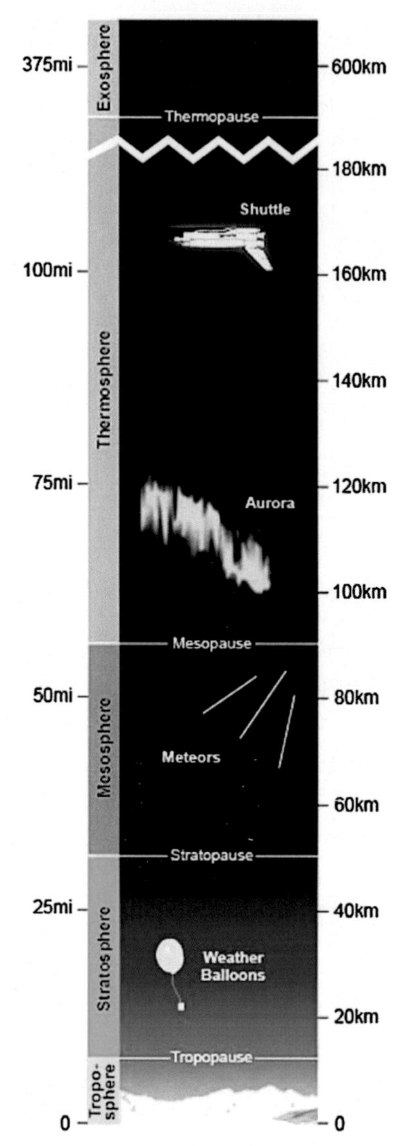

FIGURE 3.1.2
Layers of the Earth's atmosphere. *Source: NOAA.*

2. Natural features often play an important role in the quality of our air. Pollutants released into a very windy area may have fewer local impacts than the same amount released into a sheltered area like a valley. The role of winds isn't always positive. For example, the same winds that may help move pollutants away from one area can contribute to regional issues like acid deposition.

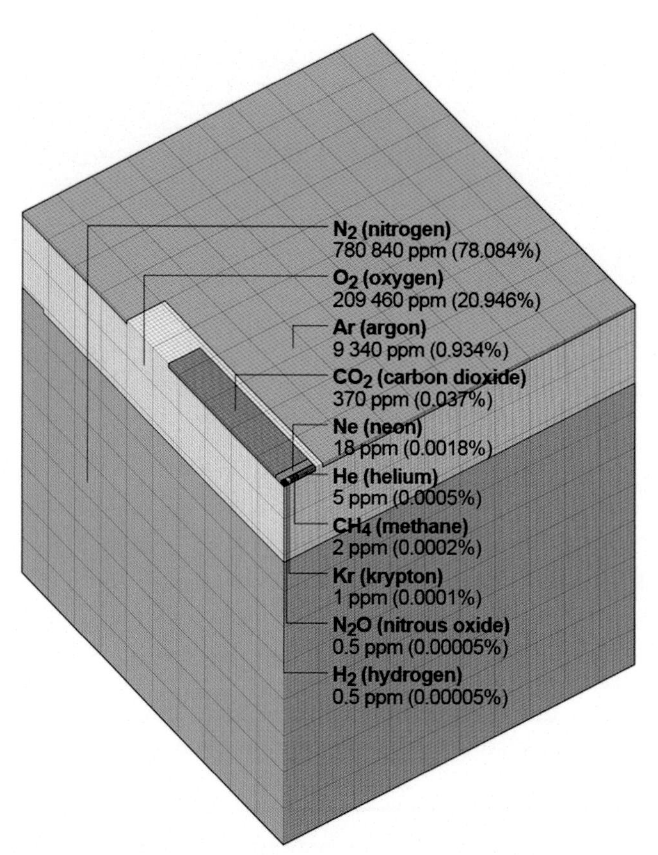

FIGURE 3.1.3
Composition of the Earth's atmosphere. *By Cmglee [CC BY-SA 3.0], via Wikimedia Commons.*

Local features like lakes and mountains may also create microclimates such as rain shadows, desert-like conditions that occur on the lee sides of mountain ranges. Because water holds its heat longer than the atmosphere, locations near large water bodies may have more moderate temperatures in the fall and winter (until ice forms) than in surrounding areas. However, cold winds blowing across the open waters of large lakes like Erie can bring very heavy, sometimes disabling, lake-effect snowstorms (Fig. 3.1.4).

3. The severity and impacts of air pollution episodes depend not only on the amount and type of air pollutants released, but also where they are released. For example, the problem of brown smog formation in Los Angeles has historically been severe because of (1) the many sources of smog ingredients (e.g., tailpipes on vehicles using Los Angeles freeways); (2) the climatology, with lots of sunlight creating the perfect environment

FIGURE 3.1.4
NASA satellites capture a building "lake effect" storm. *Source: NASA.*

for smog formation; (3) the geography, with the city's location between mountains to the east and the Pacific Ocean to the west creating ideal conditions for smog formation; and (4) the millions of residents in the Los Angeles area who can potentially be affected by the smog. Brown smog follows a daily progression, with pollutants like nitrogen oxides (NO_x) and ozone gradually shifting eastward from downtown Los Angeles to the suburbs.

4. The effects of air pollutants on ecosystems and human health often remain poorly understood because many different pollutants may coexist in an area at the same time, and their concentrations may fluctuate rapidly over time. We'll look at the challenges involved in establishing cause–effect relationships in the case of airborne PM and its impacts on human health.

5. Climate and weather are different. Weather refers to short-term changes in parameters like temperature and precipitation, while climate refers to conditions over periods of 30 years or more. Remember that with climate change, the focus isn't on daily fluctuations, it's the long-term alterations in atmospheric conditions.

Sources and Types of Air Pollutants

Sources: Air pollutants can be either primary, those emitted directly into the atmosphere, or secondary, those forming in the atmosphere as a result of the reactions among other chemicals driven by energy from the sun. There are two general types of sources: point and nonpoint.

> *Point* sources are those released at a single point. A smokestack at a coal-fired power plant or the tailpipe of a vehicle are examples of point sources. *Nonpoint* sources are those not released at discrete points. For instance, dust blown from a field by the wind or given off by a wildfire (Fig. 3.1.5) are examples of nonpoint sources.

Types of pollutants: There are many types of air pollutants that affect human health and the environment. We'll focus some attention on those that are regulated by the Clean Air Act originally passed by the US Congress in 1972. These are commonly known as "criteria" pollutants because numerical standards establishing safe environmental levels are set for them, and they are also important concerns globally. The first five are primary pollutants, while the sixth, ozone, is a secondary pollutant.

a. ***PM***: PM includes small solid particles and liquid droplets suspended in the atmosphere. PM may contain various constituents, including metals, nitrates and sulfates, organic compounds, and soil and dust particles. Of greatest concern are particles with a diameter of 2.5 μm or less (PM 2.5). Particles of this size not only are typically the most difficult to control at point sources, but they also pose the greatest risk to human health, as they are able to penetrate deeply into the lungs.

b. ***Sulfur dioxide (SO_2)***: Primarily released from the burning of fossil fuels, SO_2 is a respiratory irritant and one of the two main ingredients involved in the formation of acid deposition.

FIGURE 3.1.5
Wildfires are a common source of nonpoint air pollution. *Source: Nerval (Public Domain), via Wikimedia Commons.*

c. ***Nitrogen dioxide (NO_2)***: Also a product of fossil fuel combustion, NO_2 comes from both stationary sources like power plant smokestacks and mobile sources like tailpipes. Like SO_2, it is a respiratory irritant and another important ingredient in the formation of acid deposition. It also plays a key role in photochemical smog formation.

d. ***Carbon monoxide (CO)***: A colorless, odorless gas given off by combustion, CO comes mostly from mobile sources. It is primarily a human health threat, reducing oxygen delivery in the body and, at high concentrations, causing death.

e. ***Lead***: Once a major pollutant present in tailpipe emissions, lead in the air now comes mostly from ore smelters and metal processing operations. The banning of leaded gasoline in the 1970s led to a dramatic decline in environmental levels of this highly toxic element, which can cause a host of human health problems, including damage to the nervous system.

f. ***Ozone (O_3)***: A secondary pollutant formed when the sun's energy interacts with the oxides of nitrogen and volatile organic compounds (VOCs) in the atmosphere, O_3 is a strong oxidant that can damage vegetation. Ground-level or "bad" O_3 can also cause respiratory irritation in humans and is a particular threat to those with asthma. Remember, however, that O_3 in the stratosphere protects us from harmful UVR (Fig. 3.1.6).

Other Important Air Pollutants: there are several other pollutants that are important to consider for a variety of reasons.

a. ***Carbon dioxide (CO_2)***: CO_2 is the leading cause of human-related climate change. One of the challenges, of course, with this gas is that it is released by so many activities that are considered vital to many nations' economies.

b. ***Methane (CH_4)***: A reduced carbon gas, this is also important because of its role in climate change. Of particular concern are the many natural sources of CH_4, including livestock, landfills, and termites.

c. ***Nitrous oxide (N_2O)***: N_2O is another gas that contributes to climate change; release from fertilized agricultural fields is a common source.

d. ***Persistent organic pollutants (POPs)***: Including many toxic compounds such as PCBs, DDT, and others, POPs persist in the environment and tend to build up to high levels in top members of food webs, potentially causing a variety of harmful effects. Although many of these have been

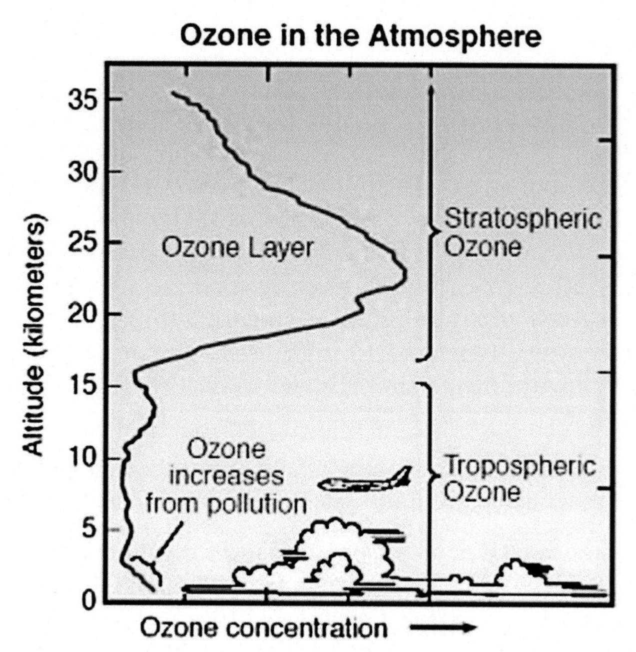

FIGURE 3.1.6
The complex role of ozone in our atmosphere. *Source: www.airnow.gov.*

banned, POPs are still a threat to the environment and are the subject of one of our case studies.

e. ***VOCs***: Volatile organic compounds (VOCs) are those chemicals that readily move from solution into the atmosphere. VOCs are widely used for a variety of industrial and commercial purposes. Some, like benzene, are known or suspected carcinogens. VOCs also pose a substantial threat to groundwater because they move readily through most subsurface systems, where they avoid exposure to the atmosphere.

f. ***Chlorofluorocarbons (CFCs)***: CFCs are industrial compounds used for a variety of purposes, such as propellants and cooling agents. Identified as major culprits in the depletion of stratospheric ozone, CFCs were banned by the Montreal Protocol. However, CFCs still persist in the atmosphere because they break down very slowly over time.

g. ***Household air pollution (HAP)***: Also known as indoor air pollutants (IAPs), this category includes a wide range of substances, including PM, mold, and asbestos, which may pose a risk to humans exposed indoors. We'll cover this group of pollutants in one of our case studies.

AIR QUALITY AND ATMOSPHERIC SCIENCE CASE STUDIES

Results of efforts to protect air quality have been mixed. We've had some successes: the use of catalytic converters has substantially reduced tail-pipe emissions, cap-and-trade approaches have been used to reduce the emission of acid deposition precursors like SO_2, and an international treaty banned the manufacture and use of CFCs, a main culprit in ozone depletion. In other cases, however, we've made less progress. Many cities still suffer levels of air pollution deemed hazardous to human health, some workers are still exposed to high levels of carcinogenic pollutants in their work environments, and climate change remains a vexing global challenge.

In this textbook, there are six air quality and atmospheric science case studies presented. These occur at multiple scales, from global to local:

> ***Global issues***: Several of our case studies focus on atmospheric issues that are caused by daily human activities and impact social and ecological systems on a global scale. Few issues are more pressing than climate change. We'll look at how this phenomenon is already affecting several iconic animal species around the globe. Fig. 3.1.7 shows global temperature changes from 1900 to 2012.

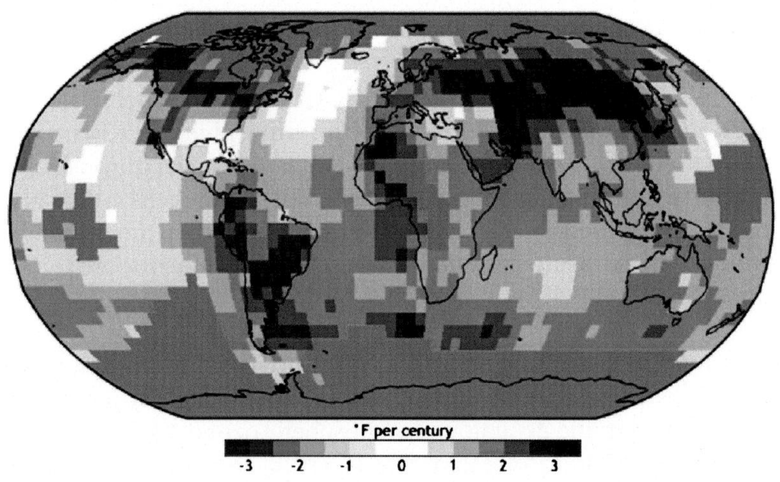

FIGURE 3.1.7

Observed trend in temperature from 1900 to 2012; yellow to red indicates warming, while shades of blue indicate cooling. Gray indicates areas for which there are no data. There are substantial regional variations in trends across the planet, although the overall trend is warming. *Map from FAQ appendix of the 2014 National Climate Assessment. Originally provided by NOAA NCDC.*

Another important global issue is the depletion of the ozone layer over the Antarctic. Unlike the case with climate change, we have taken significant steps to slow ozone depletion, and there has been some improvement as a result of these efforts. We'll examine the impacts of ozone depletion on skin cancer rates and on primary producers in marine ecosystems. Finally, we'll consider the transport of POPs from the tropics to the Arctic, where they can harm both human health and Arctic wildlife.
Regional issues: Important regional air quality issues include brown smog and acid deposition. We'll focus on airborne PM, a problem that typically plagues large cities that have high levels of traffic, power plant emissions, and other sources of small particles. We'll explore sources of PM and how it affects human health.
Local issues: An issue still challenging many parts of the developing world is the presence of large point sources of pollution like smokestacks. We'll examine the Met-Mex metals smelter in Mexico and consider the effects of pollutants like lead on the surrounding community. We'll also look at issues of transnational pollution and environmental justice.
Household issues: Historically, indoor air pollution was rarely considered a serious environmental threat. We now know better. Anyone who smokes or lives with a smoker is likely exposed to dozens of potentially carcinogenic air pollutants on a daily basis. Other sources of IAPs include chemicals used in various household products, particulates like asbestos, molds and harmful microbes, and even odorless, tasteless radon gas, which is capable of causing lung cancer (Fig. 3.1.8).
A study by the US Environmental Protection Agency (EPA) found that concentrations of several important air pollutants were higher indoors than they were outdoors. Our case study on HAP will focus on the plight of women who are forced to rely on peat, animal dung, and other vegetable material as fuels for cooking and heating indoors in developing nations like India.

These case studies will allow you to work with real data to better understand environmental problems and how to solve them. In addition to taking an in-depth look at the science behind these air quality and atmospheric science issues, we'll work on some skills important for the practicing environmental scientist. In particular, we'll:

- review the steps necessary to insure that high-quality data are generated;
- consider the challenges faced when considering laboratory versus field approaches;
- evaluate the strengths and weaknesses of studies attempting to establish cause–effect relationships between a pollutant like PM and human health;

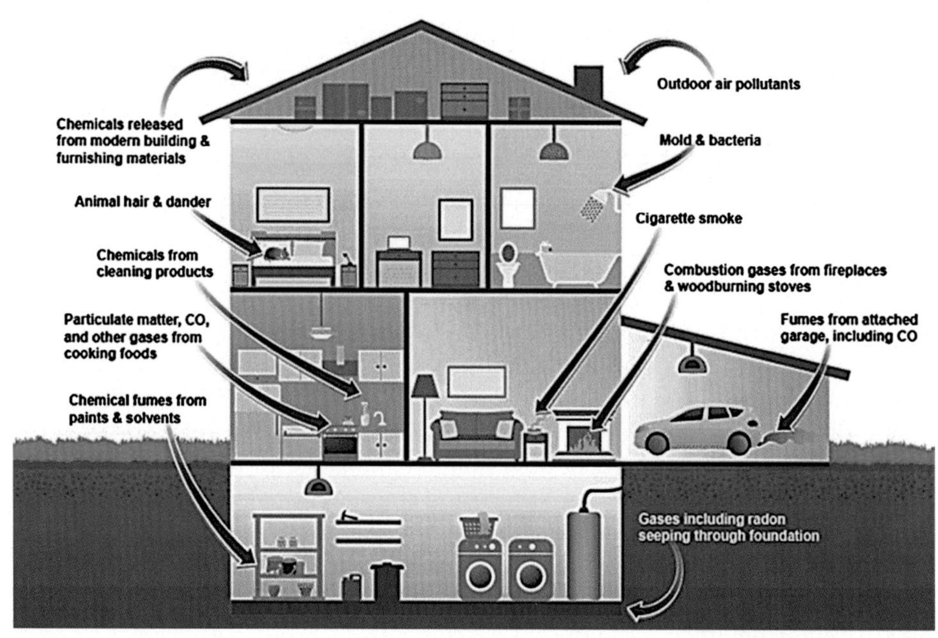

FIGURE 3.1.8

Indoor air pollutants can enter buildings from the outside or be generated inside from common building materials and household items. *Source: US EPA.*

- look at several of the ethical issues surrounding air pollution; and
- conduct a cost–benefit analysis of alternate sources of home energy in developing nations.

A reminder: On your Resource Page, we've listed some general websites that will help you review the basics. In addition to these general resources provided, each case study will include specific resources chosen to help you understand the material covered by that issue.

Chapter 3.2

Ozone Depletion

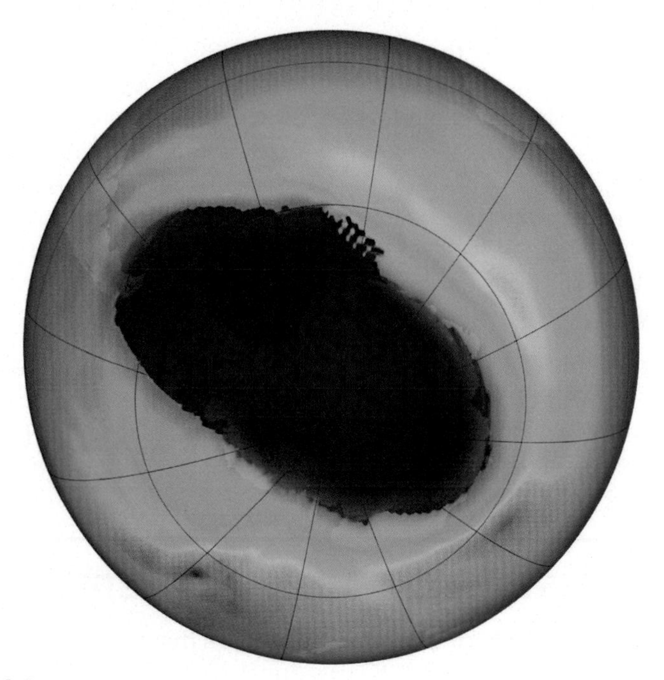

FIGURE 3.2.1
The image above is the first one publicly released from the Aura mission. Acquired by the mission's ozone monitoring instrument (OMI) on September 22, 2004, the image shows dramatically depleted levels of ozone in the stratosphere over Antarctica. Purple shows areas with very low ozone concentrations (as low as 125 Dobson units), while turquoise, green, and yellow show progressively higher ozone concentrations. To learn more about the mission, please read the Aura Fact Sheet at earthobservatory.nasa.gov/Library/Aura/, or visit the Aura website at aura.gsfc.nasa.gov/. NASA Identifier: AntOzone_AOM20040922.

Ozone Depletion: We've known for decades that the ozone layer in our stratosphere is being depleted by human-made compounds like CFCs (Fig. 3.2.1). In this case study, we review how this process occurs and consider the impacts of the resulting increase in UVR levels on humans and on marine primary producers. We'll conclude by considering the Montreal Protocol, which has made strides toward reducing this threat, and we'll look at obstacles threatening the continued recovery of the ozone layer.

Background: Ozone depletion is one of those unusual scientific phenomena that was predicted as a possibility based on scientific theory before it was ever observed. In 1974, two chemists, Sherwood Rowland and Anthony Molina, hypothesized that CFCs could lead to substantial depletion of the stratospheric ozone layer. Their work demonstrated not only how the chemical structure of CFCs could lead to ozone degradation, but also how CFCs released at the Earth's surface could reach the vulnerable ozone layer in the stratosphere. The two scientists were awarded the Nobel Prize for Chemistry for their efforts.

It wasn't until 11 years later that Farman et al. (1985) published a paper in the journal Nature first documenting a decrease in ozone levels in the stratosphere over the South Pole. The combination of sound science and documented impacts led the nations of the world to sign the Montreal Protocol in 1987 restricting the production of CFCs. The mean ozone hole size (OHS) increased steadily during the 1980s (Fig. 3.2.2), yet, thanks to the Montreal Protocol, the rate of increase began to level off by the 1990s. Still, it is important to underscore that there has been no documentation of a consistent reduction in the size of the yearly OHS despite these efforts.

Because of the rapid chemical interactions that are driven by both the changing concentration of CFCs and the temperature of the atmosphere, the size and location of the ozone hole can vary dramatically from day to day (Fig. 3.2.3). The dark blue colors in this figure indicate the areas of ozone thinning over the Antarctic in October 2008.

THE CHEMISTRY OF FIGURE OZONE DEPLETION

Before considering the effects of ozone depletion in the stratosphere on human health and on ecosystems, we'll review the chemistry of this phenomenon

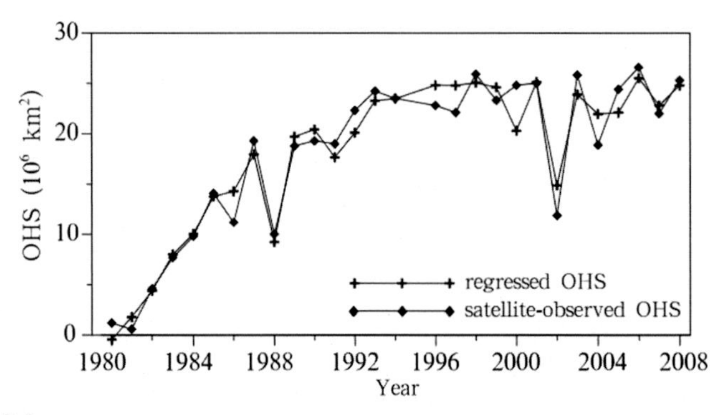

FIGURE 3.2.2
Time series of regressed ozone hole size (OHS) and satellite-observed OHS during 1980–2008 (Lin-Gen et al., 2012).

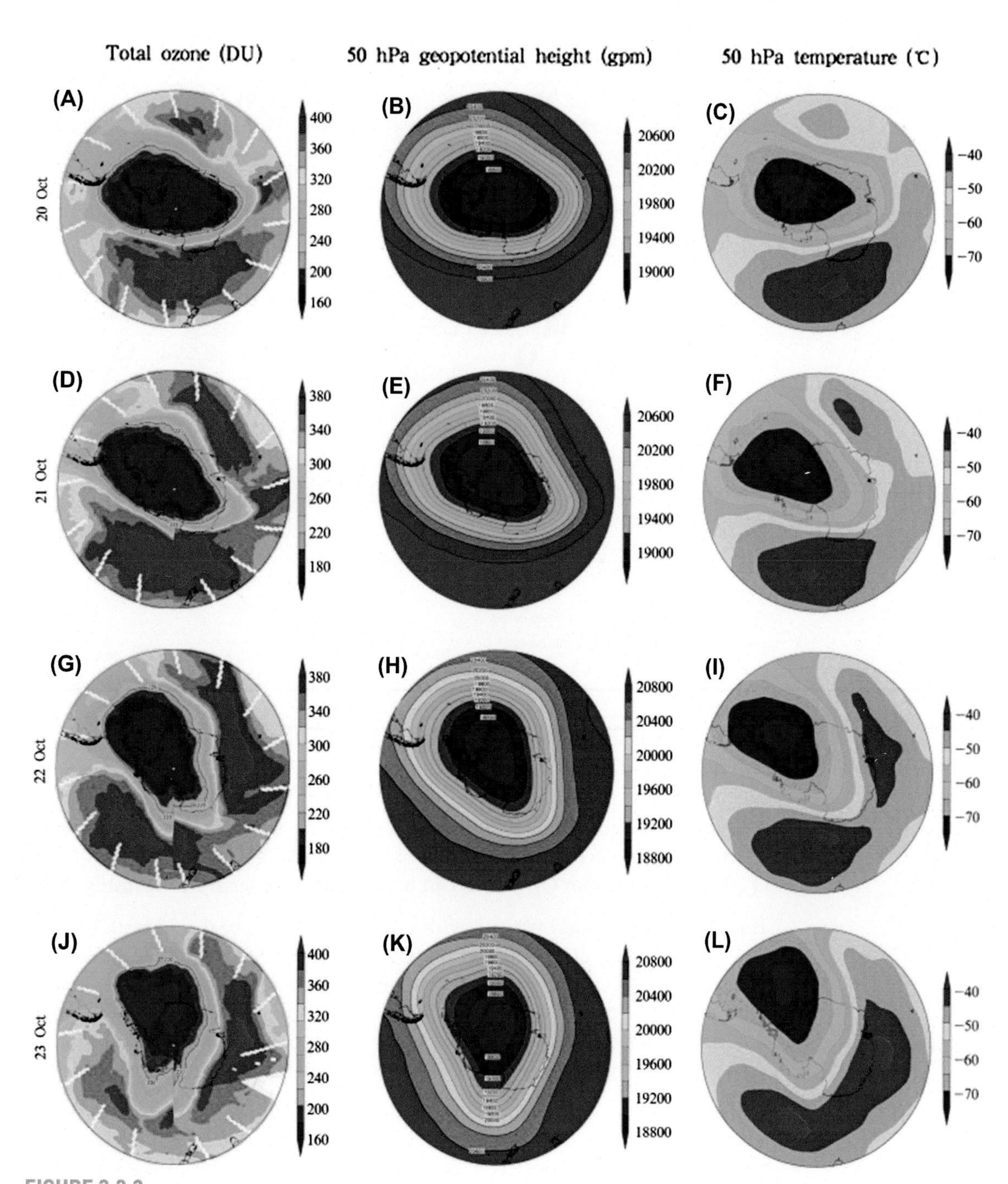

FIGURE 3.2.3

Column ozone, 50 hPa geopotential height, and temperature fields over Antarctica from October 20–25, 2008 (Lin-Gen et al., 2012).

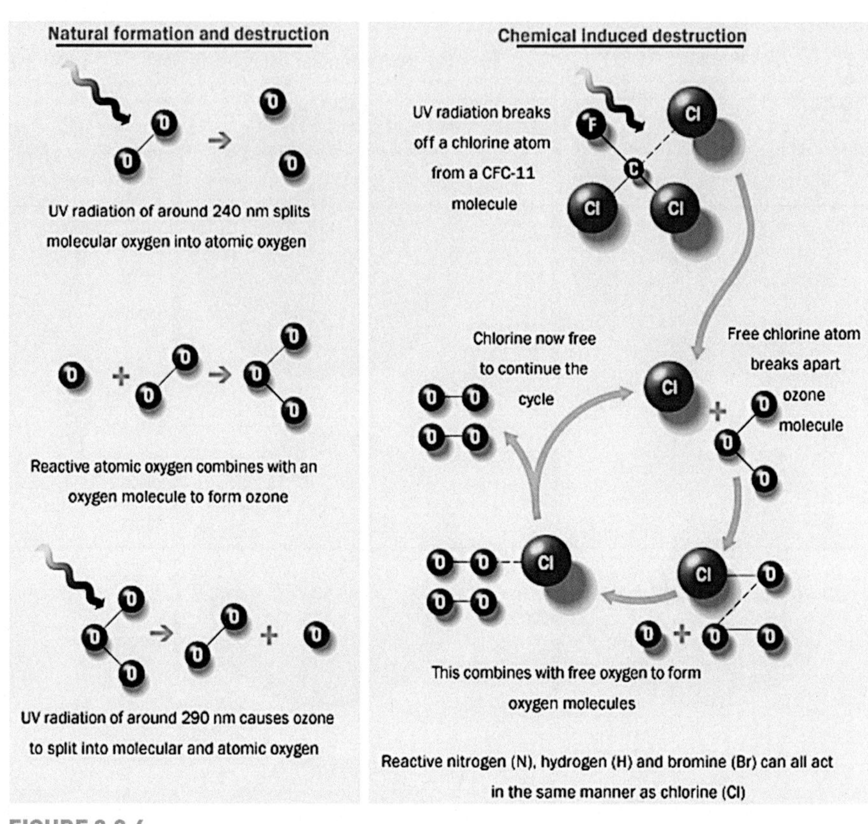

FIGURE 3.2.4
Formation and destruction of ozone in the stratosphere. *Courtesy of NASA GSFC.*

(Fig. 3.2.4). Ozone is naturally formed in the Earth's upper atmosphere as O_2 molecules interact with incoming UVR from the sun. This radiation breaks apart the O_2 molecules into reactive oxygen atoms that then bind with other O_2 molecules to form O_3, which can be broken apart again by incoming UVR.

This cycle continuously repeats, with O_2 and O_3 cycling back and forth as reactive oxygen atoms are broken off by UVR. Even though individual molecules are split and then recombined, the overall amount of O_3 in the stratosphere is relatively stable under natural conditions. It is through this continuous process that certain wavelengths of UVR are absorbed by ozone molecules, preventing the arrival of large quantities of this harmful radiation at the Earth's surface.

While a number of different compounds can break down ozone molecules in the stratosphere (we'll look at a few of these at the end of this case study), most attention has focused on CFCs, considered to be most responsible for ozone depletion. CFCs and other depleting compounds move slowly into the stratosphere following their release from the Earth's surface. When radiant energy from the sun interacts with these molecules, the CFCs are split, releasing a highly reactive chlorine atom, which breaks apart the ozone molecule, producing O_2.

During winter months over the South Pole, when incoming solar radiation is low, CFCs are bound to ice crystals that make up the polar stratospheric clouds that are common during the coldest winter months in the Antarctic atmosphere. During these colder months, the ozone hole can be quite small to nonexistent. However, with the warmer temperatures and increased solar radiation in spring, more chlorine atoms are split off CFCs and attack O_3 molecules, ultimately increasing the size of the ozone hole.

Things to remember about CFCs and these reactions:

- CFCs can persist in the stratosphere for decades and have the potential to break down ozone for long periods of time.
- The reactions described in Fig. 3.2.4 are called the "chlorine cycle." Each atom of chlorine produced can be recycled back to react with new O_3 molecules. It is estimated that one chlorine atom can destroy more than 100,000 ozone molecules before it is removed from the stratosphere.
- CFCs are also important greenhouse gases. Although their concentrations in the atmosphere aren't high, their structure makes them highly efficient at trapping heat.

UV RADIATION IN A NUTSHELL

UVR comes in three varieties: UVA, UVB, and UVC, each with successively shorter wavelengths and higher associated energy. As sunlight passes through the Earth's atmosphere, all of the highest energy UVC and 90% of the moderately energetic UVB are absorbed by ozone, water vapor, oxygen, and carbon dioxide. UVA radiation is less affected by the atmosphere, but this lower energy radiation has less of an impact on humans or ecosystem health.

The Problem: Perhaps the most well-known impact of the thinning of the ozone layer is the increased penetration of UVB to the Earth's surface. One estimate is that for each 2% of ozone depletion, an additional 1% of UVB reaches the surface. Two areas of concern have emerged: human health and ecosystem impacts.

Human Health: UVB is a biologically energetic form of radiation. If UVB strikes human skin, it can alter the structural integrity of cellular DNA and lead to the formation of skin cancers called squamous cell carcinomas (Fig. 3.2.5). Experts have estimated that for each 1% increase in UVB reaching the Earth's surface, a 1.5% increase in the number of skin cancers will occur.

CRUNCHING THE NUMBERS

Exercise 1: Few areas of the world receive more hours of sun annually than Australia. In addition, stratospheric ozone levels over the country can decrease dramatically in the late spring as the ozone hole shifts northward from the South Pole. Armed with the above statistics and population figures that can be accessed online, calculate the number of Australians expected to develop skin cancer if the ozone layer over Australia is thinned by 10%. Assume that 10% is an average annual figure.

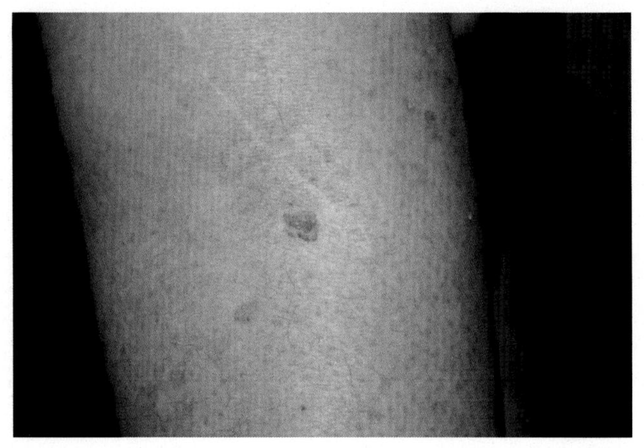

FIGURE 3.2.5
Squamous cell carcinoma typically displays as a pink, raised lesion on the skin. *By Kelly Nelson (photographer), via Wikimedia Commons.*

Ecosystem Health: A less well-publicized concern about the increased UVR reaching the Earth's surface is its impacts on ecological systems. UVB impairs photosynthesis, increases susceptibility to disease, and reduces size, productivity, and quality in many plant species. In particular, scientists are concerned that marine primary producers may be harmed by increased radiation levels, as UVB is capable of damaging components of cellular DNA, leading to mutations that can damage or even kill the organisms.

Scientists have also found that UVB impairs cellular division in developing sea urchin eggs, may alter the movement and orientation of plankton, and contributes to coral bleaching (Fig. 3.2.6). Because UVB may affect organisms that cycle nutrients and energy through the biosphere, we can expect changes in their populations to alter uptake and storage of carbon in the world's oceans, perhaps with the potential for cascading impacts through ocean food webs.

Science in the Field and in the Laboratory: One of the challenges when trying to determine the effects of increased radiation on primary producers is to design

FIGURE 3.2.6
Marine organisms living in shallow water can experience damaging levels of UV radiation, as demonstrated by the bleached coral here. *NOAA photos by Bernardo Vargas-Ángel.*

and carry out an experiment that accurately mimics conditions in nature while controlling for a host of other confounding factors. Duplicating conditions that occur in the field can be very challenging.

We'll analyze several experiments designed to help determine the severity of the threat that increased UVR poses to marine primary producers. In doing so, we'll consider several problems that scientists face when trying to design the ideal experiment.

Study #1: ***Dealing with multiple stressors***. Remember that ecosystems are exposed to a combination of natural and anthropogenic factors. For example, at the same time that nutrient levels may be elevated in spring runoff entering a lake, temperatures of lake water are also increasing. Populations of primary producers may respond to both of these changes, as well as many others, in complex and hard to predict ways.

In this study, Cabrerizo et al. looked at how various types of marine algae, including a dinoflagellate, a diatom, a chlorophyte, and a haplophyte, responded to changing levels of UVR, temperature, and nutrient concentrations in the water.

Let's take a second to examine their results. Because they were varying so many different experimental treatments simultaneously (UVR, temperature, and nutrient levels), their data can be difficult to interpret, so let's start with Fig. 3.2.7A. In this figure, the authors show the impacts (change in photosynthesis on the y-axis) for the four different species (various lines on the graph) under a range of temperature conditions (x-axis, with cooler temperatures to the left and warmer temperatures on the right) under low nutrient concentrations.

Their results indicated that at cooler temperatures, all four algal species experienced some decrease in photosynthesis when exposed to UVR. However, at warmer temperatures similar to those anticipated with climate change, UVR did not inhibit photosynthesis in haptophytes (***Dunaliella salina***, triangle in Fig. 3.2.7A) and did allow diatoms to maintain stable photosynthetic levels (***Alexandrium tamarense***, circle in Fig. 3.2.7A). However, UVR did inhibit photosynthesis in

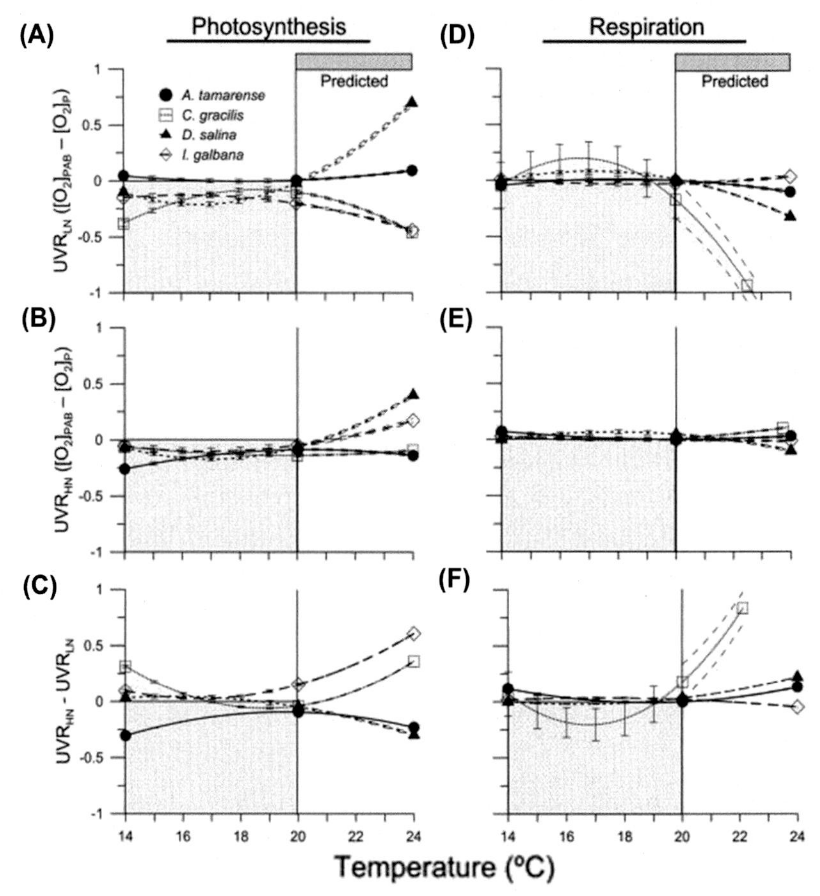

FIGURE 3.2.7

UV radiation (UVR) effect, evaluated as the difference in photosynthesis (A,B,C) and respiration (D,E,F), between typical and increased UVR levels across a range of water temperatures. Samples were grown and incubated in high-nutrient (HN) and low-nutrient (LN) conditions, and at three temperatures, 14, 17, and 20°C. The lines (*solid and broken*) represent the best fit using a polynomial function, while the vertical lines represent 95% confidence intervals. *Source: (Cabrerizo et al., 2014).*

larger species of chlorophytes and dinoflagellates (squares and diamonds in Fig. 3.2.7A).

Fig. 3.2.7B shows the same photosynthetic response across temperatures for the four species, but this time under high nutrient concentrations. Note that by changing this one additional variable, the negative impacts on the larger chlorophytes and dinoflagellates that were apparent under low nutrient concentrations (squares and diamonds in Fig. 3.2.7B) were mitigated.

Fig. 3.2.7C allows us to visualize and isolate the impact that the different nutrient concentrations had on UVR's impacts on photosynthesis by simply looking at the difference between panels A and B for each species. This helps identify those species that may be particularly responsive to increases in nutrient concentration.

DRILLING DEEPER

Exercise 2: Consider the biological characteristics of the species compared in the Cabrerizo et al. (2014) study.

- Why might the four species shown in Fig. 3.2.7C react so differently?
- What does this imply about which environmental conditions measured in this study control productivity in this marine environment?
- Are there other environmental conditions that might impact productivity that were not included in this study?
- What does this experiment tell you about the combined, and potentially interacting, effects of UVR, temperature, and nutrient concentration on respiration (Fig. 3.2.7D–F) of the four species?

Like the marine species considered in this study, all organisms face certain requirements if they are to maintain their health, and they are also often exposed to various stressors that can harm them. These environmental conditions rarely occur in isolation. For example, increased light may stimulate terrestrial plant growth rates, but only under conditions of sufficient moisture. Thus, both light and water availability interact to determine how the plant ultimately responds.

In this study, the ultimate response (photosynthesis) to an environmental stressor (UVR) varied widely, depending upon species, nutrient levels and temperature, demonstrating the importance of accounting for multiple factors in order to better understand how species and ecosystems respond.

Study #2: ***Field versus laboratory approaches***. K. Zacher took on another challenge faced in experimental design: linking laboratory experiments to research conducted in the field. The advantage of laboratory experiments is the ability to control many of the conditions that might influence organism response so that you can isolate the effect of one or more specific experimental treatments. The challenge is that it is hard to duplicate the conditions found in nature in laboratory experiments. One can get closer to achieving natural conditions by running experiments in the field. While it is harder to control for all conditions in the field, such research does provide an opportunity to apply experimental treatments in a realistic setting.

Zacher designed a field experiment to study of the effects of UVB radiation and PAR (photosynthetically active radiation) on the spores of three species of Antarctic seaweeds. For her field exposures, she built aluminum frames with black bottoms and UV-transparent plexiglass covers to hold Petri dishes containing the spores of seaweeds. She held them at four different depths, 1, 2, 4, and 8 m, below the ocean's surface. The number of spores germinating at each depth was measured.

Because water absorbs much of the radiation that reaches the Earth's surface, the deeper the frame, the lower the exposure of the spores to UVR. To replicate this differential exposure to radiation under the controlled conditions of a laboratory, a similar set of Petri dishes was set up with artificial radiation exposure measured after four different time intervals (1, 2, 4, and 8 h).

PUTTING SCIENCE TO WORK

Exercise 3: Any experiment must include replication. This is the only way that the researcher can calculate how much variability is normal (due to random chance) within a given treatment versus how much variability is due to the differences that result from the treatment itself.

In Zacher's field study, she had replicate Petri dishes placed into replicated aluminum frames, which were exposed at four different depths.

1. What are some of the sources of variability that would lead to different spore counts at any given depth?
2. What other factors (besides differences in the amount of UVR) might differ among the treatment depths?
3. What are some of the sources of variability that would lead to different spore counts for the Petri dishes exposed to UVR in the laboratory?
4. What is the benefit of conducting both field and laboratory experiments? How are they comparable? How are they different?

The left side of Figure 3.2.8 shows the number of germinations at different depths in the field for the three seaweed species. The right side shows the comparable laboratory results. Note that for each treatment depth (radiation duration), there are three different bars. These represent replicates of the treatments in which spores were exposed to PAR wavelengths only (P, striped bars), to PAR wavelengths with the addition of UVA radiation (PA, white bars), and to PAR wavelengths with the addition of UVA and UVB radiation (PAB, black bars).

The first species represented in the top two plots of Figure 3.2.8 (indicated by AF and AL) is ***Adenocystis utricularis***, an intertidal species that showed little negative

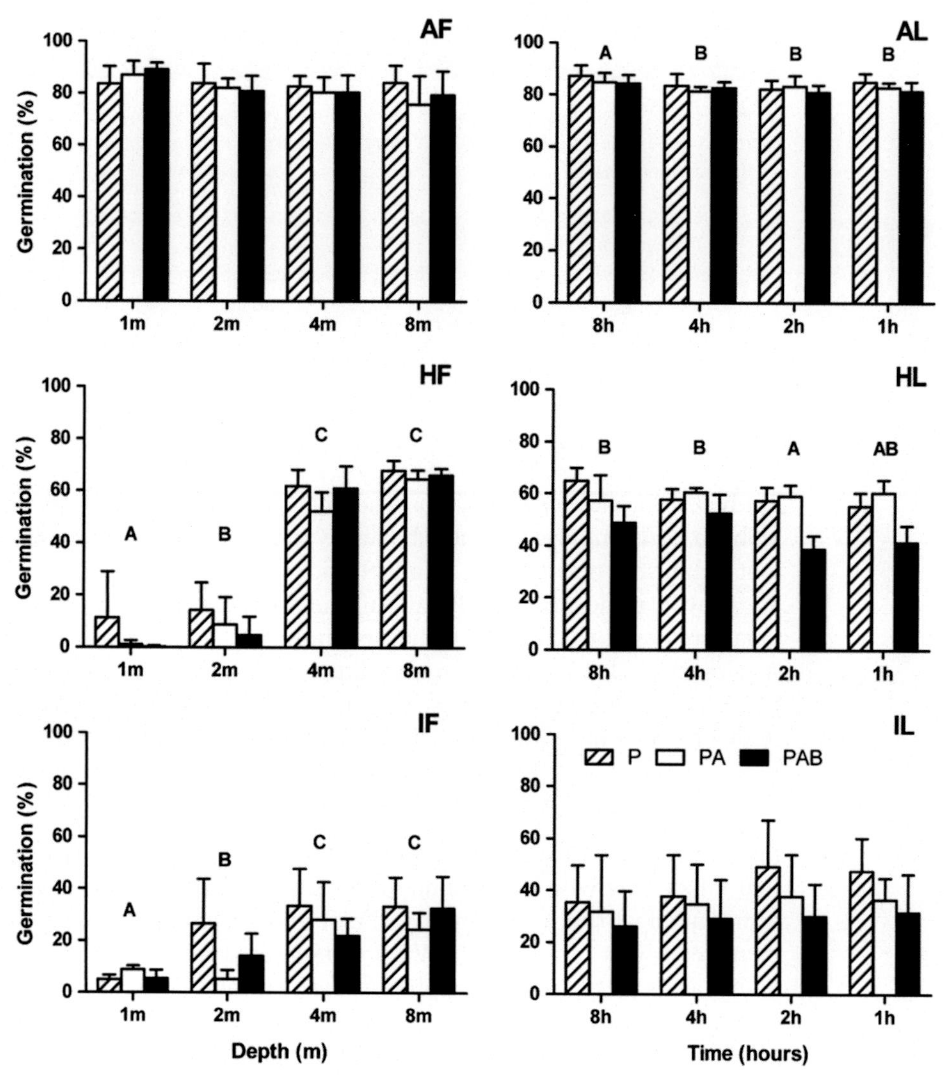

FIGURE 3.2.8

Results of field (left) and laboratory (right) exposure to UV radiation for three different species. *Source: (Zacher, 2014).*

impact, as the number of spores that germinated remained fairly constant across all treatment levels in both the field and laboratory experiments. The second and third species, ***Himantothallus grandifolius*** (HF and HL in the figure) and ***Iridaea cordata*** (IF and IL in the figure), typically live in subtidal regions where their exposure to UV radiation is less. Germination in these two species was strongly affected in the 1 and 2 m field exposures, but generally much less affected in laboratory exposures. Thus, one of the challenges of laboratory research is revealed: how closely do lab results mirror what will happen in the field?

CRITICAL THINKING

Exercise 4: Conducting research isn't always about finding the results you expect. Read through the article by Zacher (available on your Resource Page) to help you answer the following questions:

1. What might have caused the differences noted between the field and laboratory results for the two subtidal species?
2. The general pattern for the two subtidal species was sensitivity to UVR in the field but insensitivity in the laboratory. In what ways might you be able to modify and improve the laboratory portion of the study?
3. How does Zacher explain *Adenocystis utricularia's* ability to tolerate higher levels of PAR and UVB radiation?

Study #3: ***The Role of Adaptation***. The final study in this set by Hylander and Jephson confronts yet another challenge and reality of experiments: the fact that organisms may be able to adapt and alter their responses to a stressor. You may hear the terms "acclimation" and "adaptation" used to describe how organisms react to changes in their environment. Adaptation occurs when a species, over a long period of time, adjusts to changes in its environment. It is part of an evolutionary process (e.g., think of the ability of camels to survive in deserts with little water). Acclimation is a short-term response to a change in the environment (e.g., excessive sweating when exposed to high ambient temperatures or shivering in cold temperatures).

Many species are capable of modifying their physiology to tolerate conditions that might otherwise be harmful. In this case, Hylander and Jephson exposed a marine algal species (a dinoflagellate) and a zooplankter (a copepod) to increased levels of PAR and UVR and measured levels of mycosporine-like amino acids (MAA), which are photoprotective natural sunscreens, an effective defense against the effects of UVR. They exposed one group of each species to radiation and left another set of organisms untreated as a control. In this way, their experiment could effectively identify any production of MAA that might result from exposure to radiation.

There was a marked increase in MAA production over time in both the algae (left) and copepods (right) when exposed to UVR (indicated by the black circles in Fig. 3.2.9) compared to the control. This finding suggests that some species may be able to cope with increased UVR levels.

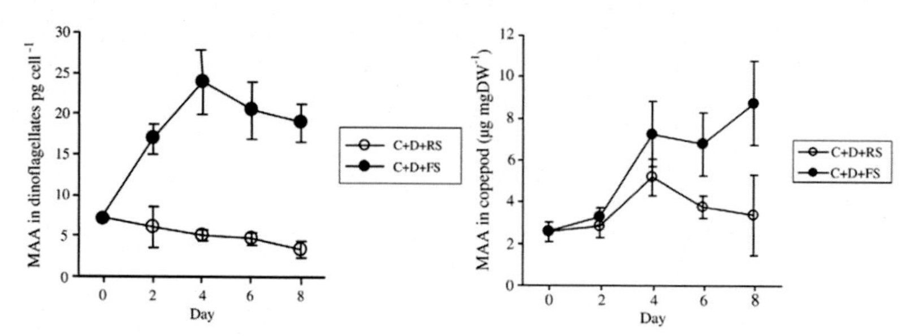

FIGURE 3.2.9

Production of mycosporine-like amino acids (MAA) by marine plankton species in response to UV radiation. *Source: (Hylander and Jephson, 2010).*

DRILLING DEEPER

Exercise 5: Consider two long-term global phenomena: climate change and the depletion of the protective ozone layer.

- Identify two species, one terrestrial and one aquatic, and describe two ways that each might adapt to the long-term changes brought about by these two phenomena.
- For each species you've chosen, identify one additional member of its food web and describe how it might be affected by changes in the first species.

CRITICAL THINKING

Exercise 6: While some might see the results of this study as an argument for disregarding the continued threat that ozone depletion poses to marine populations, a critical assessment of this work may raise more questions than comforting answers. For example, we have no evidence from these data that the increase in MAA is sufficient to prevent UV damage to these species. It is possible that even though they have a protective response, UV damage may still occur at levels that threaten the population. Also, it isn't clear whether or not there are other limitations on MAA production. In the algae, production peaks on day 4 and then begins to decline. Is this species capable of sustaining the necessary production of MAA over long periods of time? If so, at what cost?

- Read through the Hylander and Jephson article on your Resource Page.
- What additional questions do their results raise?
- How might their experiment be redesigned to answer some of these questions?

SYSTEMS CONNECTIONS

Exercise 7: It is possible that if UVR levels continue to increase over time, species like those in this study with the ability to produce natural sunscreens might have a competitive advantage over species that don't.

- Do some research on the marine food webs of the Antarctic. Which species you've discovered in your search would likely be most vulnerable to the effects of continued or increased exposure to harmful UVR? Why?
- Assume that the only response to the radiation was the production of MAA in certain plankton species. Describe how the marine food web of the Antarctic would likely be affected if this were the case.
- How might the additional threat posed by climate change further alter the Antarctic's marine food web?

Solutions: As we noted earlier, the Montreal Protocol was signed to reduce the production and use of CFCs. This was a fine example of the Precautionary Principle at work. When considering whether or not to take action to address an environmental issue, it is important to consider the "cost-benefit;" that is, how does the cost of addressing the issue (banning the use of CFCs in this case) compare to the potential risks to the environment and human health if nothing is done? In this case, experts concluded that the risk of doing nothing exceeded the cost of taking action. This is the basis of the Precautionary Principle.

SOLUTIONS

Exercise 8: Return to the Tools and Skills section and review the basics of cost-benefit analysis, and then complete a cost-benefit analysis for the issue of climate change as it relates to your hometown.

- Do you believe the costs of taking action are more or less than the costs of doing nothing? Justify your choice.
- What factors make your analysis in this case so difficult?

Not So Fast! There are a couple of twists in the CFC story that have complicated the ozone depletion issue.

1. Hydrochlorofluorocarbons (HCFCs) were to be a safer replacement for ozone-depleting CFCs. While they can degrade ozone, they do so at a much lower rate. Unfortunately, beginning around 2000 the use of air conditioners in nations like China and India began to skyrocket, increasing levels of HCFCs in the atmosphere. Efforts are now underway to restrict the use of HCFCs.

2. In the US and other developed nations, HFC coolant 410a, which doesn't react with ozone, has been the replacement gas of choice. Unfortunately, while this gas doesn't affect the ozone, it is 2100 times more powerful than CO_2 as a greenhouse gas, and it cannot be regulated under the Montreal Protocol, as this treaty deals only with ozone-depleting gases!

CONSIDER THIS

Exercise 9: There clearly are "winners and losers" in the climate change picture. Those nations free to burn unlimited amounts of fossil fuels to achieve economic growth have been "winners," while those nations which haven't can be seen as the "losers." While perhaps not as obvious as the case with climate change, there are aspects of environmental justice associated with the issue of ozone depletion.

- Who are the winners in the case of ozone depletion? Who are the losers?
- What human population might be seen as the victim of an environmental injustice in the case of ozone depletion? Why?

REFERENCES

Cabrerizo, M.J., Cabrillo, P., Villafane, V.E., Helbling, E.W., 2014. Current and predicted global change impacts of UVR, temperature and nutrient levels on photosynthesis and respiration of key marine phytoplankton groups. J. Exp. Mar. Biol. Ecol. 461, 371–380.

Farman, J.C., Gardiner, B.G., Shanklin, J.D., 1985. Large losses of total ozone in Antarctica reveal seasonal ClO_x/NO_x interaction. Nature 315 (6016), 207–210.

Hylander, S., Jephson, T., 2010. UV protective compounds transferred from a marine dinoflagellate to its copepod predator. J. Exp. Mar. Biol. Ecol. 389, 38–44.

Lin-Gen, B., Zhong, L., Xiang-Dong, Z., Yong-Feng, M., Long-Hua, L., 2012. Trend of Antarctic ozone hole and its influencing factors. Adv. Clim. Change Res. 3 (2), 68–75.

Zacher, K., 2014. The susceptibility of spores and propagules of Antarctic seaweeds to UV and photosynthetically active radiation—field versus laboratory experiments. J. Exp. Mar. Biol. Ecol. 458, 57–63.

Chapter 3.3

Persistent Organic Pollutants (POPs)

FIGURE 3.3.1

This cross-section of the atmosphere over Africa shows clouds, dust, and smoke from fires returned by the Cloud-Aerosol Transport System (CATS) instrument aboard the International Space Station. *Source: NASA.*

Overview: POPs are toxic chemical compounds that, when released into the environment, do not degrade or break down for long periods of time. As such, they are able to travel long distances in the wind (Fig. 3.3.1) and water and potentially impact the environment far from where they are released. These harmful substances can accumulate in organisms and pass from one species to another up the food web.

To address this global concern, 90 countries came together to sign the Stockholm Convention in 2001, agreeing to reduce or eliminate the production, use, and release of 12 key POPs (Fig. 3.3.2). However, the long-lived nature of these pollutants means that, even with the treaty in place, many areas are still exposed to these toxic chemicals.

A major impetus for the Stockholm Convention was the finding of POPs contamination in relatively pristine Arctic regions, thousands of kilometers from any known source. In this case, certain pesticides and PCBs present in the environment at lower latitudes have moved poleward and entered the Arctic ecosystem. During this case study, we'll look at what these POPs are, where they originate, how they move to the Arctic, how they are introduced into Arctic food webs and affect various species, and how they can affect the health of

The "Dirty Dozen"

aldrin [1]
chlordane [1]
dichlorodiphenyl
trichloroethane (DDT)[1]
dieldrin[1]
endrin[1]
heptachlor[1]
hexachlorobenzene [1,2]
mirex[1]
toxaphene[1]
polychlorinated biphenyls
(PCBs) [1,2]
polychlorinated dibenzo-p-
dioxins[2](dioxins)
polychlorinated
dibenzofurans[2] (furans)

1-Intentionally Produced.
2-Unintentionally Produced -
Result from some industrial
processes and combustion.

FIGURE 3.3.2
The original toxins included in the Stockholm Convention banned list include several pesticides, industrial chemicals, and by-products. *Source: US EPA. Since the creation of the list in 2001, many more chemicals have been added (see* www.pops.int *for the complete list).*

indigenous peoples like the Inuit. We'll conclude by considering steps that might reduce the severity of this problem.

Background: POPs are mostly large, carbon-based chlorinated molecules, including pesticides like DDT and chlordane, long-banned industrial compounds like PCBs, and "accidental" pollutants like dioxin, which are not intentionally made, but are released to the environment as by-products from the burning of plastics, the chlorination of certain types of industrial waste, and other processes. These compounds are considered to be "legacy" POPs because they have caused environmental concerns for decades in spite of reductions in their production and use.

A second group of newer POPs, considered to be "emerging" pollutants, is also being measured at elevated levels in the Arctic ecosystem. These include brominated flame retardants like PBDEs (polybrominated diphenyl ethers) and perfluorinated compounds like perfluorooctane sulfonates (PFOS), which are used in and on various products to help resist heat and stains.

Important properties of "legacy" POPs include the following:

- Because they are chlorinated hydrocarbons and their bonds (e.g., C—C, C—Cl) are very resistant to both oxidation and hydrolysis, POPs break down in the environment very slowly. Under certain conditions, they can remain in the environment for decades.

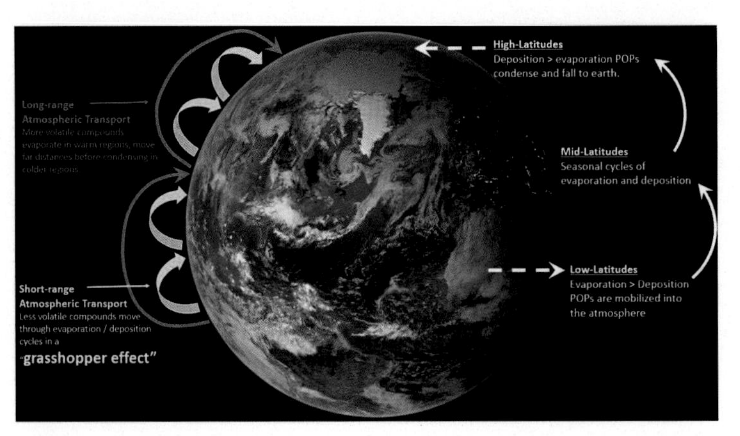

FIGURE 3.3.3

The "grasshopper" effect: pathways for the long-range transport of persistent organic pollutants. *Source: Jennifer Pontius*

- POPs tend to be semivolatile and evaporate slowly. Thus, they can enter the atmosphere and be transported long distances by prevailing winds and then be deposited into environments like the Arctic far from their source.

- They are sparingly soluble in water and highly soluble in fats. This property results in a tendency for POPs to be found at higher concentrations in biota as one moves up exposed food webs.

- Some POPs can interfere with biological systems at very low levels, and they may pose a threat, not only to fish and wildlife, but also to humans who consume them in their diets.

Term to Learn: The phrase "grasshopper effect" is used to describe the cycles of evaporation and deposition that may occur multiple times as POPs move from sources in the tropics to the Arctic, with the movement resembling the hopping of a grasshopper (Fig. 3.3.3). Wania (2003) labeled most POPs as "multiple hoppers" that will be carried to the poles in repeated evaporation and condensation reactions.

THE PROBLEM

For years, scientific research has demonstrated that concentrations of POPs in various components of Arctic food webs are elevated to levels that one might not expect for an ecosystem so far removed from obvious sources of these pollutants. In addition, studies of indigenous populations of the far north like the Inuit have shown surprisingly high levels of PCBs and other POPs in blood

FIGURE 3.3.4
Melting Arctic ice could release long-dormant POPs back into Arctic ecosystems. *Source: USGS.*

samples, and there are even suggestions that the health of children in these populations may have been affected by exposure to these POPs. In this case study, we'll take investigate how this unusual case of transglobal pollution has occurred.

There are several potential ways that these pollutants can enter the Arctic environment:

Wind currents: Tracing the movement of most POPs in the environment is complicated because these compounds can exist in different phases (e.g., as a gas or attached to airborne particles) and can be exchanged among environmental media. For example, some POPs can be carried for many kilometers when they evaporate from water or land surfaces into the air or when they are adsorbed by airborne particles. Then, they can return to Earth on particles or in snow, rain, or mist.

Ocean currents: POPs may enter the world's oceans in rivers draining landscapes where POPs are used (e.g., pesticides) or released in industrial effluents. Once in the marine environment, these contaminants can be carried by currents to the waters of the far north, where they may enter aquatic food webs.

Remobilization: POPs buried long ago in Arctic ice may now be reappearing in the atmosphere and entering Arctic food webs. A team of scientists led by J. Ma et al. (2011) compared data on low volatility POPs in the Arctic atmosphere to model projections and concluded that, in fact, alterations brought about climate change, including sea ice melt, have allowed long-dormant POPs to reenter the atmosphere (Fig. 3.3.4).

Biovectors: As northern latitudes continue to warm, another potential route for POPs into the Arctic is via migrating salmon. Having accumulated POPs during their time in the Pacific Ocean, the salmon, seeking cooler waters, might move into Arctic rivers, carrying their POPs burdens with them. Predators or humans feeding on the salmon might then take in the fish's POPs burden. We'll consider an even more interesting example of biovectors when we discuss how POPs move up Arctic food webs.

SYSTEMS CONNECTIONS

Exercise 1: The behavior of POPs in the Arctic is likely to get more complicated with a changing climate. Dig into the literature and answer the following:

- In addition to the melting of ice and salmon migration, what is one additional way that climate change might increase the activity of POPs in the Arctic?
- What's one way that climate change might decrease activity of POPs in the Arctic?
- Overall, do you think that climate change will make the problem of POPs in the Arctic better or worse?

Movement of Persistent Organic Pollutants in Arctic Food Webs

To help keep track of trends in levels of POPs in Arctic wildlife that might pose a risk to human consumers or other wildlife species, the Canadian government began a POPs monitoring program in 1991. Figure 3.3.5 shows the types of samples taken and the frequency of monitoring POPs, while Figure 3.3.6 shows the POPs being sampled, including a number of "emerging" chemicals listed at the bottom of the table. In 2003–2004, the biological monitoring program was changed so that scientists could detect a "10% annual change in contaminant concentration over a period of 10–15 years with a power of 80% and confidence level of 95%." This change meant that fewer locations would be sampled.

CRUNCHING THE NUMBERS

Exercise 2: Based on what you know about experimental design and statistical power:

- How many observations would be required if the goal was to detect a 5% change instead of a 10% change (maintain power and confidence intervals)?
- What if you think that a 1% change should be detectable?
- Why do the requirements for the number of observations change, depending on how small a change you want to be able to confidently detect?
- Why might the government make such a change in a long-running monitoring program?

CRITICAL THINKING

Exercise 3: Compare this monitoring program to the US Geological Survey monitoring program discussed in the Everglades case study in the Water Resources section.

- What are the pros and cons of each program's monitoring design?
- Which program do you think is stronger from a scientific standpoint? Why?
- How might you improve the Canadian POPs monitoring program?

The results of the Canadian POPs monitoring program from the early 1990s to 2011 represent a good news–bad news scenario (Fig. 3.3.7). Most of the "legacy" POPs, like PCBs and DDT, have shown a decreasing trend during this time period, although in some species, like seals and belugas, there has been little change over time. The story for "emerging" POPs is similar, with peak levels occurring in the early 2000s. Restrictions and bans on the use of some of the "emerging" POPs may be responsible. However, there is still a suite of POPs near the bottom of the table that show either significant increasing trends or mixed results across locations sampled.

Media	Locations	Sampling years[1]	Frequency
Air - hi volume	Alert	1992-2010	7 day continuous
Air - passives	Up to 7 arctic/sub-arctic locations	2005-2011	Quarterly
Arctic char (searun)	Cambridge Bay, Pond Inlet , Nain	2004-2011	Annual
Arctic char (landlocked)	Lakes Resolute, Char, Amituk and Hazen	2004-2011	Annual
Burbot	Fort Good Hope, Great Slave Lake West Basin and East Arm	2004-2011	Annual
Lake trout	Lake Laberge, Kusawa Lake, Great Slave Lake West Basin and East Arm	2004-2011	Annual
Caribou	Northern Yukon and Southwestern Nunavut (Porcupine, Qamanirjuaq herds)	2006, 2008	Single study
Ringed seals	Arviat, Resolute, Sachs Harbour and other locations to 2009	2004-2011	Annual
Beluga	South Beaufort, Cumberland Sound	2004-2011	Annual
Polar bears	West Hudson Bay and other locations	2004-2011	Annual
Seabirds (thickbilled murre, blacklegged kittiwakes	Prince Leopold Island, Coats Island	2004-2011	Annual

[1]All programs include data from earlier years based on existing data or reanalysis of archived samples

FIGURE 3.3.5

Overview of the Northern Contaminants Program (NCP) persistent organic pollutants monitoring program media 2003–2011. *Source: (Government of Canada, 2013).*

Major groups of POPs and other persistent organics	NCP I (1991-1996)	NCP II (1997-2002)	NCP III (2003-2011)
PCBs[1]	Air, snow, sediment, seawater, biota	Air, seawater, sediment, biota	Air, snow, seawater, biota
OC pesticides[2]	Air, snow, sediment, seawater, biota	Air, seawater, sediment, biota	Air, snow, biota
Chlorobenzenes	Air, snow, sediment, seawater, biota	Air, seawater, sediment, biota	Air, snow, biota
Chlorinated dioxins/furans	Biota	Air, sediment, biota	Biota
Chlorinated naphthalenes (PCNs)		Air, biota	Air, biota
Chlorinated paraffins		Air, sediment, biota	Biota
Endosulfan		Air, seawater, biota	Air, seawater, biota
Polybrominated diphenyl ethers (PBDEs)		Sediment, biota	Air, snow, seawater, sediment, biota
Hexabromocyclododecane (HBCDD)			Air, snow, seawater, biota
Other Brominated and chlorinated flame retardants			Air, snow, seawater, biota
Penta and hexabromobiphenyls			Air, biota
Current use pesticides[3]			Air, snow, seawater, lake water, biota
Perfluorooctane sulfonate (PFOS) and other perfluoro-alkyl acids and alcohols			Air, snow, seawater, lake water, sediment, biota
Siloxanes			Air

FIGURE 3.3.6
Major groups of persistent organic pollutants and other persistent organics in environmental compartments of the Canadian Arctic determined by the NCP core monitoring and research programs. *Source: (Government of Canada, 2013).*

SYSTEMS CONNECTIONS

Exercise 4: Go to your Resource Page and click on the link to the Canadian Arctic Contaminant Assessment Report III; review the discussion on page 392 about why they may be witnessing an increase in polar bear POPs levels.

- Why would trends in polar bears differ from trends in other Arctic species?
- What specifically makes polar bears more susceptible?

In some locations, POPs levels have remained stable or actually shown opposite trends for the same variable measured in other locations.

- What are some of the possible reasons for this geographic variability in POP trends?

TABLE 3. Overview of time trends of selected POPs and persistent organics in Canadian arctic air and biota. Estimated for all results from early 1990s to 2011

	Air	Burbot[1]		Lake trout[2]			Landlocked char[3]		Sea-birds[4]	seals[5]			Beluga[6]		Polar bears[7]
		FGH	GSL	KW	LL	GSL	H	A	PLI	SBS	LS	HB	SBS	CS	HB
PCBs															
ΣCBz															
ΣHCH															
ΣCHL															
ΣDDT															
toxaphene															
endosulfan															
SCCPs															
PCNs															
PCDD/Fs															
ΣPBDEs															
HBCDD															
PFOS and precursors															
PFCAs and precursors															

Legend:
- Limited or no results to assess trends
- No statistically significant change (typically < 3%/yr)
- Significant declining trend (typically > -5%/yr)
- Significant increasing trend (typically > +5%/yr)
- Significant increase in the early 2000s, currently stable or declining

[1]FGH = Fort Good Hope, GSL = Great Slave Lake - East Arm and West Basin
[2]LL & KW = Lake Laberge and Kusawa; GSL = Great Slave Lake - East Arm and West Basin
[3]H = Lake Hazen. A= Amituk Lake
[4]PLI = Prince Leopold Island, Lancaster Sound
[5]SBS = Southern Beaufort Sea (Ulukhaktok, Sachs Harbour), LS = Lancaster Sound (Resolute, Arctic Bay and Grise Fiord), HB = Hudson Bay (Arviat and Inukjuaq)
[6]SBS = Southern Beaufort Sea (Hendrickson Is); CS= Cumberland Sound (Pangnirtung)
[7]HB = Hudson Bay – for all compounds except PFOS and PFCAs where results for N. Baffin Island and Baffin Bay were used

FIGURE 3.3.7
Overview of time trends of selected POPs and persistent organics in Canadian Arctic air and biota. Estimated for all results from early 1990s to 2011. *Source: (Government of Canada, 2013).*

EXPLORE IT

Exercise 5: In addition to long-term monitoring programs, the behavior of POPs is also tracked by satellites designed to detect the movement of aerosols. This information allows scientists to model the long-range atmospheric transport of POPs from high-release areas to the Arctic.

Ozone Mapping and Profiler Suite (OMPS) is a series of backscattered UV radiation sensors designed to measure atmospheric ozone and aerosol levels (Fig. 3.3.8). The Total

continued

EXPLORE IT—Cont'd

Ozone Mapping Spectrometer (TOMS) aerosol index is a measure of how much UVR is backscattered from an atmosphere containing aerosols (with Mie scattering, Rayleigh scattering, and absorption) compared to that of a pure molecular atmosphere (pure Rayleigh scattering). Determining which wavelengths of UVR are backscattered helps scientists determine what kind of aerosols are present and in what quantity across the globe.

Explore NASA's OMPS webpage (**https://ozoneaq.gsfc.nasa.gov/**) to learn more about this instrument.

- Browse through the science highlights. What kind of information does the OMPS sensor provide and what does it tell you *about the common sources of POPs?*
- How do scientists know they are tracking probable aerosol pollutants and not simply clouds moving across the globe?

NASA also has an interactive portal that allows us to track aerosol levels recorded at different stations across the globe. Visit the MAPSS: Multi-sensor Aerosol Products Sampling System website at **http://giovanni.gsfc.nasa.gov/mapss/**. You will view a time series of aerosol deposition data for the Tiksi station in the Arctic. You will need to specify the Tiksi station, the sensor, and the variables you would like to see plotted after making the following selections for data visualizations:

Select Sation: Tiksi
Select Plot: Time Series
Select Measurements: Basic
Product: OMI aerosols L2, Ver 003
Parameter: Best AOD
Layer: Best AOD at 354nm
Measurement: Median
No. of pixels with value: All
Final algorithm flag: All
Select Date Range: Date Picker
Date: Enter 2001-01-01 to Enter current date

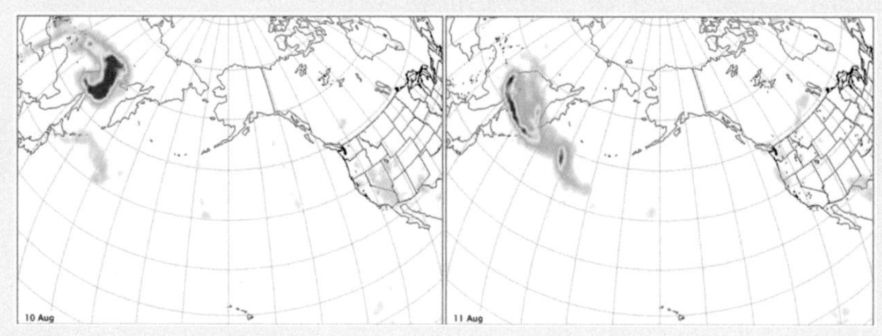

FIGURE 3.3.8

An aerosol index measured by the OMPS sensor tracks smoke from fires over Lake Baikal as it begins to cross the Pacific Ocean. *Source: NASA.*

EXPLORE IT—Cont'd

This includes specifying the Tiksi station and the OMI aerosol values for the median value of the best aerosol optical depth (AOD) at 354 nm. This gives us an idea of the typical amount of small particulates in the atmosphere over Tiksi over the default date range. Be sure to select the ADD button so the interface knows to plot this selection of measurements. Now you are ready to PLOT DATA.

- What do you notice about long-term trends in aerosol concentrations at this Arctic station?
- Are there any seasonal patterns as well?
- What does this information tell you about the continued transport of POPs to the Arctic?

Environmental Sciences in Action: Guano as a Biovector

E. Choy of Canada's University of Ottawa and her colleagues have studied how fulmars, a common Arctic bird species, might transport POPs taken in from their marine-based diet into nearby terrestrial food webs through their waste, or guano. Their study focused on Cape Vera on Devon Island, Nunavut (Fig. 3.3.9), where a colony of northern fulmars nests on the sea cliffs. Previous research has shown that the guano of the birds has contaminated nearby ponds with POPs. Choy and her colleagues analyzed POPs, including PCBs and DDT, in the nearby food web, measuring concentrations in lichens, snow buntings, collared lemmings, and ermine.

Choy and her colleagues found that snow buntings were the most contaminated species, with PCB and DDT concentrations surpassing environmental guidelines for protecting wildlife. This bird species feeds on midge larva, an aquatic species known to accumulate POPs. Levels were also elevated in ermine, which prey on lemmings and snow buntings. Concentrations across all species were significantly inversely correlated with distance from the seabird cliffs and also increased with higher levels of $\delta^{15}N$ (the higher the $\delta^{15}N$, the higher a species is in the food web) (Fig. 3.3.10). The work of Choy et al. was instrumental in identifying biovector transport as a source of organic contaminants to certain components of the terrestrial food web.

Choy et al. (2010b) cited several lines of evidence to support their hypothesis that the fulmars acted as biovectors transferring POPs into nearby terrestrial food webs. While this is based primarily on simple correlations, the scientists were able to make a strong case for causation. Their justification included three primary lines of evidence:

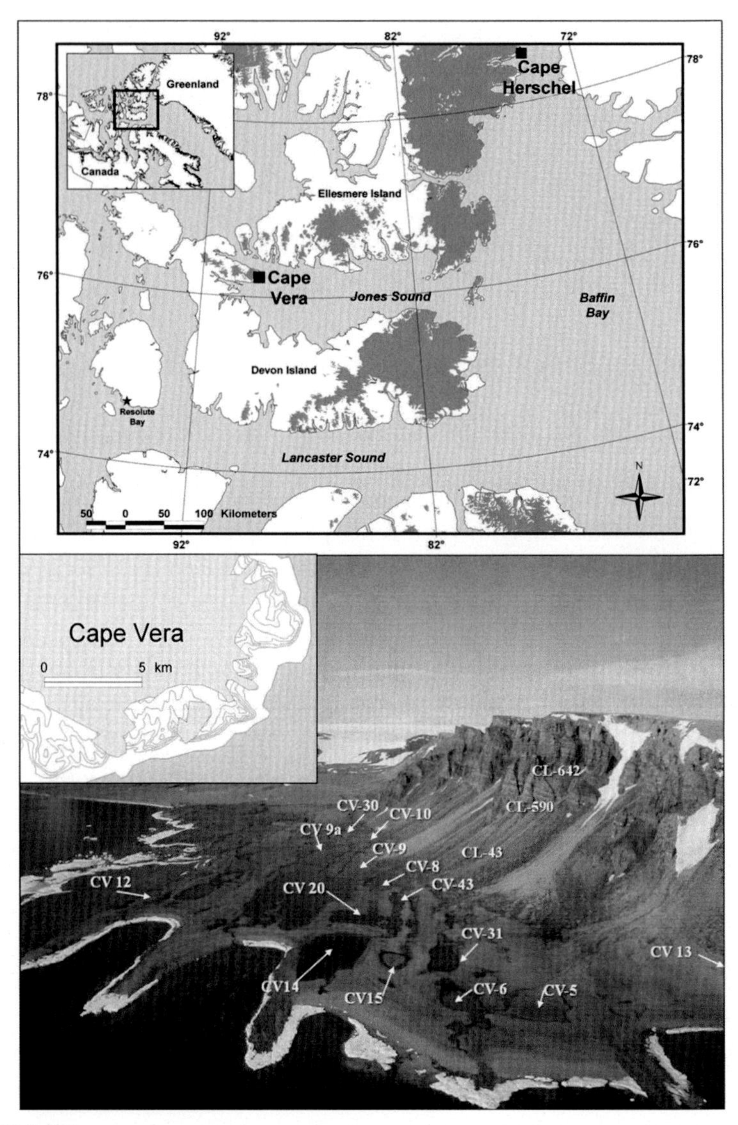

FIGURE 3.3.9

Main study site at Cape Vera, Devon Island, Nunavut. *Arrows and labels* indicate ponds where samples were collected at the base of the seabird cliffs. *Source: Choy et al. 2010b.*

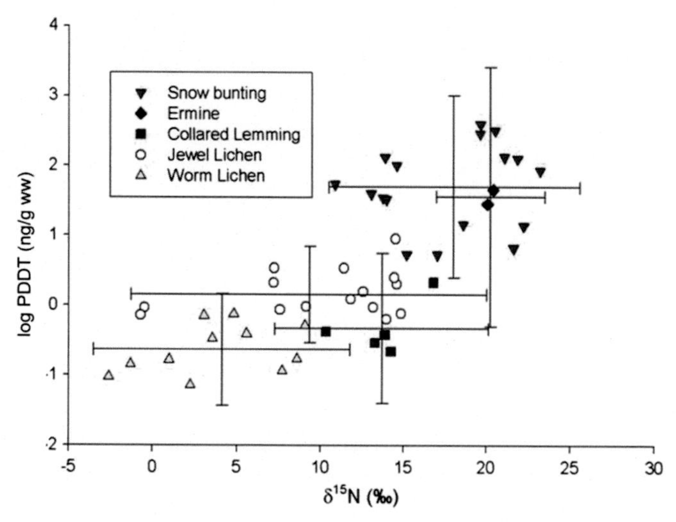

FIGURE 3.3.10

The relationship between log PDDT (ng/g ww) and $\delta^{15}N$ values among members of the ecosystem at Cape Vera, Devon Island, Nunavut. *Error bars* represent the 95% confidence intervals surrounding the species means of log PDDT (ng/g ww) and $\delta^{15}N$ values for each species. *Source: Choy et al, 2010b.*

1. *Establish that there must be an additional source beyond atmospheric transport (i.e., POPs contamination was higher in biovector regions than anywhere else)*: POPs levels in snow buntings at Cape Vera were greater than values reported elsewhere, often exceeding available guidelines for safe consumption.

2. *Show that the closer to the biovector source, the higher the contamination levels*: Jewel lichen samples taken at greater distances from the bird's nesting areas contained lower levels of POPs than those taken near the colonies.

3. *Demonstrate the potential mechanism for bioaccumulation in populations*: Species higher up in the food web exhibited higher levels of contamination; there was a positive correlation between $\delta^{15}N$ and POPs contamination.

PUTTING SCIENCE TO WORK

Exercise 6: The golden rule of statistics is that correlation does not equal causation. Yet in Choy et al.'s study, the scientists present a strong case that fulmar guano is causing greater POPs contamination in nearby terrestrial ecosystems.

- When is it acceptable for scientists to infer causation from relationships?
- What else could the scientists have done to test this causal relationship?

DRILLING DEEPER

Exercise 7: In an earlier study, Choy et al. (2010a) looked at the movement of mercury via guano. Read this article posted at the link on your Resource Page.

- How did mercury differ from the POPs in its behavior?
- How can you explain the different response of POPs and mercury?
- Based on this, which poses a more significant threat to Arctic ecosystems?

Environmental Sciences in Action: Polar Bears as a Keystone Species

Alaska's tundra and its proximity to the Bering Sea and Arctic Ocean make it a home for a wide variety of wildlife, including the iconic polar bear. During the long northern winters, mammals like polar bears metabolize fat to survive, releasing POPs that have built up in their fat into their bloodstream. In the spring, a critical period of reproduction for Alaskan wildlife, additional POPs that have accumulated in the ice and snow are also released into the environment and the food web. While atmospheric concentrations of many POPs have decreased in recent years, many of these compounds are still at or above levels that may threaten polar bear reproduction and survival due to these additional sources.

A team of scientists led by B. Jenssen of the Norwegian University of Science and Technology recently reviewed the combined effects of climate change and POPs on polar bears (Fig. 3.3.11). In addition to melting ice contributing to POPs contamination, the changing climate in the Arctic has other direct implications for polar bears. The additional stress of climate change makes isolating the effects of POPs on species like polar bears difficult. Possible interactions include:

- ...More open water and less sea ice mean fewer ringed seals in the bears' diet; as bears metabolize fat stores to compensate, POPs like PCBs stored in the fat reenter the blood.
- ...As polar bears are forced to travel greater distances to find ice, their food/energy requirements increase, requiring them to consume more prey, leading to a greater uptake of POPs.
- ...Increased reliance on food sources like seabird eggs and migrating cold water fish like salmon may also increase the levels of POPs in the diets of the bears.

Jenssen et al.'s work shows that prolonged fasting periods due to climate change and decreased sea ice may result in increases in the tissue concentrations of POPs in polar bears. The scientists concluded that fasting-induced increases of POPs would increase mortality rates and decrease reproductive success beyond effects caused by loss of habitat alone. However, few studies have addressed this possibility.

PUTTING SCIENCE TO WORK

Exercise 8: Jenssen et al. (2015) cited the need for additional studies on the effects of POP exposure on polar bear populations and how such impacts relate to the effects of climate change-induced habitat loss.

- Design an experiment that you might perform to determine how POPs and the changing climate will affect polar bear populations.
- How might you account for the combined effects of habitat loss and POP concentrations?

FIGURE 3.3.11
Climate change induced fasting and exposure to persistent organic pollutants (POPs) are anthropogenic stressors that have the potential to affect the individual health, reproduction, and mortality of polar bears. Prolonged fasting may increase tissue concentrations of POPs and thus enhance negative fitness effects, whereas high levels of POPs may cause metabolic effects that interfere with fasting tolerance. Effects on the individual level propagate to population levels. *Photo: Rune Dietz. Source: Jenssen et al. 2015.*

Effects of Persistent Organic Pollutants on Humans in the Arctic

Scientists have long known that the Inuit of Nunavik in Arctic Quebec (Fig. 3.3.12) are exposed to elevated levels of PCBs and methyl mercury because of their diet of fish and the fat of sea mammals like Beluga whales and seals. Scientists have investigated the issue of elevated pollutants in young Inuit children. Universite Laval researcher H.T. O'Brien et al. (2012) measured concentrations of mercury, lead, PCBs, pesticides, PBCEs, and PFOS in the blood of 155 Inuit children attending childcare centers in Nunavik, Canada. Lead, PCB-153, a PBDE, and PFOS were detected in all samples, and mercury was found in 97% of the samples. PBDE levels were higher than those found in similar samples globally.

A second study, conducted by O. Boucher and his colleagues (Boucher et al., 2014), measured prenatal exposure of 94 Inuit to PCBs, mercury, and lead. The scientists measured various types of cognitive impairment in the Inuit infants and related them to levels of the three pollutants. Results of their study suggested that each of the three neurotoxic pollutants affects a different brain function, with, for example, prenatal PCB exposure resulting in impaired memory.

FIGURE 3.3.12
Photographer Hans Blohm with Inuit children in Kuujjuaq, Nunavik (Quebec), Canada. *Source: By Harald Finkler (CC BY-SA 3.0).*

CRITICAL THINKING

Exercise 9: We've provided links to each of these studies on your Resource Page. Read the two studies and answer the following:

- Are you convinced by these studies that POPs represent a serious threat to Inuit populations? Why or why not?
- Which study presents a stronger case for POPs as a threat to Inuit health? Justify your choice.

SOLUTIONS

Exercise 10: Clearly the issue of POPs in the Arctic is a complicated one. Much has already been done to address this problem, including the signing of the Stockholm Convention and ongoing monitoring efforts across the Arctic.

- What are the strengths of these ongoing efforts?
- What additional approaches do you think the Canadian government should adopt to address the issue of POPs?
- What strategies could be implemented globally to reduce the amount of POPs introduced to the Arctic environment?

CONSIDER THIS

Exercise 11: Inuit populations contribute very little to POPs in the environment, yet they suffer disproportionately from these toxic contaminants. Why should we be concerned about these contaminants in small populations so far away? What steps could we personally take to help? Do you think that this is an issue of environmental justice? Why or why not?

REFERENCES

Boucher, O., Muckle, G., Jacobson, J.L., Carter, R.C., Kaplan-Estrin, M., Ayotte, P., Dewailly, E., Jacobson, S.W., 2014. Domain-specific effects of prenatal exposure to PCBs, mercury, and lead on infant cognition: results from the Environmental Contaminants and Child Development Study in Nunavik. Environ. Health Perspect. 122 (3), 310–316.

Choy, E.S., Gauthier, M., Mallory, M.L., Smol, J.P., Douglas, M.S.V., Lean, D., Blais, J.M., 2010a. An isotopic investigation of mercury accumulation in terrestrial food webs adjacent to an Arctic seabird colony. Sci. Total Environ. 408 (8), 1858–1867.

Choy, E.S., Kimpe, L.E., Mallory, M.L., Smol, J.P., Blais, J.M., 2010b. Contamination of an Arctic terrestrial food web with marine-derived persistent organic pollutants transported by breeding seabirds. Environ. Pollut. 158, 3431–3438.

Government of Canada, 2013. Canadian Arctic Contaminants Assessment Report III (2013): Persistent Organic Contaminants in Canada's North.

Jenssen, B.M., Villanger, G.D., Gabrielsen, K.M., Bytingsvik, J., Bechshoft, T., Ciesielski, T.M., Sonne, C., Dietz, R., 2015. Anthropogenic flank attack on polar bears: interacting consequences of climate warming and pollutant exposure. Front. Ecol. Evol. http://dx.doi.org/10.3389/fevo.2015.00016.

Ma, J., Hung, H., Tian, C., Kallenborn, R., 2011. Revolatilization of persistent organic pollutants in the Arctic induced by climate change. Nat. Clim. Change 1, 255–260.

O'Brien, H.T., Blanchet, R., Gagne, D., Lauziere, J., Vezina, C., Vaissiere, E., Ayotte, P., Dery, S., 2012. Exposure to toxic metals and persistent organic pollutants in Inuit children attending childcare centers in Nunavik, Canada. Environ. Sci. Technol. 46 (8), 4614–4623.

Wania, F., 2003. Assessing the potential of persistent organic chemicals for long-range transport and accumulation in polar regions. Environ. Sci. Technol. 37 (7), 1344–1351.

Chapter 3.4

Particulate Matter (PM) Pollution

FIGURE 3.4.1
Dust pollution during bore well drilling in Munnekolala. *Photo By Amol.Gaitonde (CC BY-SA 3.0), via Wikimedia Commons.*

Overview: PM consists of a mixture of solid and liquid particles suspended in the air. The mixture may include acids, organic chemicals, metals and dust particles, as seen in Fig. 3.4.1. PM is regarded as a significant air pollutant, affecting the respiratory and circulatory systems of humans who breathe in the tiny particles. PM also represents a substantial challenge to regulators because of its many, often hard to control, sources.

In this case study, we'll focus on a challenge confronting scientists and regulators attempting to understand and solve environmental problems: how to establish cause–effect relationships. While it is possible to assess the health of people and the composition and density of particulate pollution in the air, establishing the link between the two is much more difficult. We'll look in depth at several studies designed to help understand the effects of PM exposure on human health, and, in the process, evaluate the strengths and weaknesses of each approach.

Background: Most attention on PM has focused on smaller particles. PM 10 refers to particles with a diameter of 10 μm or less, and PM 2.5 refer to particles

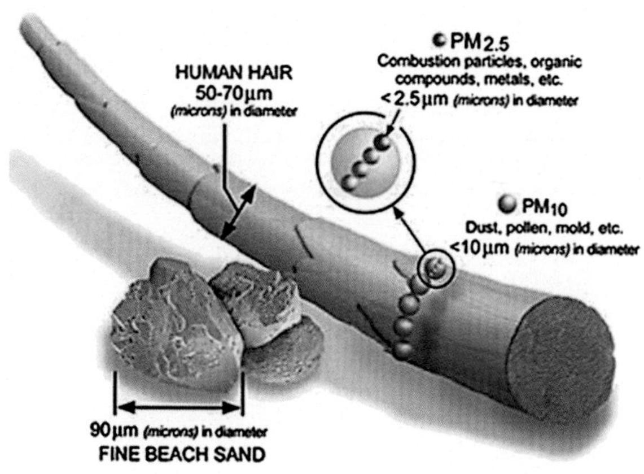

FIGURE 3.4.2
Relative sizes of particle pollution. *Source: US EPA.*

2.5 μm or less (Fig. 3.4.2). Anything less than 10 μm is considered to be inhalable, but the smaller a particle is, the more easily it can enter the lungs, where it can cause serious health effects. The smallest particles, less than 1 μm in diameter, can remain in the atmosphere for long periods and be transported great distances by the wind, potentially increasing the number of humans exposed. These smaller particles are also responsible for reduced visibility (haze) in many locations.

EXPLORE IT

Exercise 1: While we can't see particles smaller than PM 2.5 with the unaided eye, these tiny pollutants have a large impact on visibility because of their tendency to scatter visible light. Because they are small, they can also travel far from their source, impacting visibility in many scenic locations like national parks.

The CAMNET network of "haze" cameras provides live photos and air quality conditions from scenic vistas across the eastern US (Fig. 3.4.3). Visit the CAMNET website at **http://www.hazecam.net/** linked on your Resource Page. Click the tab for the gallery of good and bad days and examine how much visibility (and hence air quality) can change for the same location from day to day.

Pick one of the hazecam sites and familiarize yourself with what a good and bad day from this haze camera look like. Choose that same location from the left menu of live cameras.

- How does the current level of haze compare to your standard good and bad days for this location?
- What is the current level of fine particles (low, medium, or high) based on the provided air pollution scale?
- Does your visual assessment of the level of haze match the reported concentration of fine particles?

continued

EXPLORE IT—Cont'd

Now click on the Real Time Pollution Maps and Forecasts tab. Select the current AQI (air quality index) tab and examine air quality across the US.

- Does the current AQI for your selected site on this modeled map match what you saw from the site specific live camera?
- Where is the current region of most unhealthy air quality, according to this map (unhealthy to hazardous)?
- Based on these locations and current conditions, what do you suspect might be the source of the PM and ozone for this region? In the example shown in Figure 3.4.4, there were many fires in the Pacific Northwest leading to hazardous levels of PM and ozone at the time of assessment.
- We typically think of cities as having the highest consistent levels of air pollution. Does your map of current AQI match this expectation? If not, why might this be?

FIGURE 3.4.3
An example of a good and bad haze day at the Burlington, Vermont CAMNET hazecam.
Source: CAMNET.

EXPLORE IT—Cont'd

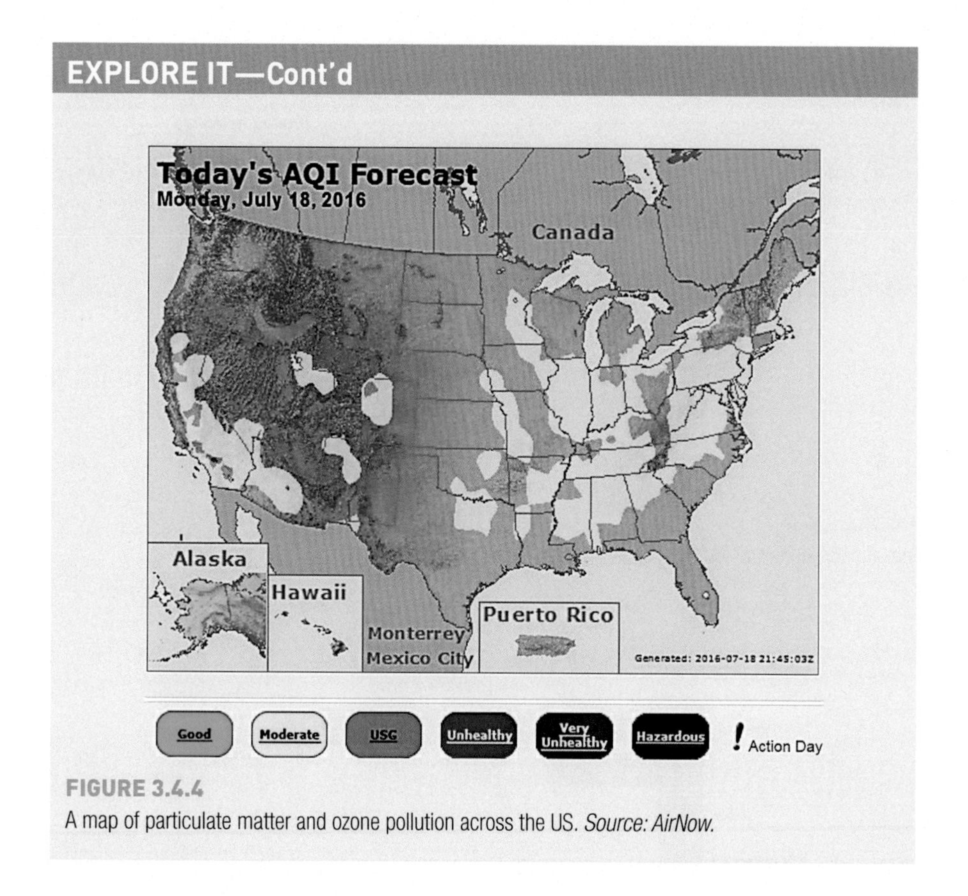

FIGURE 3.4.4
A map of particulate matter and ozone pollution across the US. *Source: AirNow.*

SOURCES OF PARTICULATE MATTER

PM can be emitted directly from sources (primary particles) such as fires, smokestacks, or fields, or can form during secondary reactions among chemicals in the atmosphere (secondary particles). These secondary particles make up most of the fine particle pollution in developed nations.

Both primary PM pollution and the chemicals involved in the development of secondary PM pollution can come from human or natural sources. Across the entire US, fires are the primary source of PM 2.5 pollution (Fig. 3.4.5). However, this varies from location to location, depending on proximity to primary sources, land cover and land use, fire frequency, and weather patterns. For example, Fig. 3.4.6 shows the primary sources of PM 2.5 in a relatively rural state, Vermont, with comparatively little industry and a predominantly forested landscape. In Vermont, total emissions are relatively low, and the primary source of PM 2.5 is fuel combustion.

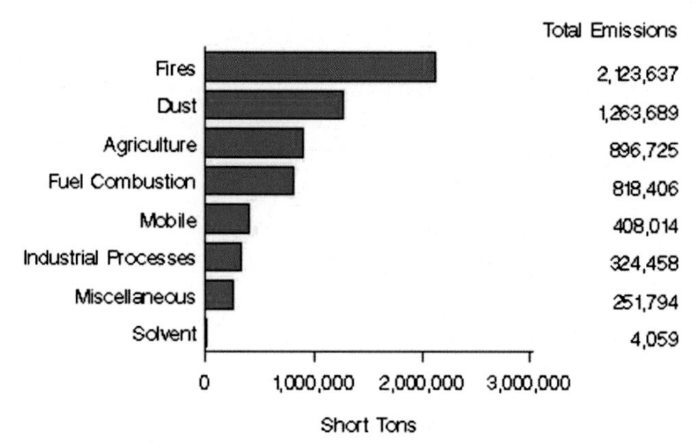

FIGURE 3.4.5

The primary sources of PM 2.5 pollution across the US in 2011. *Source: NEI 2011 US EPA.*

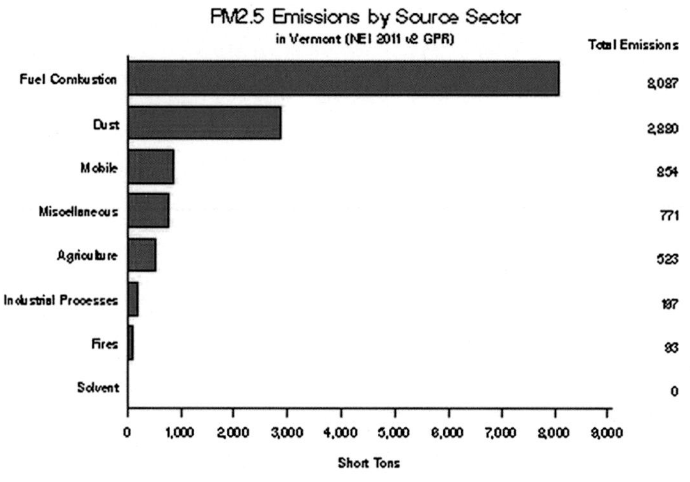

FIGURE 3.4.6

A breakdown of the primary sources of PM pollution for Vermont, a relatively rural state with low fire frequency. *Source: NEI 2011 US EPA.*

Anthropogenic sources: Important human-related sources of PM include combustion engines, fossil fuel–burning power plants, various industries (e.g., mining, gravel pits), erosion of pavement by traffic, and abrasion of brakes and tires. Depending on their origin, particles released from these sources may contain a variety of hazardous pollutants, including heavy metals like cadmium and lead, black carbon, and polycyclic aromatic hydrocarbons (PAHs), which, if inhaled, can be very bad for human health.

Natural sources: Soil and dust released by agricultural activities, traffic on unpaved roads, sand from deserts, and smoke from fires are important natural sources of PM.

DRILLING DEEPER

Exercise 2: Visit the US EPA's Air Emissions Sources website for PM linked from your Resource Page (http://www3.epa.gov/cgi-bin/broker?_service=data&_debug=0&_program=dataprog.national_1.sas&polchoice=PM#pmloc).

Here, you can view emissions summaries at the national or state level.

Select the link for state and local summaries, and select two states from the map of PM 2.5 sources to examine; include one that you would expect to be very high in PM emissions and one that you would expect to be very low.

- Which did you pick as your high and low emissions states? Why?
- Click on your chosen "high" and "low" emissions states. How does the total quantity of PM 2.5 emissions differ between the two? Does this agree with what you expected? Why or why not?
- Do the primary sources of PM 2.5 emissions differ between the two states? How can you explain this?

Measuring PM: A thorough assessment of PM concentrations requires continuous monitoring. Particles in the air are commonly collected by high-volume samplers, which are glorified vacuum cleaners (Fig. 3.4.7) that suck a known volume of air through a clean, preweighed filter pad for a fixed period of time, typically 24 h. Filters may have a variety of pore sizes to facilitate the sampling of different sizes of PM. After the sampling period ends, the filter's weight can be compared to the initial value to determine the mass of PM collected during the fixed time period. In addition, the material on the filter pad can be analyzed for contaminants, like heavy metals and PAHs.

CRUNCHING THE NUMBERS

Exercise 3: PM levels measured by high-volume samplers can be expressed on a weight per volume basis as $\mu g/m^3$. Assume your filter pad weighed 63.5 g before sampling and 78.9 g after 24 h. During this period of time, 600 m^3 of air were drawn through the filter pad.

- What is the concentration of PM in your sample?
- Assume that you determine that a filter contains a total of 20 g of PM, and you measure the lead in the sample at 4 mg. What is the concentration of lead in the PM in $\mu g/g$?

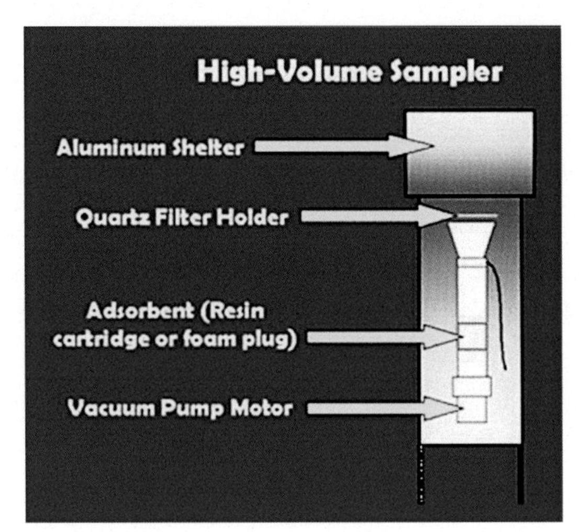

FIGURE 3.4.7
Structure of a high-volume PM sampler. *Source: US EPA.*

Collecting PM samples with a high-volume sampler is fairly straightforward. Sampling other types of air pollutants isn't always so easy. For example, greenhouse gases like methane are being emitted from oil and gas fields where fracking and other types of fuel extraction are ongoing activities. A team of National Oceanic and Atmosphere Administration (NOAA) scientists, led by Dr. Joost de Gouw, is surveying levels of methane, ozone, and other gases over these fields to evaluate the significance of this pollutant source.

Flying low in Miss Piggy, a turboprop plane converted into a flying lab (Fig. 3.4.8), the team sucks air samples from over the fields into the plane through a tube. The air sample goes directly into equipment designed to measure each pollutant, and then the sample is vented out via the plane's exhaust. The team is trying to determine the levels of greenhouse gases being given off in areas where fracking and related activities are occurring.

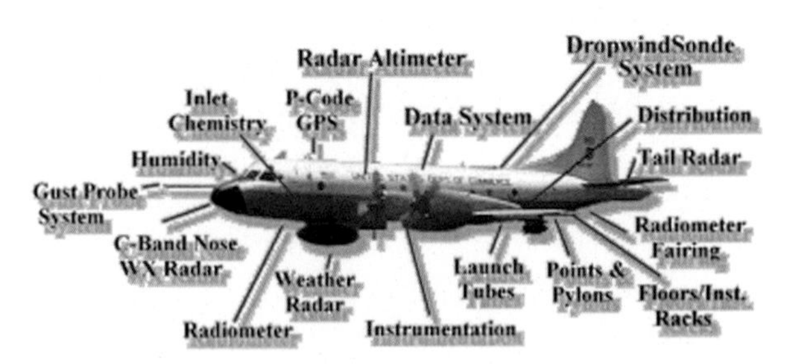

FIGURE 3.4.8
The NOAA flying laboratory collects a wealth of data as it samples the atmosphere. *Source: NOAA.*

To help make up for the scarcity of ground-level PM monitoring stations, remote sensing, combined with modeling and existing surface measurements, have recently been used to assess human exposure to PM at global scales. The resulting assessment shows that elevated PM levels are an issue around the world (Fig. 3.4.9).

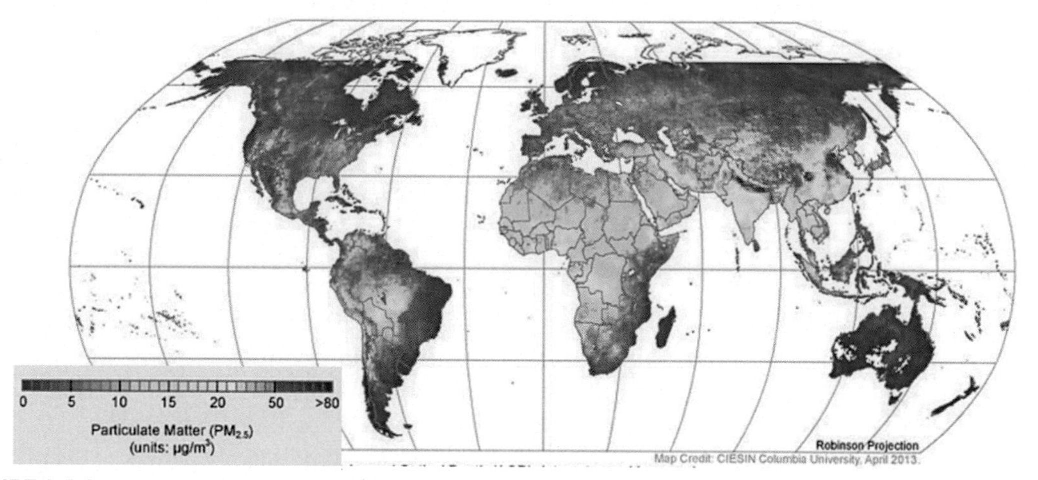

FIGURE 3.4.9

Global annual PM 2.5 concentrations from MODIS and MISR Aerosol Optical Depth data sets provide annual "snapshots" of particulate matter concentrations. *Source: Battelle Memorial Institute and Center for International Earth Science Information Network/ Columbia University (CC BY-SA 3.00).*

DRILLING DEEPER

Exercise 4. Fig. 3.4.10 shows the World Health Organization's (WHO, 2005) interim target levels for PM 2.5. In addition to the air quality guideline values required to avoid human health impacts, interim targets are given for each pollutant. These are proposed as incremental steps in a progressive reduction of air pollution and are intended for use in areas where pollution is high. These targets differ from thresholds set by the US EPA. Compare these standards to the PM 2.5 concentrations recorded across the globe in Fig. 3.4.9.

- Look up the latest US EPA primary annual standard for PM 2.5. A link can be found on your Resource Page.
- Which standard (WHO or US EPA) is more stringent?
- Which regions shown in Fig. 3.4.9 exceed those values?
- What might account for these differences in PM 2.5 levels across regions?
- It's not uncommon for states and nations to set different standards for the same pollutants. Why might this happen?

DRILLING DEEPER—Cont'd

WHOair quality guidelines and interim targets for particulate matter: annual mean concentrations[a]

	PM_{10} ($\mu g/m^3$)	$PM_{2.5}$ ($\mu g/m^3$)	Basis for the selected level
Iinterim target-1 (IT-1)	70	35	These levels are associated with about a 15% higher long-term mortality risk relative to the AQG level.
Interim target-2 (IT-2)	50	25	In addition to other health benefits, these levels lower the risk of premature mortality by approximately 6% [2–11%] relative to theIT-1 level.
Interim target-3 (IT-3)	30	15	In addition to other health benefits, these levels reduce the mortality risk by approximately 6% [2-11%] relative to the -IT-2 level.
Air quality guideline (AQG)	**20**	**10**	These are the lowest levels at which total, cardiopulmonary and lung cancer mortality have been shown to increase with more than 95% confidence in response to long-term exposure to $PM_{2.5}$

FIGURE 3.4.10
Air Quality Guidelines. *Source: WHO, 2005*

The Problem: While elevated PM levels can lead to environmental problems like decreased visibility and discoloration of masonry, the major concern with PM is its impact on human health. Much attention is focused on PM 2.5. Because of their small size, these particles can most easily penetrate deeply into the lungs of exposed humans. According to the US EPA (2014), the following health effects have been linked to particle pollution:

- premature death in people with heart or lung disease
- nonfatal heart attacks
- irregular heartbeat
- aggravated asthma
- decreased lung function
- respiratory symptoms, such as irritation of the airways, coughing, or difficulty breathing

People with heart or lung disease are especially vulnerable to adverse impacts of PM pollution. But any physical activity that causes people to breathe more rapidly and deeply can also increase the risks of health impacts if PM is present in the air. The WHO estimates that 6–13% more cardiopulmonary deaths

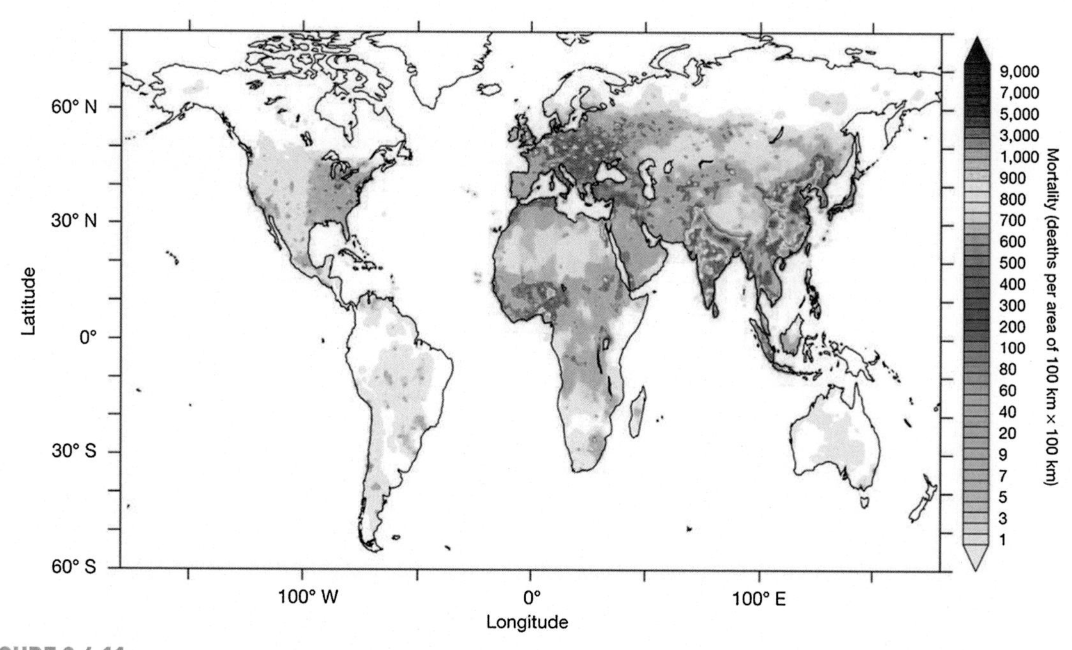

FIGURE 3.4.11

Units of mortality = deaths per area of 100 × 100 km (color-coded). In the *white areas*, annual mean PM 2.5 and O_3 are below the concentration–response thresholds at which no excess mortality is expected. *Source: (Lelieveld et al., 2015).*

occur among individuals who are exposed to levels of 10 μg/m³ PM 2.5 in the air they breathe. Scientists are also studying the potential link between PM exposure and low infant birth weight, preterm deliveries, and infant death.

Environmental Sciences in Action: Recent studies (Lelieveld et al., 2015; Thurston et al., 2015) shed additional light on the severity of the PM problem facing regulators. Lelieveld and his colleagues used a global atmospheric chemistry model to look at the link between premature death and seven PM emission categories in both rural and urban areas around the globe (Fig. 3.4.11). Using 2010 data, the investigators estimated that outdoor air pollution, mostly PM 2.5, led to 3.3 million premature deaths globally, with a doubling of that number expected by 2050 if steps are not taken to reduce emissions. Thurston et al. (2015) focused on data from the US and found that even small increases in PM pollution could lead to a 3% increase in overall deaths and an estimated 10% increase in the risk of death from heart disease.

The most important sources of PM vary among areas of the globe (Fig. 3.4.12), with residential energy use the greatest contributor in India and China, where total mortality from PM is greatest. Emissions from traffic and power generation are the most important sources in much of the US, while in other parts of the world, including Russia and Europe, agriculture is the greatest source of

PM. Natural sources, like desert dust, are important in northern Africa and the Middle East. It is sobering to note that estimated global deaths from outdoor air pollution, mostly PM 2.5, exceed the total from malaria and HIV combined.

CRITICAL THINKING

Exercise 5: Think about the major sources of PM described above.

Which do you think are easy to control? Which are more difficult to control? Which are virtually impossible to control?

Consider three common air pollutants and their sources:

- SOx emitted from a power plant smokestack;
- CO released from a tailpipe; and
- PM 2.5 given off in smoke from backyard burning and as dust carried by the wind from farmers' fields.

As a regulator, which would you rather try to control? Why?

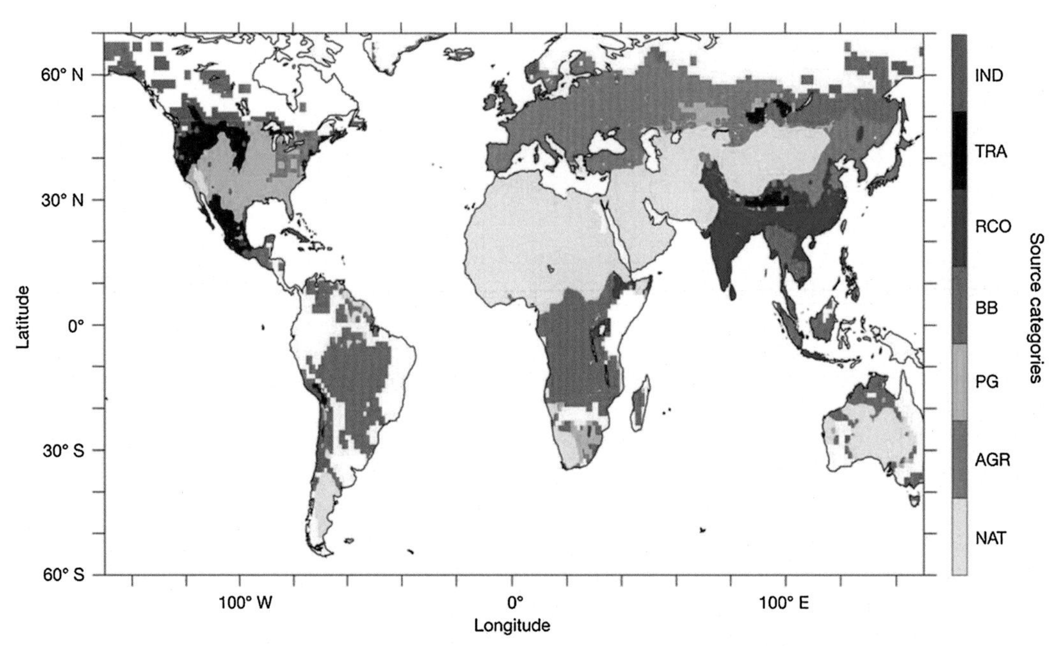

FIGURE 3.4.12

Source categories (color-coded): *IND*, industry; *TRA*, land traffic; *RCO*, residential and commercial energy use (for example, heating, cooking); *BB*, biomass burning; *PG*, power generation; *AGR*, agriculture; and *NAT*, natural. In the *white areas*, annual mean PM 2.5 is below the concentration–response threshold. *Source: (Lelieveld et al., 2015).*

For more information about how PM can impact your health and about ways to minimize your exposure, see the US EPA's pamphlet on Particle Pollution and Your Health linked on your Resource Page.

Establishing the Link: But how can any single study "prove" that PM 2.5 can cause serious health issues? You might suspect that a relationship exists between PM concentrations and health risk, but how do you demonstrate cause–effect? After all, it could be some other confounding factor, such as another air pollutant, that causes the effect you detect.

A traditional experiment in which various subjects are intentionally exposed to different levels of PM pollution and their health tracked is one way, but using humans as test subjects raises serious ethical concerns. So how do scientists confirm causal relationships when so many confounding factors are present? We'll examine four studies focused on the relationship between human health, exposure to PM, and the hazardous pollutants associated with these particles. Below is a brief description of each.

Study #1: Kunzli et al. (2010) assessed the relationship between hardening of the arteries (arteriosclerosis) and exposure to PM 2.5 in residents living along Los Angeles freeways.
Study #2: Health implications of long-term exposure of a large cohort of women to PM 2.5 were examined by Miller et al. (2007) in a large-scale nationwide assessment in the US.
Study #3: University of Cincinnati scientists led by F. Perera et al. (2009) monitored the exposure of pregnant New York City women to pollutants from the incomplete combustion of fuels and then looked for genetic changes in the cord blood after delivery of the infants.
Study #4: Mills et al. (2007) exposed human volunteers who had had previous cardiovascular episodes to diesel exhaust and measured their responses after exercising on stationary bikes.

CRITICAL THINKING

Exercise 6: Go to your Resource Page to find the link to each of the four articles connecting PM pollution to human health effects. One rapid approach to gathering information from scientific literature is to read the abstract, examine the figures and tables (captions should explain the information presented in the figures), and then skim the results and conclusions. This allows you to digest the key information from many papers in the same amount of time it might take you to read through just one more thoroughly.

Take this rapid review approach for each article listed on your Research Page and, based on the information you glean, complete the following:

- Make a list of the strengths and weaknesses of each study.
- Which study do you think makes the strongest case for cause and effect between exposure to PM and cardiopulmonary illness? Why? The weakest case? Why?
- How do the results of the study you chose as the strongest substantiate the cause and effect relationship between PM pollution and human health?

Long-Term Worries: A recent study by Peterson et al. (2015) suggests that PAHs associated with some types of PM (e.g., burning of diesel fuel, home heating oil, and coal) may have long-term effects on children. Researchers found that prenatal exposure to PAHs was linked to changes in the structure of the brain and to intellectual deficits and behavioral problems in childhood.

Scientists measured PAH levels in the air and in the blood and urine of 40 mothers in late pregnancy and followed the children until the ages of 7 to 9, performing imaging tests on their brains. The greater the exposure of the children to PAHs, the greater was the loss of white matter surface on the left sides of their brains. This loss correlated well with behavioral problems in the exposed children. Again, correlation isn't causation, but, given the potential for damage, such evidence merits careful attention on the part of those responsible for regulating PM levels.

PUTTING SCIENCE TO WORK

Exercise 7: Because of the direct risk to human health, there have been many studies focused on the impacts of PM pollution. Often, instead of conducting new research, leading scientists will write synthesis papers that summarize the body of scientific literature already published. This allows them to look for patterns and consistencies in study results that together could be used to build a strong case for the cause and effect relationship between PM and human health.

For this exercise, you will take on the role of the synthesis scientist. Use the Google Scholar database search to find more articles on human health and PM pollution.

- List at least four articles that you feel collectively build a strong case for the cause and effect relationship between PM and human health. To keep this down to a manageable time frame, focus on reviewing the abstract, figures, and conclusions only.
- After reading these articles, describe how the authors attempt to establish a cause and effect relationship between PM pollution and human health.
- Identify any gaps or flaws in the cause–effect connection from this group of studies.
- Could you design another study to address these gaps? What would this entail?

Solutions

While reducing total PM emissions is challenging, there are some straightforward ways to manage some sources of particle pollution. Commonsense steps to reduce particles contained in smoke include banning cigarette smoking, mulching refuse instead of burning it, limiting the use of fireplaces and wood stoves and making certain they're burning efficiently, and encouraging care when lighting campfires to reduce wildfire risk.

Similarly, vehicle emissions, particularly from diesel-powered cars and trucks, are important PM sources. Replacing older engines with newer, cleaner-burning ones and using mass transit, walking, and cycling are ways to reduce PM emissions from transportation sources.

An important tool for improving air quality is establishing and enforcing regulations. The Clean Air Act requires the US EPA to set national air quality standards for PM. The law also requires that the US EPA periodically review the standards to ensure that they provide adequate health and environmental protection and to update those standards as necessary. Working directly with state environmental agencies, the US EPA helps develop a State Implementation Plan (SIP) that typically includes the following:

- a monitoring program to determine whether or not an area is meeting the air quality standards, and, if not, how much of a reduction is needed to meet those standards;
- computer modeling to predict future trends and the effects of emissions reduction strategies;
- emissions inventories to quantify the sources and categories of emissions;
- control strategy studies to help identify the best way to reduce emissions;
- formal adoption of regulations to enforce reductions deemed necessary in the planning process; and
- periodic review to evaluate whether those needed reductions were achieved.

The results of these efforts have paid off with a modest but steady decrease in mean PM concentrations nationwide (Fig. 3.4.13). However, as is the case with PM concentrations, trends also vary by region, depending on the type and density of PM sources and enforcement of regulatory actions.

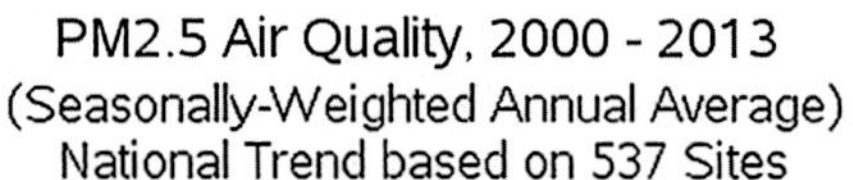

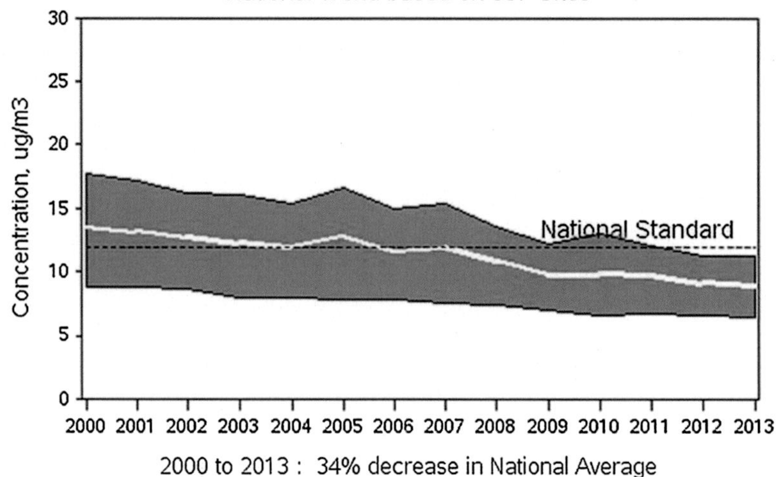

FIGURE 3.4.13

Using a nationwide network of monitoring sites, US EPA has developed ambient air quality trends for particle pollution that show a 34% decrease in PM concentrations since 2000. *Source: US EPA.*

PUTTING SCIENCE TO WORK

Exercise 8: While the decreasing trend in PM 2.5 levels is encouraging, consider two important challenges regulators face when dealing with PM. First, not all particles have the same degree of toxicity. Imagine how difficult it would be to categorize the toxicity of all the different types of particles in the air around the globe.

- Suggest an approach that might help scientists and regulators identify which types of PM pose the greatest threat to human health and, therefore, are most important to control.
- How do you balance the need to understand impacts of PM on humans with the moral responsibility to protect experimental subjects from harm?

CRUNCHING THE NUMBERS

Exercise 9: The US EPA's national trend data can be downloaded from your Resource Page (PM25National.csv).

- Using these data, calculate the slope of the trend line for the mean, 10th percentile, and 90th percentile of sites.
- Knowing that the 10th percentile represents the cleanest monitoring sites and the 90th percentile represents the most polluted monitoring sites, how do trends in PM concentrations differ between polluted and less-polluted locations?

SOLUTIONS

Exercise 10: Clearly, PM represents a threat to human health. The US EPA works closely with state agencies to come up with an SIP plan. But only a small part of the SIP includes direct actions intended to reduce emissions.

Assume you're the mayor of a large city like Chicago and that you're developing an action plan to directly reduce PM pollution in your city.

- List five steps you might take to begin to better manage the problem of PM in Chicago.
- What resistance to implementation might you expect from stakeholders?
- What partnerships could be created to ensure success of your plan?

CONSIDER THIS

Exercise 11: Often, populations in developing countries and poorer cities suffer higher rates of asthma and lung disease stemming from exposure to PM pollution.

- Why is it that these vulnerable populations are often exposed to the highest risk?
- How might the health effects of PM pollution perpetuate poverty?
- What can be done to minimize the risk of health effects from PM pollution in impoverished regions?

REFERENCES

Kunzli, N., Jerrett, M., Garcia-Esteban, R., Basagana, X., Beckermann, B., Gilliland, F., Medina, M., Peters, J., Hodis, H.N., Mack, W.J., 2010. Ambient air pollution and the progression of atherosclerosis in adults. PLoS One. 5 (2). http://dx.doi.org/10.1371/annotation/21f6b02b-e533-46ca-9356-86a0eef8434e. http://www.ncbi.nlm.nih.gov/pmc/articles/PMC2817007/pdf/pone.0009096.pdf.

Lelieveld, L., Evans, J.S., Fnais, M., Giannadaki, D., Pozzer, A., 2015. The contribution of outdoor air pollution sources to premature mortality on a global scale. Nature 525, 367–371.

Miller, K.A., Siscovick, D.S., Sheppard, L., Shepherd, K., Sullivan, J.H., Anderson, G.L., Kaufman, J.D., 2007. Long-term exposure to air pollution and incidence of cardiovascular events in women. N. Engl. J. Med. 356, 447–458. http://www.nejm.org.ezproxy.uvm.edu/doi/pdf/10.1056/NEJMoa054409.

Mills, N.L., Tornqvist, H., Gonzalez, M.C., Vink, E., Robinson, S.D., Soderberg, S., Boon, N.A., Donaldson, K., Sandstrom, T., Blomberg, A., Newby, D.E., 2007. Ischemic and thrombotic effects of dilute diesel-exhaust inhalation in men with coronary heart disease. N. Engl. J. Med. 357, 1075–1082. http://www.nejm.org/doi/pdf/10.1056/NEJMoa066314.

Perera, F., Tang, W.-Y., Herbstman, J., Tang, D., Levin, L., Miller, R., Ho, S.-M., 2009. Relation of DNA methylation of 5′-CpG island of *ACSL3* to transplacental exposure to airborne polycyclic aromatic hydrocarbons and childhood asthma. PLoS One. 4 (2), e4488. http://www.ncbi.nlm.nih.gov/pmc/articles/PMC2637989/pdf/pone.0004488.pdf.

Peterson, B.S., Rauh, V.A., Bansal, R., Hao, X., Toth, Z., Nati, G., Walsh, K., Miller, R.L., Arias, F., Semanek, D., Perera, F., 2015. Effects of prenatal exposure to air pollutants (polycyclic aromatic hydrocarbons) on the development of brain white matter, cognition, and behavior in later childhood. JAMA Psychiatry 72 (6), 531–540.

Thurston, G.D., Ahn, J., Cromar, K.R., Shao, Y., Reynolds, H.R., Jerrett, M., Lim, C.C., Shanley, R., Park, Y., Hayes, R.B., 2015. Ambient particulate matter air pollution exposure and mortality in the NIH-AARP diet and health cohort. Environ. Health Perspect. http://dx.doi.org/10.1289/ehp.1509676.

USEPA. 2014. https://www.epa.gov/pm-pollution.

World Health Organization, 2005. WHO Air quality guidelines for particulate matter, ozone, nitrogen dioxide and sulfur dioxide. Global Update. 2005.

World Health Organization, 2013. Health Effects of Particulate Matter: Policy Implications for Countries in Eastern Europe, Caucasus and Central Asia. WHO Regional Office for Europe. 13 pp. http://www.euro.who.int/en/health-topics/environment-and-health/air-quality/publications/2013/health-effects-of-particulate-matter.-policy-implications-for-countries-in-eastern-europe,-caucasus-and-central-asia-2013.

Chapter 3.5

Industrial Smokestack Pollution

FIGURE 3.5.1
A copper smelting plant in Karabash, Chelyabinsk region, Russia. *Source: Photo courtesy of By Mir76-ghost Wikimedia Commons.*

Overview: We've covered some very challenging air pollutants in our case studies. Managing PM is difficult because there are so many diffuse sources to consider. Reducing ozone pollution is hard because it's a secondary pollutant forming in the atmosphere. One partial success story in the air quality arena is the reduction of smokestack pollution.

Technologies like scrubbers that remove gases and particles from smokestacks have reduced pollutants like SO_x, NO_x, and PM released by power plants and industries. There are still challenges, however. We'll first look at continuing problems caused by inadequate controls placed on industrial sources in some parts of the world. The issues here are twofold. First, lax regulations in some nations may result in the release of harmful amounts of pollutants, including heavy metals, into the atmosphere. Second, some of the pollutants released, like lead, don't break down and may represent an ongoing problem. We'll then turn our attention to the condition of ecosystems, where there has been substantial progress. We'll focus on the impacts of acid deposition on terrestrial and aquatic ecosystems.

Background: Metal smelters (Fig. 3.5.1), which separate valuable metals like gold, copper, zinc, and cadmium from the ores they are found in, can have substantial impacts on environmental quality. Smelting, which involves heating mineral ores with chemicals that act as reducing agents, can release large quantities of metals and other pollutants into the atmosphere as well as producing solid waste in the form of tailings (Fig. 3.5.2). As a result, smelters that lack effective pollutant removal technologies can have significant effects on human health and surrounding ecosystems. Among the important pollutants released is SO_2, which can react in the atmosphere to form acid deposition. This process has contributed to widespread acidification of aquatic and terrestrial ecosystems in some parts of the world.

While the technologies that can remove pollutants like SO_2 and particulates from smokestacks of smelters exist, globally the use of pollution controls at these facilities has been spotty, often depending on effective enforcement by local environmental officials. As a result, in nations with limited or poorly enforced regulations, air pollutants released by smelters may have serious effects on the environment.

FIGURE 3.5.2
Aerial view of the NARA smelter in Gila County, Arizona, US. *By Keyes, Cornelius M., via Wikimedia Commons.*

Historically, poorly regulated smelter emissions have made international headlines with their impacts on the environment. Release of pollutants like SO_2 from a metals smelter in Sudbury, Ontario, created a virtual moonscape of 20,000 ha of stripped vegetation in the mid-20th century, with an estimated 7,000 lakes surrounding the smelter acidified by the pollution. In this case, a greening program adding tons of limestone to neutralize the soil and the replanting of thousands of trees have improved conditions.

In other cases, however, pollution from the smelters has continued, often with minimal controls. We'll focus on the Met-Mex smelter in Torreon, Mexico (Fig. 3.5.3). This plant has been the largest producer of refined silver in the world and is among the globe's leading producers of lead, gold, refined zinc, and cadmium (Soto-Jimenez and Flegal, 2011).

The Problem: At the Met-Mex smelter, large quantities of lead have been released into the atmosphere during the smelting process. Humans are exposed to this lead by breathing in lead dust, particles, or exhaust from the plant. Particles and ash containing lead can also be blown into nearby towns or onto agricultural fields where they can contaminate livestock and crops. Skin contact with contaminated soil can also expose people to lead.

Health problems associated with lead poisoning include reduced IQ, neurological damage, physical growth impairment, memory loss, seizures, and in severe cases, death. Impacts are especially serious in children, with even small amounts of ingested lead capable of causing lifelong developmental and cognitive problems.

FIGURE 3.5.3

Aerial view of the Peñoles Met-Mex metal smelter and surrounding communities, c.2011. *Source: Google Earth.*

Past sampling of lead levels in the blood of children living around the Met-Mex smelter indicated that many had unacceptably high values. As a result, in 1999, the Met-Mex smelter was ordered to install pollution control devices and clean up contaminated dust and soil around the smelter. Nearly 120 tons of polluted materials around the plant were removed. In 2000, Mexico's environmental agency, PROFEPA, lifted its restrictions on the plant.

EXPLORE IT

Exercise 1: Find the Peñoles Met-Mex metal smelter in Google Earth. Spend some time using your Google Earth skills to zoom in and around the smelter plant and answer the following questions:

- What is the predominant land use surrounding the Met-Mex smelter? What photogrammetry clues helped you determine this land use?
- Examine the historical collection of images available for this location in Google Earth. What is the time span for imagery of the smelter?
- What changes are evident over the range of image dates? (Have waste piles changed location or size? Has the plant expanded? Has surrounding development changed?)
- Explore the wider landscape. Besides impacts on human populations near the smelter, what other environmental impacts might you expect, based on the location of this smelter?

Environmental Science in Action: Soto-Jimenez and Flegal measured lead levels in the blood of 34 children living within a 6-km radius of the smelter (Fig. 3.5.4) in addition to a variety of environmental samples, including air, soil, and dust around the plant. Before we look at their results, let's review some analytical issues of importance in this type of work.

Quality Assurance/Quality Control (QA/QC): Crucial to any investigation in the environmental sciences are accurate data. The conclusions from any study are only as good as the data they're based on. Because of this reliance on high-quality numbers, many organizations that do environmental research, monitoring, and enforcement must adhere to stringent methodological standards for data collection. Even when following these protocols, those collecting the data must have a plan to check for errors and verify that the data collected are sound.

The case of the Met-Mex smelter gives us a good chance to consider how scientists ensure data quality. In the Soto-Jimenez and Flegal study, lead levels were measured in both blood and environmental samples. Two important items of note:

1. Several different measurements of human and environmental materials were included in this study. While they all determine lead

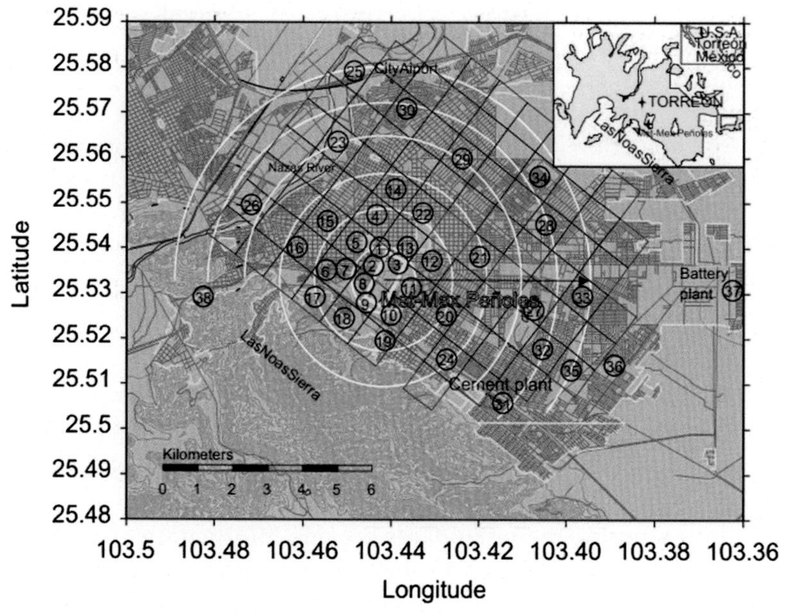

FIGURE 3.5.4

Sampling locations used by Soto-Jimenez and Flegal in their 2011 study of lead levels in children.
Source: Soto-Jimenez and Flegal , 2011.

concentrations, the units of measurement for each type of sample depend on what's being sampled. In this study, lead levels were determined in three different media:

a. blood (wt/vol): μg Pb/dL of blood (a dL is 100 mL)

b. soil and dust (wt/wt): μg Pb/g of soil or dust

c. air (wt/vol): μg Pb/m^3 of air

When reporting environmental data, make certain that the units you use are appropriate for the type of samples you've analyzed. When interpreting environmental data, make sure that you understand the units being used (particularly if any conversions were made in the data preparation). For example, contaminant concentrations expressed as weight per unit volume may have a different value if expressed in parts per million (ppm). Different studies or monitoring networks may use different units, making it hard to compare the relative contaminant values. In such cases, conversions will need to be made in order to compare the data.

2. An environmental scientist needs to ensure that all samples have been properly collected and processed, that measurements were made correctly and consistently, and that data were properly recorded. From collection to data analysis, there are many steps during which errors can be introduced, from contamination in sample collection vials to incorrect data entry. In the smelter study, the following steps were taken to ensure that adequate QA/QC was maintained in the work.

   a. Trace metal "clean" laboratories were used for analyses (Fig. 3.5.5), and extraordinary steps were taken to eliminate possible lead contamination of the samples collected.
   b. High purity reagents were used to extract lead from the samples.
   c. Standard Reference Materials from the National Institute of Standards and Testing containing known amounts of metals were used to verify the accuracy (QA) of the instruments. Run as random samples, these standards provide information about how accurate the chemical analyses have been for each run on the instrument.

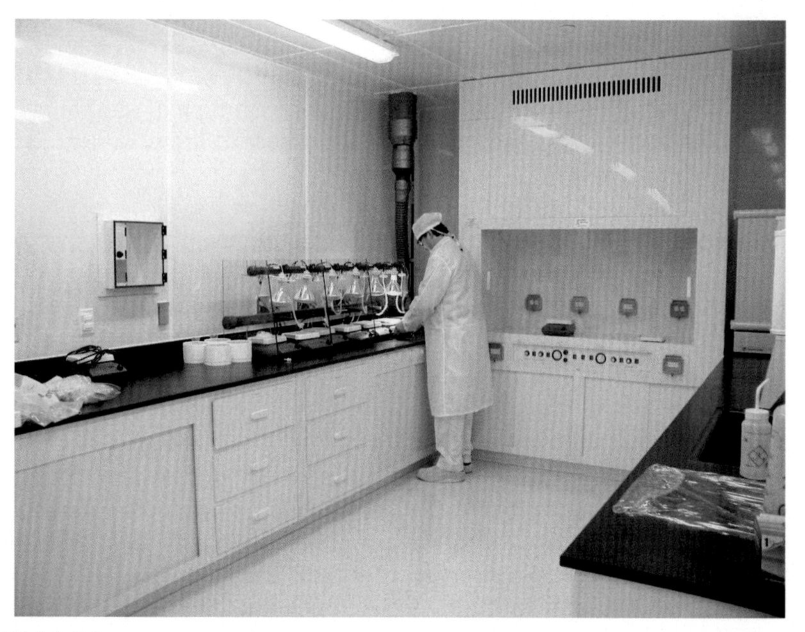

FIGURE 3.5.5
Aside from being tidy, "clean labs" have controlled air circulation to minimize dust, purified water, and acid baths to clean all equipment, and strict protocols are used to minimize the risk of contamination by technicians working in the lab. *Source: NASA.*

d. Blanks containing no samples were analyzed to insure the absence of spurious contamination on a continual basis. Similar to the standards, blanks run randomly provide information about contamination that may be coming from the equipment or reactants used to extract metals.

e. Replicates of random samples were run to insure consistent analytical results (QC). Running replicates has the added bonus of providing some detail about how accurate measurements are. This insight into the confidence intervals for measurements is important when determining if differences among samples are significant or simple variability introduced during sampling and analysis.

PUTTING SCIENCE TO WORK

Exercise 2: Here we'll consider some of the QA/QC measures for the laboratory analyses completed during the Soto-Jimenez and Flegal study. The QA/QC steps outlined above helped ensure that laboratory measurements were accurate. But there are many other activities associated with collecting and analyzing these samples.

For each of the following activities also included in this study, specify a series of QA/QC steps similar to what is outlined above that might be followed to assure high-quality data that match the objectives of the Soto-Jimenez and Flegal study:

- collection of blood samples
- collection of soil samples
- collection of air samples
- sample storage
- data entry and analysis

The Soto-Jimenez and Flegal study, occurring 6 to 7 years after Met-Mex was ordered to reduce pollution, concluded that lead emissions had been reduced by 95% during that time interval. However, although lead levels in samples of air, soil, and indoor and outdoor dust were lower than in previous years, concentrations still exceeded public health guidelines in a number of cases.

The frequency of hazardous concentrations was particularly high in samples collected nearest the smelter (Fig. 3.5.6). Blood lead levels (PbB) remained elevated (Fig. 3.5.7) in older children as well as younger ones, with 44% of the values exceeding 10 μg/dL, the level of concern for human health in the US and Mexico. This finding indicates that, despite a reduction in the release of lead pollution from the smelter, contamination from this source still remains a concern.

CRUNCHING THE NUMBERS

Exercise 3: Examine the results from the Soto-Jimenez and Flegal study presented in Figs. 3.5.6 and 3.5.7 and answer the following:

- How would you describe the relationship between lead pollution and distance from the smelter? What does this tell you about how the pollution travels and the region of greatest impact?
- If you were to set a buffer zone around the smelter, how close to the smelter would you allow households with children? Justify your response.
- Examine the R (correlation) relationships between blood lead levels and concentrations in soils and in indoor and outdoor dust. From which source does it appear that children are getting most of their lead?
- How might you explain the wide range in blood lead concentrations in children living close to the smelter?
- In Fig. 3.5.7, use the error bars for each age group to determine if there are significant differences in lead poisoning for children of different ages. What does this tell you about:
    - the sampling methods used in the study?
    - how lead enters children's blood?
    - the persistence of lead in children's blood?

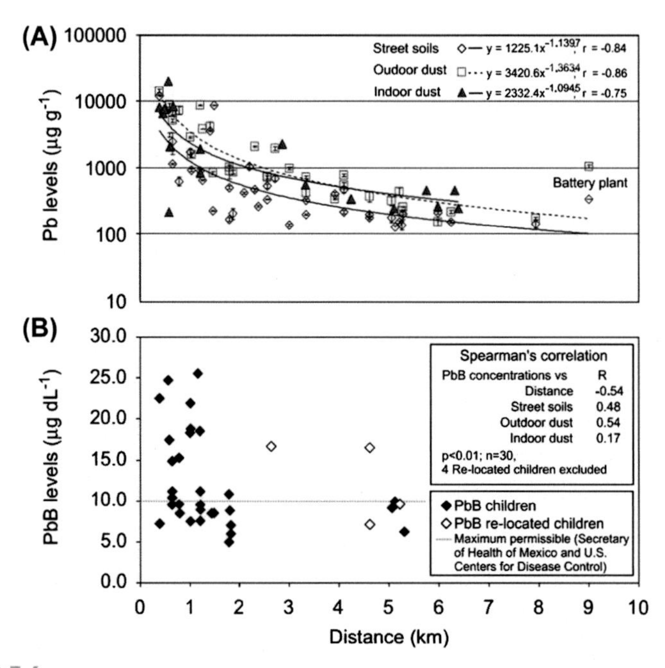

FIGURE 3.5.6

Lead levels in (A) environmental samples (µg/g) and (B) in children's blood samples (µg/dL) as a function of distance from the pollution source. *Source: Soto-Jimenez and Flegal 2011.*

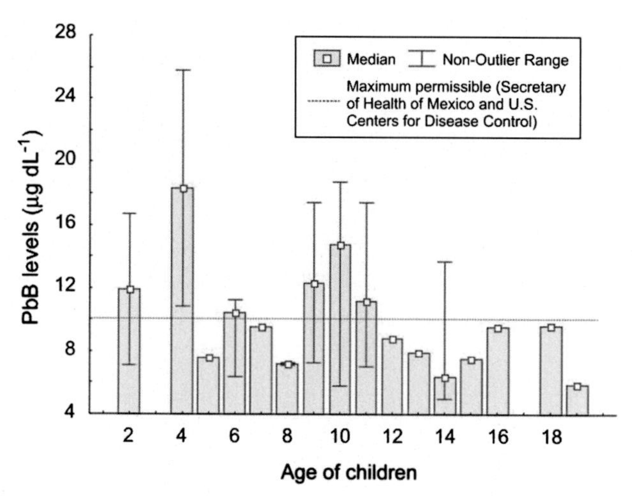

FIGURE 3.5.7
Lead levels in the blood of children of different ages. *Source: Soto-Jimenez and Flegal, 2011.*

Social Implications: The substantial levels of pollutants emitted by some metal smelters highlight two important concepts that environmental scientists should be familiar with: (a) transnational pollution and (b) environmental justice.

Transnational Pollution: Regulatory policies vary widely around the globe. If a metals smelter is required to install costly technological improvements for a plant located in one country, the owners might be tempted to move their operations to another country, which has less stringent regulations. Also, workers in the new country, which historically may have very low wages, may be attracted to jobs at the relocated smelter, thus increasing their chances of being exposed to dangerous pollutants. Similarly, many companies will outsource their smelting activities to developing nations to avoid costly regulations and negative publicity within their own borders.

DRILLING DEEPER

Exercise 4: Doe Run was the largest lead producer in the western world. It owned two primary lead smelters, one in southeastern Missouri, US, and another in La Oroya, Peru. In the US, Doe Run was cited regularly by the US EPA for exceeding emission limits, resulting in a reduction of the permitted capacity of the smelter. The company was also ordered by the US EPA to spend US $10.4 million to purchase 160 residential properties close to the smelter that were contaminated and no longer habitable. The Missouri operation ceased in 2013.

Doe Run moved a large portion of its smelting activities to La Oroya in 1997 after they were guaranteed protection by the Peruvian government from any environmental liability arising

Continued

DRILLING DEEPER—Cont'd

from the plant's prior operations. Emissions at the Doe Run smelter in Peru were still well above limits set by the WHO. The Oroya smelter closed in 2009, reopened in 2012, but, as this is written, is for sale.

Go to your Resource Page and click on the link to the Doe Run smelter operations. Read more about their smelter in Missouri and their operations in Peru.

- What were the environmental, social, and economic impacts of Doe Run switching their smelting operations from the US to Peru?
- Do you think this sort of transnational pollution should be legal? Why or why not?
- Are there any global efforts to regulate transnational pollution? Do some digging to see what these efforts might look like. Start with the Duke Law School paper on transboundary pollution found on your Resource Page.

Environmental Justice: according to the US EPA, environmental justice is the "fair treatment and meaningful involvement of all people…with respect to the development, implementation, and enforcement of environmental laws, regulations and policies." While that is a mouthful, it really boils down to this: All people have a right to a clean environment, and all people have a right to accessible natural resources. When human populations (either because of their socioeconomic status, nationality, religion, culture, etc.) are subjected to a disproportionate exposure to toxins or limited access to needed resources, we consider this an issue of environmental justice.

Because poor and underprivileged populations typically do not have the resources (either economic or political) to fight industrial development or the disposal of waste in their neighborhoods, there are often disproportionate environmental and health impacts felt by these vulnerable populations. Often in these cases, home and property values around an industrial facility like a smelter will be very low. These may be the only properties poor families can afford. Often these communities lack the strong political voice to demand improvements. With few resources to relocate, these vulnerable communities are often forced to endure, and try to deal with, the consequences of ongoing exposure to pollutants.

Smelters have been linked to issues of environmental justice. In the US, a notable example was the West Dallas (TX) RSR Corporations' used battery recycling operation, a secondary lead smelter, located in a predominantly Latino and African American neighborhood. Beginning in the 1930s and ending in 1984, this plant polluted local neighborhoods, including an elementary school and the West Dallas Boys Club in the immediate vicinity of the smelter, with elevated levels of lead (Fig. 3.5.8). Making the problem worse, because most residents living in the neighborhood surrounding the smelter were poor and couldn't afford air conditioning, most left their windows open during the summer, increasing their exposure to contamination.

Regulation of emissions was mandated by a strong lead emission ordinance passed in 1968 by the Dallas City Council, but city officials didn't enforce the standards. Blood levels in West Dallas children were 36% higher than those in control areas where children were not exposed to pollutants from the smelter. Finally, after the community organized itself and filed a lawsuit, a settlement resulted in a $4 million cleanup that removed tons of lead-contaminated soils from around 400 homes and 300 commercial operations.

However, even today, more than 30 years after the closure of the plant, soil samples in the neighborhood still show elevated lead levels (Wigglesworth, 2012). This case gives ample evidence of the long-term nature of human health and environmental concerns associated with the release of persistent pollutants like lead. Fortunately, passage of the Clean Air Act by the US Congress substantially reduced the release of most pollutants from smokestacks in the US.

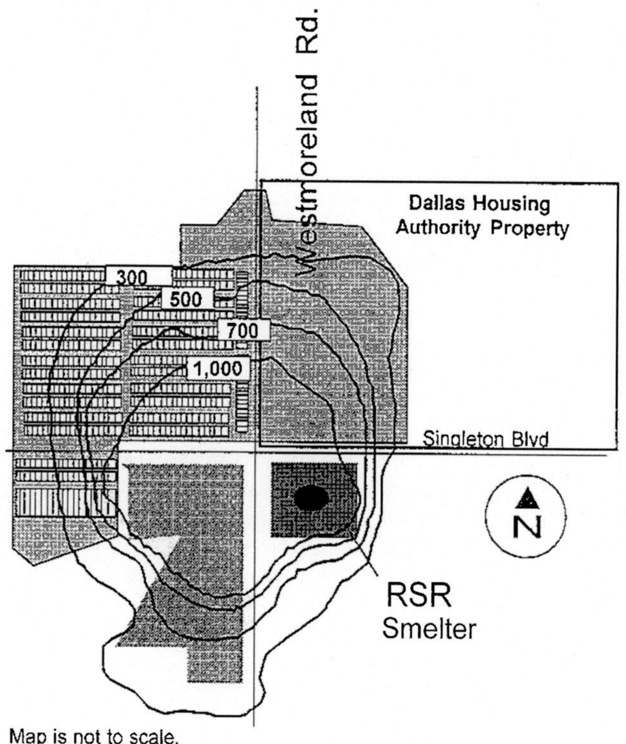

FIGURE 3.5.8

Soil lead concentrations around the West Dallas smelter. Note the proximity of the public housing development and the high levels of lead found within its soils. *Source: Agency for Toxic Substances and Disease Registry.*

SOLUTIONS

Exercise 5: As is the case for many environmental problems, there are often many smaller steps that all contribute to a larger solution.

- Brainstorm in a group to identify a plan to minimize human population exposure to smelter pollution. Be sure to consider how we might make progress in reducing the demand for the metal products, minimizing pollution from the process, cleaning existing sites, and working to help impacted populations.
- Do you think your plan might have worked to reduce the severity of the West Dallas case? Why or why not?

CONSIDER THIS

Exercise 6: The RSR secondary lead smelter is just one example of an environmental justice issue. After reading all the information about the West Dallas site linked on your Resource Page, including the US EPA cleanup progress reports, answer the following:

- Consider what makes one population more vulnerable to environmental injustice than others, using this smelter as an example. What is it about this local community that may have made it easier for officials to ignore the lead contamination for so many years?
- What finally allowed the community to successfully fight for the cleanup of the neighborhoods around the RSR smelter?
- Do you believe the regulatory agencies involved in the case (Texas and the US federal government) did enough? Why or why not?
- Do you believe something like the West Dallas lead case would have happened in a wealthy, predominantly white suburb? Why or why not?
- If you're not familiar with the phrase, look up "institutional racism." Do you think the West Dallas case is a good example of institutional racism? Why or why not?
- The federal Brownfields program is designed to convert abandoned hazardous waste sites in the US into useful properties. Find and discuss several examples demonstrating how this program has turned cases like the RSR site into useful properties.
- Do some research and find an instance where environmental pollution threatens human health close to you. Describe this problem and identify the key players and the impacted population. Does this case fit the description of environmental injustice? Has anything been done to resolve the problem? What steps might you take to help solve this problem?

CRITICAL THINKING

Exercise 7: So far in this case study, we've focused on the impacts of metal smelting operations on air quality and human health. Over the past several decades, a great deal of attention has been paid to another "smokestack problem": the formation of acid deposition, mostly from SO_x and NO_x released by fossil fuel-burning power plants, and its effects on downwind aquatic and terrestrial ecosystems.

In the 1970s and 1980s, the widespread decline in spruce forests in both the US and Europe provided a highly visible symptom of chronic pollution. However, in the early stages, it was difficult for scientists to prove a causal relationship between acid deposition and the symptoms of decline witnessed in the field.

On your Resource Page, we have provided three articles about red spruce decline in the northeastern US, starting with (1) a 1983 paper by Johnson and Siccama that provides

FIGURE 3.5.9
Acid deposition can leach calcium (Ca) from forest ecosystems and mobilize potentially toxic aluminum (Al) from forest soils. Calcium plays a unique role in the response of plant cells to environmental stress, and its depletion impairs basic stress recognition and response systems and predisposes trees to exaggerated injury following exposure to other environmental stressors. *Source: Paul Schaberg, US Forest Service.*

Continued

CRITICAL THINKING—Cont'd

insufficient evidence to link acid deposition to spruce decline, (2) a 1999 study by DeHayes et al. of the effects of acid deposition on forest health, and (3) a 2012 study by Lawrence et al. presenting evidence of a potential recovery of spruce-dominated ecosystems following enactment of the Clean Air Act by the US Congress. See also Fig. 3.5.9.

Review these three articles, and discuss the lines of evidence (and various study approaches) that allowed scientists to shift from a position of uncertainty about the role of acid deposition in spruce decline to one of acceptance and action (evidence of widespread spruce decline was the impetus for creating and enforcing the Clean Air Act).

Aquatic ecosystems are also affected by acid deposition. Do some research into the acidified clear water lakes of the Adirondacks in upstate New York.

- Why were these lakes so vulnerable to acid deposition?
- What happened to the fish in these lakes?
- Describe some of the attempts to remediate these lakes. Have these efforts been successful?
- How did the Clean Air Act lead to some improvement in the condition of these lakes?

REFERENCES

DeHayes, D.H., Schaberg, P.G., Hawley, G.J., Strimbeck, G.R., 1999. Acid rain impacts on calcium nutrition and forest health alteration of membrane-associated calcium leads to membrane destabilization and foliar injury in red spruce. BioScience 49 (10), 789–800.

Johnson, A.H., Siccama, T.G., 1983. Acid deposition and forest decline. Environ. Sci. Tech. 17 (7), 294A–305A.

Lawrence, G.B., Shortle, W.C., David, M.B., Smith, K.T., Warby, R.A.F., Lapenis, A.G., 2012. Early indications of soil recovery from acidic deposition in US red spruce forests. Soil Sci. Soc. Amer. J. 76 (4), 1407–1417.

Soto-Jimenez, M.F., Flegal, A.R., 2011. Childhood lead poisoning from the smelter in Torreon, Mexico. Environ. Res. 111 (4), 590–596.

Wigglesworth, V., 2012. The Burden of Lead: West Dallas Deals with Contamination Decades Later. Dallas Morning News. http://www.dallasnews.com/burdenoflead/20121214-the-burden-of-lead-west-dallas-deals-with-contamination-decades-later.ece.

Chapter 3.6

Household Air Pollution (HAP)

FIGURE 3.6.1
Kitchen smoke in a Tigray, Ethiopia, home. *Source: Photo by R. Waddington (CC BY-SA 2.0), via Wikimedia Commons*

Overview: While household air pollution (HAP) (Fig. 3.6.1) may not get as much media attention as other global and regional air pollution issues, the quality of air in homes has a direct and important effect on human health. Health impacts of cigarette smoke on nonsmokers living with smokers are well-documented and publicized. As a result, most people understand that cigarette smoke is a source of many different pollutants, including some that are carcinogenic. Less well-known are the effects of burning solid fuels indoors for heating and cooking.

We'll start by looking at the health problems caused by HAP in developing nations and consider some possible interventions. According to WHO (2014),

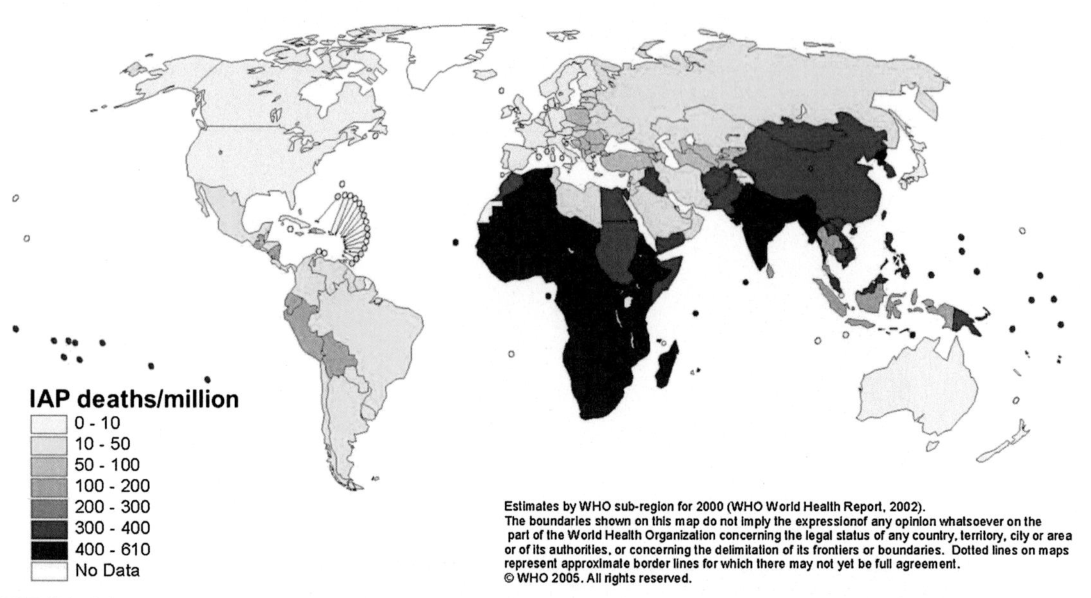

FIGURE 3.6.2
Global distribution of deaths from indoor combustion of biomass fuels. *Source: WHO, 2015. The Health and Environment Linkages (HELI). Policy Brief. Indoor air pollution and household energy.*

about three billion of the world's population heat and cook with solid fuels, including wood, animal dung, plant biomass, and coal, using open fires and leaky stoves. However, even in nations where cooking over fires is rare, indoor air quality is still an important human health issue. We'll close the case study by considering some of the indoor air quality issues common in more developed nations.

The Problem: Using data from 2012, WHO (2014) linked 4.3 million deaths globally to HAP, with many deaths occurring among children under 5 years of age. Particularly hard hit are locations in Africa, India, and Southeast Asia, where rural residents often have to rely on sources of solid fuels like cow dung, charcoal, and wood for cooking indoors (Fig. 3.6.2) (WHO, 2005). Smoke from solid fuels contains many hazardous substances, including particles smaller than 10 μm in diameter (PM 10), which penetrate deeply into the lungs and may lead to the development of respiratory and pulmonary diseases, cancers, and other illnesses.

According to the WHO (2014), about 60% of the deaths caused by HAP are the result of stroke and heart disease, with another 22% attributable to chronic obstructive pulmonary disease (COPD). The WHO estimates that HAP is

Age	Typical household[b]		236 monitored households[c]	
	Female	Male	Female	Male
0–1	216	214	209	195
1–5	212	212	199	192
6–8	173	172	156	163
9–19	207	174	196	194
20–60	227	116	221	118
60+	220	161	264	188

[a]Outdoor $PM_{10}=50$.
[b]PM_{10} concentrations: cooking area 260; living area 210.
[c]Averages for 236 separate calculations using monitored PM_{10} levels of respective households.

FIGURE 3.6.3
Exposure to particulate matter in household air pollution in Bangladesh. *Source: (Dasgupta et al., 2006).*

responsible for 2.7% of the annual global disease burden. The risk of health impacts from HAP depends on two factors: the level of pollutants occurring in the indoor air and the amount of time a particular individual is exposed.

A grim reality of HAP is that women are typically most at risk. Data shown in Fig. 3.6.3 are from a survey of household PM 10 levels in Bangladesh (Dasgupta et al., 2006). Results from 236 monitored households showed that women older than the age of 20 were exposed to the highest PM levels, presumably because they do the majority of the cooking and are exposed to the harmful pollutants for a longer period of time than males. High levels of exposure were also documented for children and adolescents of both sexes, with particularly serious exposure for children under 5. Among prime-age adults, men had half the exposure of women.

CRUNCHING THE NUMBERS

Exercise 1: The scientists who conducted the HAP study in Bangladesh monitored air quality in the kitchen, living quarters and directly outside the 236 homes they surveyed. Fig. 3.6.4 shows the mean concentration of PM10 at three different locations throughout the day.

- Graph a time series for PM 10 concentrations with overlays for each of these locations (kitchen, living quarters, ambient) throughout the course of the day.
- Is there an obvious difference in PM 10 concentrations for the three locations evident in your figure? Did the time of day impact these differences?
- Describe the temporal patterns you see for indoor air pollution.
- How does this compare to the temporal patterns in ambient pollution levels?
- Based on this information and the fact that children are often at high risk of exposure, what recommendations might you make to reduce a child's exposure to HAP?

CRUNCHING THE NUMBERS—Cont'd

Time	Kitchen	Living area	Ambient
12 a.m.	78	63	40
1 a.m.	78	63	35
2 a.m.	78	63	37
3 a.m.	78	63	39
4 a.m.	78	63	41
5 a.m.	142	114	42
6 a.m.	257	207	47
7 a.m.	466	376	57
8 a.m.	845	683	44
9 a.m.	568	459	33
10 a.m.	382	308	28
11 a.m.	257	207	33
12 p.m.	173	139	39
1 p.m.	116	94	37
2 p.m.	78	63	34
3 p.m.	78	63	40
4 p.m.	78	63	37
5 p.m.	257	207	36
6 p.m.	845	683	66
7 p.m.	525	424	101
8 p.m.	326	263	88
9 p.m.	202	163	72
10 p.m.	126	101	50
11 p.m.	78	63	46

[a]Concentrations are averages over the hour, for example, 11.30–12.30 average has been reported for 12.00.
[b]Drawing on mean values from continuous, 24-hour monitoring of 27 households in Narshingdi.

FIGURE 3.6.4
Daily PM 10 concentrations ($\mu g/m^3$).
(Source: Dasgupta et al., 2006).

Another Bangladeshi study (Mehta and Shahpar, 2004) related levels of household pollution to several variables: choices of cooking fuel, cooking locations, presence of construction materials, and ventilation practices. The authors reported that these choices were significantly affected by family income and adult education levels (particularly for women). Overall, they found that the poorest, least-educated households had twice the pollution levels that relatively high-income households with highly educated adults did.

As one can see from the "energy ladder" (Fig. 3.6.5), as a nation experiences increasing prosperity and development, there is a shift from heavily polluting solid fuels like crop waste and dung to nonsolid energy sources like natural gas and electricity. Clearly, the link between poverty and environmental health is all too obvious in the case of HAP. Let's take a look at some possible ways to begin to reduce exposures to HAP.

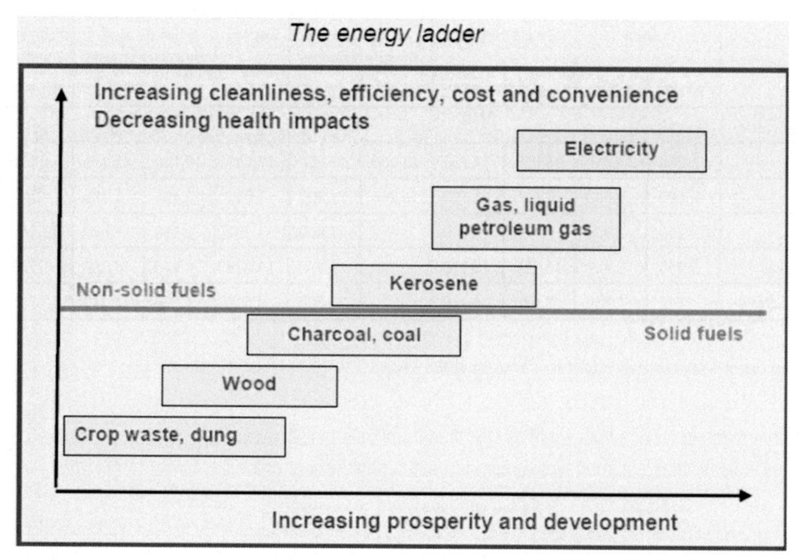

FIGURE 3.6.5
With increasing prosperity, households tend to move up the "energy ladder" and use cleaner but more expensive fuels. *Source: WHO, 2014. Indoor Air Pollution, Health and the Burden of Disease. Indoor Air Thematic Briefing 2.*

SOLUTIONS

While scientists have documented the severity of exposure to HAP in many parts of the globe, others are exploring possible ways to minimize this human health threat. We'll look at two different approaches to this issue.

Mehta and Shahpar (2004) performed a cost-effectiveness evaluation of several intervention options, including providing access to cleaner-burning fuels like liquefied petroleum gas (LPG) and kerosene and installing cleaner-burning stoves that would generate less air pollution. In addition to evaluating studies that included direct measurements of indoor air quality, the authors also looked at two health metrics, respiratory infections in young children and COPD in adults, as endpoints, and they examined epidemiological data to estimate improved health outcomes likely to result from expected reductions in HAP.

The authors calculated a cost-effectiveness ratio (CER), which compares the cost of each intervention option per healthy year added. Fig. 3.6.6 shows that a combination of improved stoves and a switch to kerosene is a cost-effective way to reduce HAP, but that the overall CER still differs markedly by region.

	Africa		The Americas		Eastern Mediterranean		Europe	South and South-east Asia		Western Pacific
	AfrD	AfrE	AmrB	AmrD	EmrB	EmrD	EurB	SearB	SearD	WprB
Cost-effectiveness ratio (CER) (I$/healthy year gained)										
LPG	6,270	11,050	14,050	7,500	24,200	11,020	17,740	15,120	7,350	1,410
Kerosene	1,000	2,000	2,410	1,180	16,200	1,800	3,010	2,450	1,380	260
Improved stoves	500	730	-	5,880	-	7,800	-	1,180	610	32,240
Combination: LPG and improved stoves	3,750	6,440	16,330	6,770	-	9,780	19,870	8,970	4,280	1,570
Combination: kerosene and improved stoves	840	1,530	8,080	3,120	-	4,500	9,510	1,950	1,040	780

Note

1. Missing values indicate that health benefits from current regional coverage exceed those of the intervention scenario.

FIGURE 3.6.6

Cost-effectiveness ratios (CERs) for interventions to reduce household air pollution from solid fuels. Lower values indicate a higher-cost effectiveness. Note that lower CER values indicate that less money is required for each healthy year gained. *Source: Mehta and Shahpar, 2004.*

CRUNCHING THE NUMBERS

Exercise 2: To explore the data further, Fig. 3.6.7 shows the cost-effectiveness uncertainty analysis for the Southeast Asia region only. Review this figure and answer the following for the South and Southeast Asia regions.

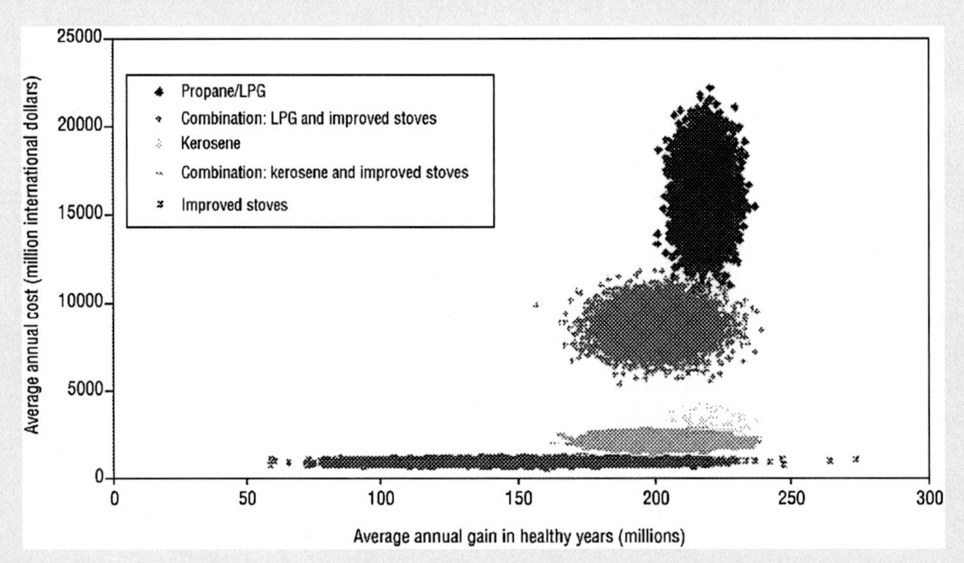

FIGURE 3.6.7

Cost-effectiveness uncertainty analysis for the Southeast Asia region. The "clouds," or uncertainty regions, illustrate the range of possible point estimates emerging from the uncertainty analysis. *Source: Mehta and Shahpar, 2004.*

CRUNCHING THE NUMBERS—Cont'd

Based on the uncertainty "clouds" plotted in Fig. 3.6.7, which intervention approach do you believe:

- Has the best chance of leading to significant health improvements?
- Is the most inexpensive to implement?
- Is the most cost-effective (e.g., unit health change per unit cost)?
- How would you rank these intervention steps in order of priority for implementation in poor, rural areas?
- How would you rank these intervention steps for more affluent and densely populated communities?
- Why would your ranking differ depending on where you are working?

Using a more direct approach, Cheng et al. (2015) selected 371 rural households in Gansu, China, to implement intervention measures, including stove improvements (Figs. 3.6.8 and 3.6.9) and health education. The research team monitored indoor air quality in 8 of the 371 homes. After the interventions, PM 4 levels dropped from an average of 455 $\mu g/m^3$ to 200 $\mu g/m^3$, and CO was reduced from 3.60 to 2.90 ppm.

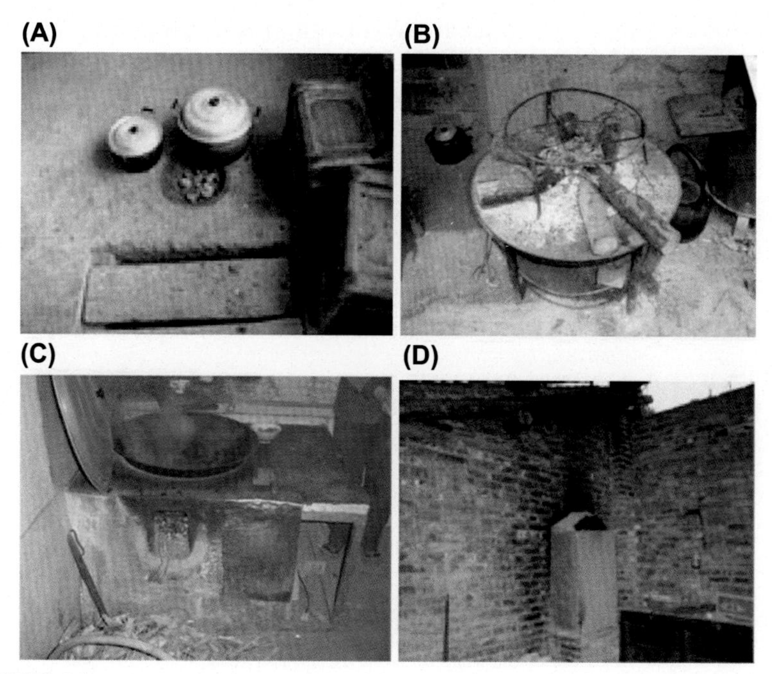

FIGURE 3.6.8
Stoves used by the households in the study area before the intervention included open pits (A) and hand stoves (B), as well as traditional stoves with either no chimney in the home (C) or a chimney that was installed too low to effectively mitigate household air pollution (D). *Source: Cheng et al. 2015.*

FIGURE 3.6.9
Improved stoves used by the households in the study area after the intervention included chimneys (A) and (B) that were specifically designed for the local residents. The improved stove was connected to a heating bed called Kang, as shown in (C).
Source: Cheng et al. 2015.

The frequency of several symptoms of poor health associated with HAP also significantly declined in the study population compared to baseline levels. However, the research team saw little improvement in lung function, indicating a need for additional evaluation of the long-term effects of HAP exposure.

PUTTING SCIENCE TO WORK

Exercise 3: Cheng et al. (2015) applied intervention strategies in eight households in order to determine if there were significant improvements following the interventions. Their air quality results are presented in Fig. 3.6.10.

Sampling spots	Time	N[a]	Mean	SD	Median	p-value[b]
Kitchen	Before	8	3.81	3.25	5.26	0.94
	After	8	3.00	1.57	3.05	
Bedroom	Before	8	2.99	2.79	2.31	0.84
	After	8	2.80	1.77	2.53	
Total	Before	8	3.40	2.96	3.81	0.72
	After	8	2.90	1.62	2.86	

[a]Sample size of households for IAP monitoring; [b]Wilcoxon signed rank test comparing data before and after intervention.

FIGURE 3.6.10
Comparison of CO levels (ppm) before and after intervention. *Source: Cheng et al. 2015.*

PUTTING SCIENCE TO WORK—Cont'd

- Compare the mean and median values for CO levels before and after intervention. What does this tell you about the distribution of the response to intervention in the eight selected households included in the study?
- Consider the reported *p*-values comparing the "before" and "after" CO levels. What do you conclude about the significance of these intervention methods?
- What key study design characteristics might influence the power of this test?
- How might you replicate this study with an improved ability to detect significant changes that resulted from intervention?

SOLUTIONS

Exercise 4: The WHO lists several possible strategies to reduce the production of and exposure to IAPs (Fig. 3.6.11). They also cite several case studies that have explored various intervention approaches on their website, available on your Resource Page at http://www.who.int/heli/risks/indoorair/indoorairdirectory/en/index2.html.

Interventions to Reduce Indoor Air Pollution Exposure

Source of pollution	Living environment	User behaviour
Improved cooking devices • Improved biomass stoves without flues • Improved stoves with flues attached **Alternative fuel-cooker combinations** • Briquettes and pellets • Charcoal • Kerosene • Liquid petroleum gas (LPG) • Biogas, producer gas • Solar cookers (thermal) • Other low smoke fuels (e.g. methanol, ethanol) • Electricity **Reduced need for fire** • Retained heat cooker (haybox) • Efficient housing design and construction • Solar water heating • Pressure cooker	**Improved ventilation** • Hoods/ fireplaces/ chimneys built into the structure of the house • Windows/ ventilation holes/ eaves spaces **Kitchen design and placement of the stove** • Kitchen separate from house reduces exposure of family (less so for cook) • Stove at waist height reduces direct exposure of the cook leaning over fire	**Reduced exposure through operation of source** • Fuel drying • Use of pot lids to conserve heat • Food preparation to reduce cooking time (e.g. soaking beans) • Good maintenance of stoves, chimneys and other appliances **Reduced exposure by avoiding smoke** • Keeping children away from smoke, e.g. in another room (if available and safe to do so)

FIGURE 3.6.11

The World Health Organization's summary of possible steps to mitigate the impact of HAP on human health. *Source: World Health Organization, 2004.*

SOLUTIONS—Cont'd

- Choose two of these case studies and summarize the intervention strategies they utilized, along with their particular effectiveness.
- Which approaches do you believe are the most likely to succeed? Why?
- Discuss the practicality of implementing these intervention strategies in different regions (considering differences in income, education, access to resources, etc.).

There's little reason to doubt that interventions such as those discussed above can help reduce the problems associated with HAP. But let's think more broadly, as there are other possible options. To do a thorough cost–benefit analysis of various approaches for improving indoor air quality, one needs to consider not only resulting health improvements, but also economic and environmental factors.

CRITICAL THINKING

Exercise 5: Assume that you are the Environmental Commissioner of a district in rural central India. Your boss asks you to prepare a detailed proposal outlining the costs and benefits for the various options available to eliminate the health hazards posed by burning biomass indoors for cooking in the 1000 homes in the district. Assume all the cooking and heating in all these homes is done by burning plant waste and cow dung.

Evaluate and critique the options listed below. Prepare a cost–benefit analysis for your boss, including your recommendation based on your analysis.

Tips to remember as you do your analysis:

- Be sure to include not only environmental and economic costs and benefits, but also those associated with human health and welfare.
- When calculating the environmental costs, consider more than the immediate improvement in HAP levels in the homes. For example, what were the environmental impacts that occurred to produce the raw materials used in your alternative? Similarly, don't just consider the benefits of improved health in the immediate family. What about the broader, long-term benefits if emissions of greenhouse gases like nitrous oxide are reduced?
- One of the challenges in doing this type of cost–benefit analysis is availability of solid numbers for your estimates. If you can't find exact numbers for some of your costs or benefits, you can use an estimate, as long as you indicate how you arrived at your number.

Options for reducing HAP in your district:

1. Electricity: hooking up to the electrical grid that is 20 km distant;
2. Setting up photovoltaic (PV) solar cells to provide electricity: OMC Power in India does this;

CRITICAL THINKING—Cont'd

3. Providing solar cookers for each home: use of box or panel cookers that concentrate the sun's rays;
4. Switching to biogas produced in biodigesters, which convert animal waste to methane;
5. Installing cleaner-burning, more efficient stoves and properly venting homes; and
6. Providing LPG for each of the district's 1000 homes.

OTHER INDOOR AIR QUALITY ISSUES

The effects of pollutants on indoor air quality are felt elsewhere around the globe. Some of the important sources in a typical home are shown in Fig. 3.6.12. Note that HAPs can be biological (molds, bacteria, viruses), physical (asbestos, smoke), chemical (cleaning fluids, formaldehyde), and even radiological (radon).

What makes these contaminants particularly dangerous is that often families don't even know they are being exposed. This is compounded by the fact that

FIGURE 3.6.12
Take the interactive tour of indoor air pollution sources at https://www.epa.gov/indoor-air-quality-iaq/interactive-tour-indoor-air-quality-demo-house. *Source: US EPA.*

many affluent populations spend much of their time indoors. According to Chapter 16 of US EPA's *Exposure Factors Handbook: 2011 Edition*, a US adult aged 18 to 64 years spends on average more than 19 h indoors per day. An average child younger than 18 spends even more time indoors per day (US EPA, 2011).

Indoor air pollution can result from the following:

- combustion of oil, gas, kerosene, coal, wood, and tobacco products;
- building materials and furnishings;
- consumer products (e.g., products used for household cleaning and maintenance, personal care, or hobbies);
- central heating and cooling systems and humidification devices; or
- penetration of outdoor pollutants (radon and ozone are examples of two pollutants formed outside the home that can travel to indoor living spaces and be trapped in the indoor environment).

PUTTING SCIENCE TO WORK

Exercise 6: For this exercise, you get to be a detective. Choose a building (e.g., your home, your school) and identify all possible sources of hazardous IAPs. The US EPA has an interactive household map that can be used to explore common pollutants found in modern homes. Visit this interactive viewer linked on your Resource Page at http://www.epa.gov/indoor-air-quality-iaq/interactive-tour-indoor-air-quality-demo-house.

In addition, the link to indoor air quality found on your Resource Page at http://www.epa.gov/iaq/pubs/insidestory.html includes information about common types of IAPs, corrective strategies, and specific measures for reducing pollutant levels. With the help of these materials, complete the following:

...List each type of IAP and identify the source or sources in your building of choice.
...Describe each potential effect on human health and rank the pollutants in your building.
...Determine one or more possible steps that might be taken to reduce the pollutant.
...Choose the one IAP you think is most important to control and defend your choice.

SYSTEMS CONNECTIONS

Exercise 7: Here we have focused on the impact of household air pollution on human health. However, pollutants and emissions from home heating and cooking fuels also have other environmental impacts. We have not explored in detail how harvesting of solid fuels might impact the environments from which they are extracted or how emissions from their burning alter ambient air quality, with repercussions for the larger ecosystem.

- Consider the path that solid fuels take (including wood, dung, and charcoal), and summarize the potential environmental threats that result from these activities. Consider the impacts of both extracting and burning the fuels on the larger ecosystem in regions where solid fuels represent the primary source of energy for human populations.
- Do some digging in the literature and find an area of the globe where using wood for fuel has had serious ecological consequences in addition to human health impacts. Discuss your findings.

CONSIDER THIS

Exercise 8: This case study clearly demonstrated the link between HAP and income in developing nations. Higher quality, cleaner-burning fuels are often unavailable or not affordable to those living in poverty. But even in developing nations with plenty of access to clean cooking and heating fuels, indoor air quality remains an environmental justice issue. Many reports and studies indicate that low-income, minority, tribal, and indigenous communities may be disproportionately impacted by HAP.

- What are some of the reasons why this may be the case?
- What steps could be taken to alleviate this toxic burden on lower income communities?

REFERENCES

Cheng, Y., Kang, J., Liu, F., Bassig, B.A., Leaderer, B., He, G., Holford, T.R., Tang, N., Wang, J., He, J., Liu, Y., Liu, Y., Liu, J., Chen, X., Gu, H., Ma, X., Zheng, T., Jin, Y., 2015. Effectiveness of an indoor air pollution (IAP) intervention on reducing IAP and improving women's health status in rural areas of Gansu province, China. Open J. Air Pollut. 4, 26–37.

Dasgupta, S., Huq, M., Khaliquzzaman, M., Pandey, K., Wheeler, D., 2006. Who suffers from indoor air pollution? Evidence from Bangladesh. Health Policy Plan. 21 (6), 444–458.

Mehta, S., Shahpar, C., 2004. The health benefits of interventions to reduce indoor air pollution from solid fuel use: a cost-effectiveness analysis. Energy Sustain. Dev. 8 (3), 53–59.

U.S. EPA, 2011. Exposure Factors Handbook 2011 Edition (Final). U.S. Environmental Protection Agency, Washington, DC. EPA/600/R-09/052F, 2011.

World Health Organization, 2004. Indoor Air Pollution, Health and the Burden of Disease. Indoor Air Thematic Briefing 2.

World Health Organization, 2005. The Health and Environment Linkages Initiative (HELI): Indoor Air Pollution and Household Energy. Policy Brief.

World Health Organization, 2014. Household Air Pollution and Health. Media Centre. Fact Sheet N°292.

Chapter 3.7

Climate Change

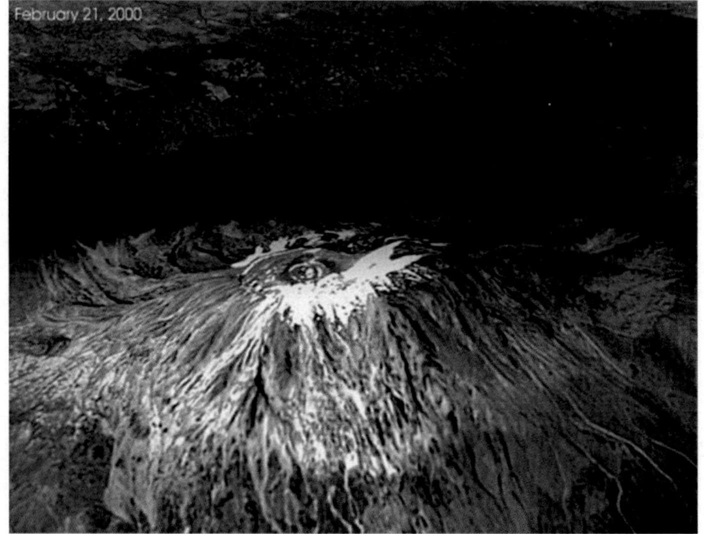

FIGURE 3.7.1

These two images taken by NASA show the changes in snow accumulation on the summit of Mount Kilimanjaro. *The top image was taken on February 17, 1993, and the lower one, on February 21, 2000. NASA.*

FIGURE 3.7.2
Boats left stranded by drought under the bridge over Sioni Reservoir in the Republic of Georgia. *Photo by Vladimer Shioshvili*
(CC BY-SA 2.0).

Overview: Perhaps there is no more far-reaching global environmental issue than climate change. While the Earth's climate has always varied over time, the current rate of change and the nature of these changes are testing the ability of ecosystems, and hence human systems, to adapt. A comprehensive review of this topic would require an entire volume, as climate change has the potential to affect the globe in so many different ways (Figs. 3.7.1 and 3.7.2).

While much attention has been paid to melting glaciers, rising sea levels, and extreme weather events, to help focus this case study, we'll look at one of the potential indirect impacts of a changing climate on biodiversity. By looking at three iconic species of wildlife on three different continents (the great panda of China, the African lion, and the kangaroos of Australia), we'll explore the complex interactions and repercussions that changes in climate patterns can have. Then we'll take an in-depth look at some of the regulatory and policy approaches being used or considered for combatting climate change and think about what an effective climate mitigation plan might look like.

Background: Before diving into this case study, be sure to review the basics of climate change from the various sources listed on your Resource Page. You should have an understanding of (1) atmospheric structure and dynamics, (2) the anthropogenic gases that contribute to climate change and (3) the range of possible impacts that might be expected with a changing climate.

Key points to remember about climate change:

1. Climate change is much more than "global warming." While the mean temperature of the Earth's land surfaces and oceans has been increasing over the past several decades and is projected to continue to rise, this change is not consistent across all parts of the globe. Climate is controlled by complex interactions between the Earth's surface and its atmosphere. For example, it is projected that temperatures in Europe may actually decrease as a result of an influx of cold water from the melting of the Greenland ice sheet that may effectively shut down the Gulf Stream that normally brings warm waters to Europe's coastline (Fig. 3.7.3).

   In addition to changes in temperature, there will also be many other types of altered weather patterns. For example, the disruption of the jet stream resulting from warming in the Arctic is projected to shift the position and slow the movement of weather systems across the globe. According to the latest United Nations Intergovernmental Panel on Climate Change (UN IPCC) climate report (IPCC, 2014), "it is likely that circulation features have moved poleward since the 1970s, involving a widening of the tropical belt, a poleward shift of storm tracks and jet streams, and a contraction of the northern polar vortex." This means that high-pressure systems may persist longer than normal, resulting in dry conditions, while low-pressure systems may "sit," resulting in intense rainfall and flooding.

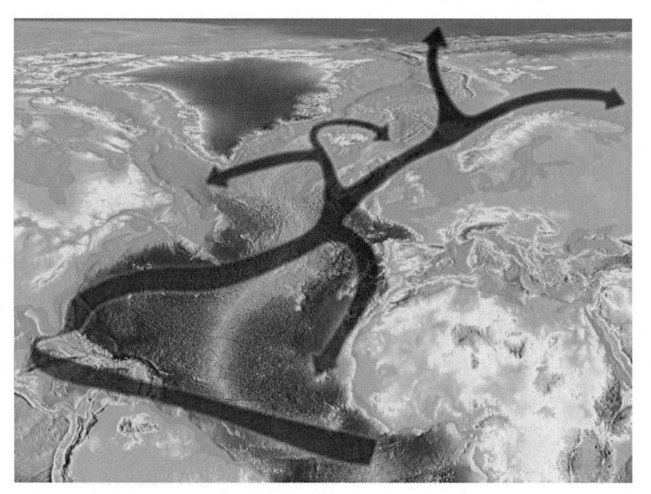

FIGURE 3.7.3
A map of the Gulf Stream highlighting the movement of warm equatorial waters toward Europe. *Source: RedAndr (CC BY-SA 4.0) via Wikimedia Commons.*

Because of increased energy stored as latent heat in the atmosphere, we may see an increased number of violent storms. But keep in mind that climate refers to long-term changes of 30 years or more in the patterns of weather conditions. A single weather event isn't necessarily related to the changing climate, but a change in the pattern of weather conditions over many years may signal larger changes in the Earth's climate system that can have significant and lasting impacts on ecosystems, which have evolved to historic norms.

2. While natural factors like the tilt of the Earth and the presence of water vapor in the atmosphere have important influences on the climate, dramatic increases in the concentration of anthropogenic greenhouse gases are primarily responsible for recent increases in global temperatures. By measuring the abundance of heat-trapping gases over time, coupled with other climate drivers like fluctuations in solar radiation, the IPCC calculated the "radiative forcing" (RF) of each climate driver, in other words, the net increase (or decrease) in energy attributable to that driver. Carbon dioxide has the highest positive RF (Fig. 3.7.4) of all the human-influenced climate drivers evaluated by the IPCC.

3. Predictions about the long-term effects of climate change are based on the outputs of computer models that incorporate the complex relationships, interactions, and feedback among climate drivers. As such, there is a high degree of variability among model projections, depending on the data that go into the models. Adding to the uncertainty are feedback mechanisms, which may speed up or slow down the rate of future change.

DRILLING DEEPER

Exercise 1: Feedback mechanisms result from changes in conditions or reactions that either speed or reduce the rate of change. A feedback that increases warming is called a "positive feedback," while one that slows warming is a "negative feedback."

For instance, the melting of highly reflective Arctic snow and ice leads to increased absorption of solar energy by dark ocean waters that replace them, thus accelerating the rate of warming. This is an example of a positive feedback mechanism.

- What are some other positive feedback mechanisms that may accelerate the rate of climate change?
- What is an example of a negative feedback mechanism related to climate change?

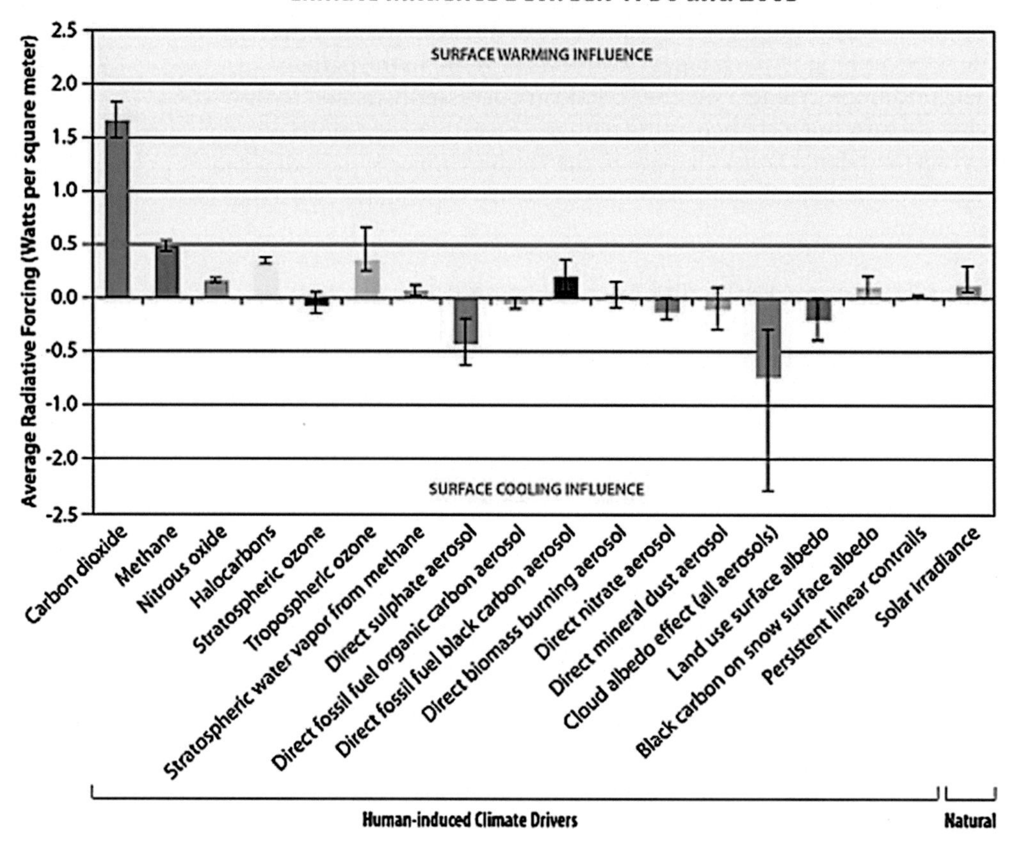

FIGURE 3.7.4

Relative contribution of the various climate drivers to surface temperature change. Positive radiative forcing values represent average surface warming, while negative values represent average surface cooling. *Source: Le Treut et al. (2007).*

4. Any successful efforts to manage climate change will require a collective effort across political borders and geographic boundaries. This is one example of an environmental issue with a global reach. Reducing emissions of gases like CO_2 is particularly challenging because burning fossil fuels has long been a key ingredient in the prosperity of developed nations. Reduction strategies are fraught with issues related to environmental justice, often pitting developed versus developing nations.

   For example, many of the countries that contribute the fewest greenhouse gas emissions will be the most likely to suffer from climate change. Tuvalu (Fig. 3.7.5), an island nation in the Pacific

Ocean, notably played an active role in the 2009 UN Climate Change Conference in Copenhagen when their representative, Ian Fry, delivered a final plea with tears in his eyes, concluding, "the fate of my country rests in your hands."

The Problem: Anthropogenic emissions of greenhouse gases are destabilizing the Earth's climate. A changing climate will likely alter both the Earth's ecosystems and its human societies. To focus on one particular type of impact, we'll consider three different megafauna in three different parts of the globe. Each animal is being or likely will be affected by climate change in different ways. Remember that ecosystems will survive climate change. They may just look very different than they do now, as some species will vanish and others will move into the system. One question is this: how committed are we to maintaining the iconic species we consider here?

FIGURE 3.7.5

Aerial view of Tuvalu's capital, Funafuti, in 2011. Tuvalu is a remote island nation of low-lying atolls, making it vulnerable to climate change. *Photo: Lily-Anne Homasi/DFAT [CC BY 2.0], via Wikimedia Commons.*

Australia: *Animal at risk*: *the kangaroo (Family*: *Macropodidae)* (Fig. 3.7.6).

While many of the kangaroo species in Australia are adapted to warm temperatures, recent research indicates that an increase in average temperature of only 2°C could dramatically impact kangaroo populations by shrinking the range of suitable habitat available to them. Scientists E.G. Ritchie and E.E. Bolitho (2008) of James Cook University used 3 years of field data and computer models to predict how expected temperature changes in Australia might affect kangaroos. Of particular concern is the effect climate change would have on the amount of available water in northern Australia. If waterholes dry up

FIGURE 3.7.6
Eastern gray kangaroo at Shadbolt in Australia's Greater Bendigo National Park. *Photo by Peterdownunder (CC BY-SA 3.0).*

and drought reduces productivity of pastures where kangaroos feed, death and reproductive failure among kangaroos may result.

Ritchie and Bolitho found that even a temperature increase as small as 0.5 °C (projected to occur by 2030) would shrink kangaroo's geographic range. An increase of 2 °C (projected by 2070) might shrink that range by 89%. The authors noted that, while kangaroos are quite mobile, the vegetation they've adapted to wouldn't likely migrate rapidly enough to keep up with the animals. Their study shows that an increase of 6 °C could spell extinction for this iconic animal.

A Special Case: the musky rat-kangaroo (***Hypsiprymnodon moschatus***) (Fig. 3.7.7) represents a special case. The smallest species of kangaroo, it's found only in a tropical rainforest in northeastern Queensland, Australia. Because this species is adapted to this one particular habitat, if changes in temperature and precipitation patterns alter the rainforest habitat as predicted by climate change models, the musky rat-kangaroo may well vanish, unable to migrate to new locations or adapt to the changing conditions (Bates et al., 2014).

The loss of this species, aside from its innate value, could have serious repercussions for the rainforest ecosystem. A fruit eater, it is one of only two seed dispersers in the forest (the other being the cassowary, a large bird). The loss of this kangaroo, given its vital role, could further upset the ecological balance in this tropical rainforest.

FIGURE 3.7.7
Hypsiprymnodon moschatus, the musky rat-kangaroo, in its native forest habitat. *By RachTHeH from Norwich, UK (CC BY 2.0) via Wikimedia Commons.*

SYSTEMS CONNECTIONS

Exercise 2: The musky rat-kangaroo is an integral component of the rainforest ecosystem it calls home. Do some research on this species and describe some of the repercussions its loss might have on the larger forested ecosystem of Queensland.

Similar to the case of the musky rat-kangaroo, the potential extinction of the much larger and well-known gray kangaroo would also have cascading impacts throughout its ecosystem. Do some research on this species as well, and describe potential repercussions of its loss on the larger outback ecosystem of Australia. Be sure to consider potential impacts on native communities and local economies.

China: *Animal at risk: the giant panda* (***Ailuropoda melanoleuca***) (Fig. 3.7.8).

The rarest member of the bear family, pandas live high in the mountains of northwestern China, where they subsist almost entirely on bamboo, with an adult panda requiring between 12 and 38 kg daily. Similar to the role of the musky rat-kangaroo in Australia's tropical rainforest described above, giant pandas also play a crucial role in the bamboo forests by spreading seeds and facilitating vegetation growth.

Since 1984, the giant panda has been listed as endangered in the International Union for Conservation of Nature and Natural Resources (IUCN)'s Red List of Threatened Species. There are only about 1,600 left in the wild. The co-location of the panda's primary habitat in the heart of an important economic region in China has resulted in the loss and fragmentation of much of the panda's

FIGURE 3.7.8
Six-month-old baby giant panda at China's Wolong Nature Reserve. *Photo by Sheila Lau (own work) (public domain) via Wikimedia Commons.*

habitat. But recent research indicates that development isn't the only threat to the panda.

Scientists from Michigan State University and the Chinese Academy of Sciences have modeled the effects of climate change on the forests of the Qinling Mountains in Shaanxi Province (Tuanmu et al., 2013). These mountains are home to 275 giant pandas, about 17% of the remaining wild population. Because of their geographic isolation and inability to migrate should their habitat become compromised, the pandas are particularly vulnerable to the effects of climate change.

The scientists' models incorporated field data on the distribution of bamboo in the forest, a number of climate projections, and historic data on the area's temperature and precipitation to evaluate how three species of bamboo would fare under changing climate regimes. Their projections indicated that between 80% and 100% of the bamboo in the giant panda's habitat would vanish by the end of this century, primarily due to elevated temperatures. Because bamboo comprises almost the entire diet of the panda population in the region, the projected changes in bamboo distribution suggest a potential shortage of food for this population.

Tuanmu et al.'s work further concluded that while some places within the region may become climatically suitable for bamboo species under a changed climate, most of these areas would likely lie outside the present network of nature reserves or far from the current bamboo stands. This fragmented landscape may separate pandas from their main food source, threatening the long-term survival of the species.

CRITICAL THINKING

Exercise 3: Panda conservation efforts have been in place for many decades, but this recent research highlights the need to expand activities to account for the threats posed to bamboo habitat by climate change. Among the steps being considered to sustain panda populations in the face of climate change is the development of heat-tolerant strains of bamboo to plant in the forest and the development of land bridges that might enable the pandas to migrate to other areas where they might find bamboo to feed on.

- Take some time to learn about some of the ongoing conservation approaches taken by the World Wildlife Fund's Giant Panda program (linked from your Resource Page), and discuss which of these may prove most effective in protecting panda populations over the next century.
- Are there other activities that might directly address the threats posed to pandas by climate change?

Africa: *Animal at risk: the African lion* (***Panthera leo***) (Fig. 3.7.9).

FIGURE 3.7.9
African lion. *Photo by Chi King (CC BY 2.0) via Wikimedia Commons.*

Of the species we have discussed so far, the "king of the beasts" is relatively well-adapted to cope with the direct impacts of climate change on its food and water supplies. According to a Heinz Center for Science, Economics, and Environment report (2012), lions should be better able to handle water shortages than many other species. They can go for days without water during droughts. They are also tolerant of high levels of solar radiation, and thus lions should be able to handle the expected warmer temperatures. Even their dietary flexibility should help them as the climate changes. They are opportunistic feeders and should be able to find food even as prey species shift their ranges.

However, the fate of the lion in a changing climate is far from assured. Lions are susceptible to many diseases. Some, like anthrax, are projected to become more common and virulent with climate change (Heinz Center or Science, Economics, and Environment, 2012). This threat is compounded by the increasing isolation of lion populations. Many of these populations are relatively small, limiting the potential for recovery after widespread disease-induced mortality.

Epidemic disease risks for lions living in fragmented habitats also become significantly higher as contacts with human and domestic animal populations become more frequent. But perhaps the greater threat is that climate extremes may increase the risk of interaction between epidemic and endemic pathogens. This is partially because of the added stress that climate extremes put on ecosystems, but also because pathogens tend to thrive in warmer environments.

Climate extremes can also change the lion's exposure to pathogens and the effect of the disease on the lions themselves. These pathogens are normally tolerated in isolation, but, in cases of co-infection (simultaneous infection by multiple pathogens), they can result in catastrophic mortality. Infection by more than one pathogen can change transmission rates and affect the virulence of a disease.

ENVIRONMENTAL SCIENCE IN ACTION: ENVIRONMENTAL FORENSICS

Two epidemics of canine distemper virus (CDV) among African lion populations in the Serengeti in 1994 and in the Ngorongoro Crater in 2001 resulted in a substantial number of deaths in each population. It is estimated that the epidemic claimed approximately 30% of the lion population of the Serengeti. Although their numbers dropped to about 2000, the Serengeti population was able to recover, but recovery following a similar outbreak among the smaller Ngorongoro population was not as successful. A team of scientists led by L. Munson of the University of California at Davis (Munson et al., 2008) was puzzled by the high mortality rates in these two cases, since CDV is not uncommon among lions and rarely leads to excessive mortality.

The team performed serological analyses on blood samples collected from more than 500 lions between 1984 and 2007. As can be seen in

Fig. 3.7.10, they found that antibodies indicating CDV infections occurred frequently during the period examined (vertical bars). The two fatal outbreaks, indicated by the red vertical bars, occurred after periods of extreme drought.

These droughts led to widespread herbivore die-offs, most notably of Cape buffalo (*Syncerus caffer*). As a consequence of heavy tick (*Babesia*) infestations on starving buffalo, the lions were infected by unusually high numbers of the ticks. Infections that resulted from tick bites were magnified by the immunosuppressive effects of the co-occurring CDV infection. In combination, this co-infection led to unprecedented mortality. Such mass mortality events may become increasingly common if climate extremes disrupt historically stable relationships between coexisting pathogens and their susceptible hosts.

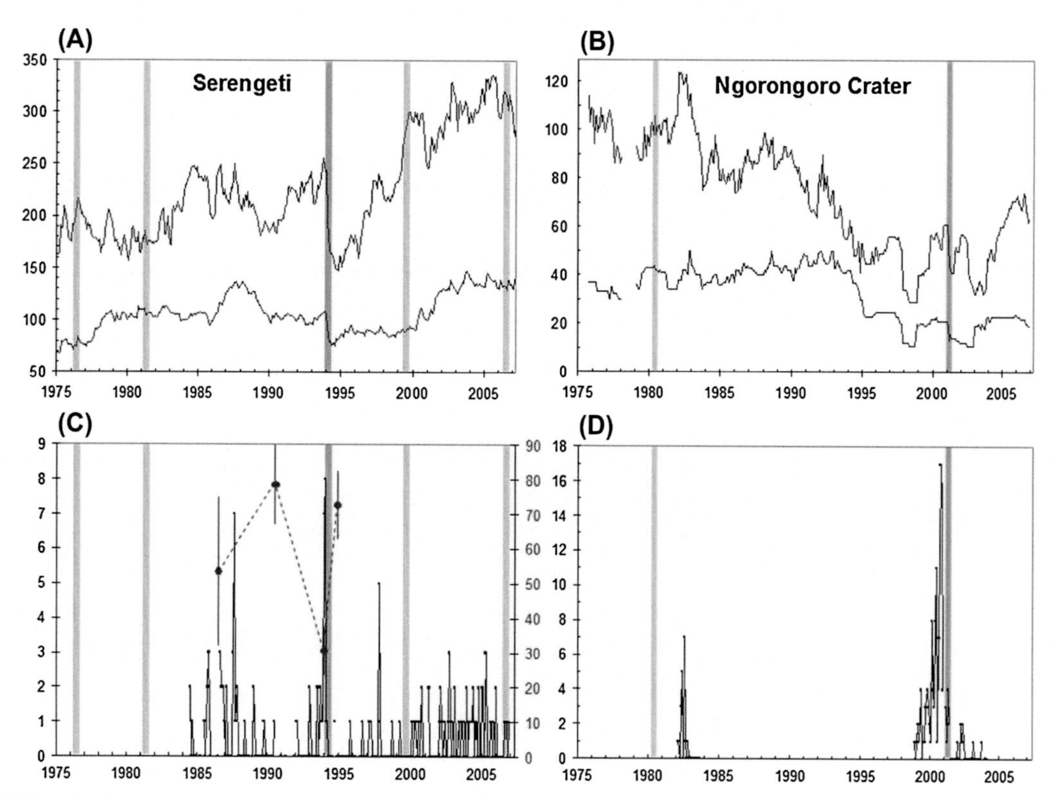

FIGURE 3.7.10

Timing and impact of canine distemper virus outbreaks in (A) the Serengeti and (B) Ngorongoro Crater. The *top lines* show the total population, and the *narrow lines* show the number of adults >4 years old. *Gray bars* indicate likely timing of "silent" outbreaks that were detected only retrospectively by serology; *pink bars* show fatal outbreaks (C,D). *Source: Munson et al. 2008.*

CRUNCHING THE NUMBERS

Exercise 4: They say one picture is worth a thousand words, and, considering the amount of data presented in Fig. 3.7.10, it might be worth much more than that.

Across the x-axis are 30 years of total lion counts (upper line), along with the number of breeding-aged individuals (lower line) for two different populations (Serengeti in A and Ngorongoro in B). The vertical bars show periods of CDV outbreaks identified by blood serum samples collected from the lions. Take some time to study these figures and answer the following:

- Describe the differences between the Serengeti and Ngorongoro Crater populations. Be sure to consider total individuals, variability, trends over time, and changes in demographics.
- Did all outbreaks of CDV (horizontal bars) result in similar mortality rates for these two populations? Explain why this might be.
- How did the mortality rate of the CDV outbreaks compare to other mortality episodes?
- Consider the resilience of the two populations after the CDV outbreaks. How did the two populations respond, and what might have driven that response?
- What are the long-term changes in the two populations?
- Based on the information presented in this figure, what do you believe will be the status of each population 30 years from now?

DRILLING DEEPER

Exercise 5: Climate change is also expected to lead to changes in disease spread and virulence and mortality rates in humans as well as in other mammalian populations.

Do some research to explore some of the greatest human health threats expected to be exacerbated by climate change. You can start with the information from the National Institute of Health (NIH) Climate Change and Human Health Program linked from your Resource Page.

- Discuss one disease that you feel may have the greatest impact on the global human population.
- Discuss one that may impact a community near you.
- How do you think the impact of climate change–induced health threats compares to the climate-induced threats to global water and food supplies?

SYSTEMS CONNECTIONS

Exercise 6: The impacts of climate change on the animals discussed in this case study don't occur in isolation. Climate change is only one threat facing these species. Go to one of the links posted on your Resource Page, and read about one of the three animals discussed above.

- What are the additional stresses that the animal you chose faces?
- How do these stresses interact to either increase or reduce the overall pressure on these populations?
- Identify an animal species in your area that may be affected by climate change. What are the likely effects on this species?

Solutions: Most of the global community of nations is now in agreement that climate change is real and that it poses a serious threat to ecological and social systems. Many groups have come together to work to solve this problem. There are a number of approaches, both in the realms of policy and technology, being considered to reduce carbon emissions, the primary driver of climate change. Below, we discuss two policy approaches already being used around the globe: cap and trade, and a carbon tax.

APPROACH 1: CAP AND TRADE

In this approach, a regulatory body, like the US EPA, sets a limit on emissions of CO_2 that can be released from a particular group of sources, like power plants. It then allocates a number of permits that give each emitter the right to release a specific amount of CO_2 (e.g., a ton). The total number of permits, and hence the total amount of emissions allowed for each group, are fixed by the cap. These permits are tradable such that if one emitter implements new technologies to reduce emissions, it can sell its permits for financial gain. Similarly, an emitter that can't or doesn't want to invest in new technologies to reduce CO_2 emissions can purchase permits from a source that has. The cap on emissions is gradually lowered, forcing emitters to continue to find ways to reduce their carbon emissions or pay for the right to continue polluting.

For an example of a cap-and-trade system already in place, follow the link on your Resource Page to learn about the Regional Greenhouse Gas Initiative (RGGI) started in 2009 by 10 northeastern and mid-Atlantic US states. You might also want to look at Europe's cap-and-trade program and some of the challenges it has faced.

APPROACH 2: CARBON TAX

A carbon tax is a levy charged for the use of fossil fuels and is based on the amount of carbon given off when a fossil fuel is burned. The tax, added to the cost of coal, oil, and natural gas, increases the price of such fuels, which encourages consumers to reduce their use. Also, because fossil fuels generate different amounts of carbon/ton burned (coal>oil>natural gas), taxes differ, depending on the fuel burned. Implementation of a carbon tax should promote increased energy efficiency and development and use of energy alternatives. Because of concerns that poorer members of society would be unfairly penalized by a carbon tax, proponents suggest that rebates could be made directly to consumers or that revenues from the taxes generated might be used to lower other types of taxes.

British Columbia, Canada, implemented a revenue-neutral carbon tax in 2008 with revenues from the tax returned to citizens of the province through reductions in other taxes. The tax rate was initially set low and has gradually increased since its inception. Opponents of a carbon tax have claimed that it would destroy jobs and limit economic growth by increasing the price of doing business. Yet the evidence from British Columbia tells a different story.

"Revenue-neutral" by law, the policy requires cuts to other taxes equal to what it raises. The result is that British Columbia now has the lowest personal income tax rate in Canada, with additional cuts helping low-income and rural residents. Since the tax came in, fuel use in the province has dropped by 16%, while in the rest of Canada, it's risen by 3% (counting all fuels covered by the tax). To put that accomplishment in perspective, Canada's Kyoto target was a 6% reduction after 20 years.

In contrast, in 2014, Australia, with a higher per capita carbon emission profile than British Columbia (Fig. 3.7.11), voted to repeal a politically divisive carbon tax. This came on the heels of the global financial crisis that slowed growth and raised unemployment in the country. Australian voters turned against the climate policies, blaming them for rising living costs.

CRITICAL THINKING

Exercise 7: Each approach, cap and trade and a carbon tax, has its pros and cons. Read the review article by Revelle (2009) linked on your Resource Page and answer the following:

a. Identify and discuss one strength and one weakness of each approach.
b. Cap and trade worked well to reduce the precursors of acid deposition. Why do carbon emissions represent a greater challenge using this approach?
c. Review both the RGGI initiative and British Columbia's carbon tax by following links on your Resource Page. Does each approach seem to be working well? Why or why not?
d. Having reviewed each of these two approaches for reducing carbon emissions, which do you favor? Why? Include both pros and cons of each in your argument.

SOLUTIONS

Exercise 8: Assume you've been elected mayor of one of the cities listed at the bottom of this exercise. Your assignment consists of several parts:

1. Present background information about the city that might be important to consider when anticipating the likely impacts of climate change (e.g., describe its current climate. How vulnerable are its infrastructure components like roads, mass transit systems, and housing to climate change? What is the current base of its economy?)
2. Look up the forecasts for what the climate will be like in your chosen city in 2050. Summarize the important features of the 2050 climate, and describe major impacts that such a climate would have on your city.
3. Develop a plan to prepare your city for the expected impacts of climate change. Identify specific elements you'd need to include in your plan and list any particular challenges you'd face in implementing it.

SOLUTIONS—Cont'd

4. Discuss any uncertainties you would face as you developed your plan.
5. What are some steps that your city could take now to help mitigate climate change?
6. How might you rally citizens around this plan?

You can choose any of the following cities to work on; these can be done individually or in small groups:

New York City, Chicago, Miami, Houston, San Francisco, London, Bangkok, Sydney, Johannesburg, Beijing.

Facing the threat of climate change will require different approaches in different cities. Major cities along coastlines may be most threatened by rising sea levels, while inland cities may face increased threats from droughts and/or flooding. Needless to say, solutions will likely be complex, expensive, and challenging to implement.

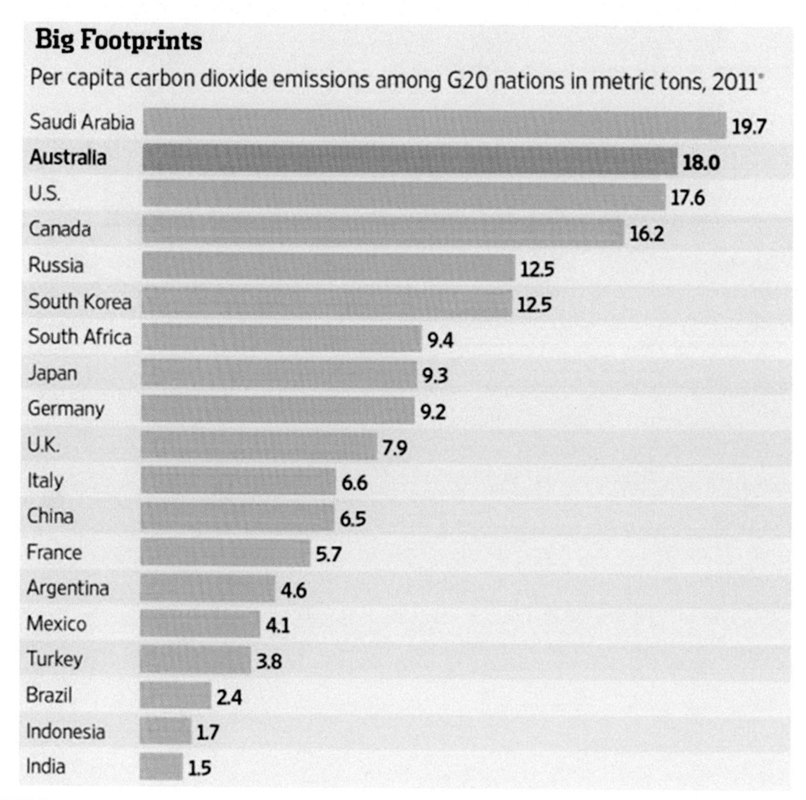

FIGURE 3.7.11

Per capita carbon emissions. *Source: US Energy Information Administration.*

CONSIDER THIS

Exercise 9: While we typically think of refugees as people fleeing from violence or persecution, recent decades have seen record numbers of people displaced by floods (Fig. 3.7.12), wind storms, droughts, and other climate-related hazards. Many find refuge within their own country, but some have to flee abroad. In the future, climate change is predicted to increase the frequency and intensity of climate-related disasters, prompting even higher levels of displacement. To date, however, such displacement has received little recognition, and refugees are often left for years in substandard camps, unable to return to their homes or create new homes in other countries.

FIGURE 3.7.12

Somali refugees flee flooding in Kenya. The refugee camps are located in areas prone to both drought and flooding, making life for refugees and delivery of supplies difficult. *Source: UNHCR/B. Bannon.*

You will find a link to the documentary Climate Refugees on your Resource Page. This documentary spanned the globe to interview some of the 25 million climate refugees now on the run. After watching this film, consider the following:

- Should climate refugees be granted similar political status to political refugees? Why or why not?
- What obligations do countries have to open their borders to climate refugees?
- What else should be done to assist climate refugees?
- What is our personal role in this crisis?

REFERENCES

Bates, H., Travouillon, K.J., Cooke, B., Beck, R.M.D., Hand, S.J., Archer, M., 2014. Three new Miocene species of musky rat-kangaroos (Hypsiprymnodontidae, Macropodoidea): description, phylogenetics and paleoecology. J. Vert. Paleon 34 (2), 383–396.

Heinz Center for Science, Economics, and Environment, 2012. Climate-Change Vulnerabilities and Adaptation Strategies for Africa's Charismatic Megafauna. Washington, D.C. 56 pp.

IPCC, 2014. Climate change: synthesis report. In: Pachauri, R.K., Mayer, L.A. (Eds.), Contribution of Working Groups I, II and III to the Fifth Assessment Report of the Intergovernmental Panel on Climate Change (Core Writing Team. IPCC, Geneva, Switzerland. 151 pp.

Le Treut, H., Somerville, R., Cubasch, U., Ding, Y., Mauritzen, C., Mokssit, A., Peterson, T., Prather, M., 2007. Historical overview of climate change. In: Solomon, S., Qin, D., Manning, M., Chen, Z., Marquis, M., Averyt, K.B., Tignor, M., Miller, H.L. (Eds.), Climate Change 2007: The Physical Science Basis. Cambridge University Press, Cambridge, United Kingdom and New York, NY, USA, pp. 95–127. Contribution of Working Group I to the Fourth Assessment Report of the Intergovernmental Panel on Climate Change.

Munson, L., Terio, K.A., Kock, R., Mlengeya, T., Roelke, M.E., Dubovi, E., Summers, B., Sinclair, A.R.E., Packer, C., 2008. Climate extremes promote fatal co-infections during canine distemper epidemics in African lions. PLoS One 3 (6), e2545. http://dx.doi.org/10.1371/journal.pone.0002545.

Revelle, E., 2009. Cap-and-Trade Versus Carbon Tax: Two Approaches to Curbing Greenhouse Gas Emissions. League of Women Voters of the United States. Climate Change Task Force: Background Paper 10 http://lwv.org/content/cap-and-trade-versus-carbon-tax-two-approaches-curbing-greenhouse-gas-emissions.

Ritchie, E.G., Bolitho, E.E., 2008. Australia's Savanna herbivores: bioclimatic distributions and an assessment of the potential impact of regional climate change. Physiol. Biochem. Zool. Ecol Evol. Approaches 81 (6), 880–890.

Tuanmu, M.-N., Vina, A., Winkler, J.A., Li, Y., Xu, W., Ouyang, Z., Liu, J., 2013. Climate-change impacts on understorey bamboo species and giant pandas in China's Qinling Mountains. Nat. Clim. Change 3, 249–253.

CHAPTER 4

Human Impacts on the Global Landscape

Science and the Global Environment. http://dx.doi.org/10.1016/B978-0-12-801712-8.00004-4

Chapter 4.1

Introduction

FIGURE 4.1.1
Farms, wildlands, towns, and cities make up the complex mosaic landscape in which we live. *Photo by Rino Peroni (CC by SA 2.0) Flickr.*

Humans have been altering the global landscape for millennia (Fig. 4.1.1), beginning with the advent of agriculture more than 10,000 years ago. As our global population continues to grow, more pressure is placed on this landscape to provide the resources necessary to support essential human needs. National Geographic's EarthPulse (1996) estimated that more than 80% of the Earth's surface had been impacted in some way by human activity, with exponential growth in global food production a dominant factor (up 170% between the 1950s and 1990s). When you consider the global nature of climate change, you might argue that human activity has affected the entire planet, including the most remote wild areas. Aside from the loss of many unique and valuable ecosystems, these impacts threaten the many ecosystem services that these natural landscapes provide (Fig. 4.1.2).

In the landscape case studies, we'll look at some of the ways that human activities have affected terrestrial ecosystems. Before tackling these case studies, be sure to review the background materials on biomes, terrestrial food webs, soil structure and function, and other relevant topics on your Resource Page.

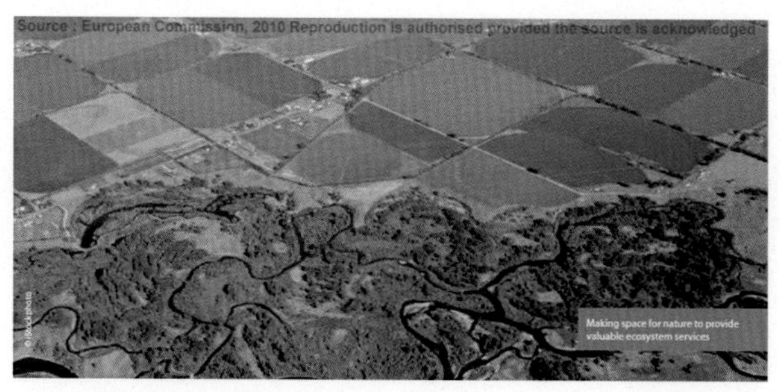

FIGURE 4.1.2
In many locations, natural landscapes are converted to food production, impacting the many ecosystem services typically provided. *By European Commission (CC BY-SA 3.0) via Wikimedia Commons.*

Stressors: Our case studies will focus on only a few of the many ways that human activities affect the landscape. While there are many types we could discuss, the following are some of the more important categories of impact.

1. ***Agriculture***: Impacts occur throughout the process of producing food and growing fiber. Clearing the landscape to grow crops or support grazing animals removes the natural habitat from the landscape. Crops often require large inputs of fertilizers and pesticides, some of which can leave the site and pollute nearby surface and groundwater. In water-stressed regions, inefficient irrigation systems can lead to additional problems, like salinization, depleted groundwater reserves, and artificially low surface water levels, which can affect aquatic ecosystems. Poor location or management of grazing animals may alter vegetation species composition, compact the soil, and contribute to desertification. Waste generated by large operations like feedlots can, if not properly treated, threaten land and water quality.

2. ***Deforestation***: Globally, forests occupy about 30% of the Earth's land surface. They are considered one of the largest sinks for carbon and account for approximately half of terrestrial plant productivity. For centuries, humans have harvested forest products. While some approaches, like selective harvesting, are sustainable and can minimize wide-scale damage to forested ecosystems, techniques like clear-cutting or conversion of forests for either development or agricultural purposes have had severe impacts on the landscape. Increased erosion and runoff from bare ground and from logging roads, particularly on steep terrain, can degrade water quality. Loss of cover on steep slopes can also increase soil instability and lead to mudslides and flooding (Fig. 4.1.3).

3. ***Mining***: While subsurface mining has comparatively little impact on surface environments, surface or strip mining requires removal of

FIGURE 4.1.3
Deforestation in Brazil leaves hillside vulnerable to runoff and erosion. *Source: Alex Rio Brazil (public domain) via Wikimedia Commons.*

overlying landscape features, with obvious deleterious effects on plants and animals. Associated issues like acid mine drainage are also important concerns in many areas. A particularly destructive type of surface mining, mountaintop removal, can cause severe damage. While reclamation of mined areas is required, restoring extensive landscapes to their original condition is a challenging, long-term process.

4. ***Urban- and suburbanization***: Conversion of natural landscapes into cities and suburbs, including all the infrastructure necessary to support them (transportation networks, shopping malls, sewage, water, and power lines), changes the fundamental character of the land, reducing biodiversity and, in many cases, degrading air and water quality as well. Urban sprawl, the outward expansion of cities into the countryside, carries with it additional impacts upon the landscape as natural habitats are developed (Fig. 4.1.4).

5. ***Waste disposal***: From littered cigarettes to illegal dumps and electronic waste (e-waste), humans produce vast amounts of trash that require careful management to avoid impacts on the environment. In addition to the obvious aesthetic concerns, improper waste disposal can destroy habitats and pollute both surface and groundwater. Secondary impacts can occur as communities are forced to transport their solid waste long distances because of a lack of disposal sites nearby.

There are many other examples of human impacts on the landscape, including the introduction of invasive species, impacts of recreational developments like golf courses and ski resorts, land application of sewage sludge, land subsidence from groundwater withdrawals, flooding of landscapes for reservoirs, and,

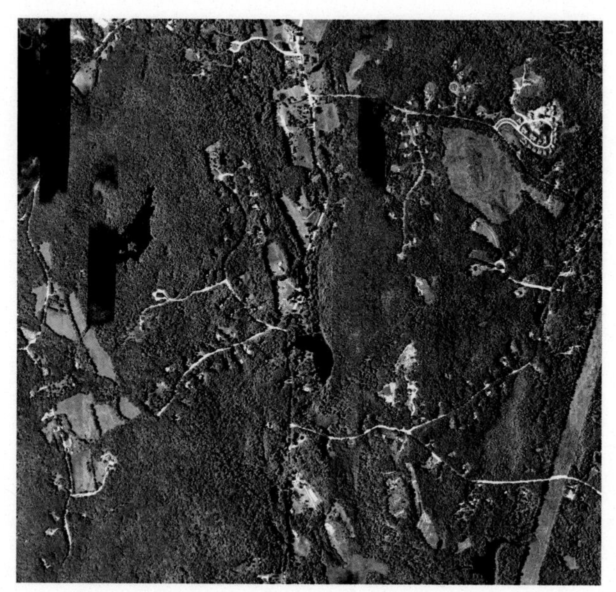

FIGURE 4.1.4
Expansion of cities to suburbs in outlying areas has dramatic impacts on ecosystems by fragmenting and isolating pockets of the natural landscape. *Source: NASA.*

perhaps most notably, climate change. In these case studies, we will consider a few of these in depth. But it is important to keep in mind that as active participants in the ecosystems in which we live, all our actions have an impact on the environment around us. While it is easy to rally behind the environmental crisis of the day, we must also consider the impacts we have on our own backyards.

Some important things to remember as you work on case studies focused on human impacts on the landscape:

1. Terrestrial landscapes vary widely across the globe. Deserts and boreal forests differ in almost all important aspects. Because of their many different temperature, precipitation, and vegetation patterns, landscapes react to stressors in different ways. For instance, ecosystems in very warm climates may have species already near their upper temperature tolerance limit. Any additional warming from climate change, and these species will migrate, adapt, or die out. A different ecosystem adapted to very dry conditions would likely fare better under extended drought conditions than one needing more precipitation.

2. Undisturbed terrestrial ecosystems are highly effective at recycling key nutrients like carbon and nitrogen, critical elements for the continued operation of the ecosystem. Human activities like highway and pipeline construction can disrupt this closed loop circulation and lead to "leakage"

FIGURE 4.1.5

Southern sea otters feed on populations of herbivorous invertebrates, such as sea urchins and large gastropods, thus playing a keystone role in preventing kelp forests from being overgrazed. *Source: US Fish and Wildlife Service.*

of valuable nutrients out of the terrestrial ecosystem and into waterways, where they may degrade water quality.

3. Keystone species (Fig. 4.1.5), organisms whose activities may affect many other species in significant ways, will be important in several of our case studies. Such species play a unique and crucial role in the way an ecosystem functions. Without these species, the ecosystem would be dramatically different or cease to exist altogether. While some keystone species are megafauna like polar bears, even tortoises and insects can play a keystone role in some settings.

4. Damage caused to terrestrial landscapes by human activities can play out in a variety of ways. In some cases, a single species may be affected, perhaps without having any major effects on ecosystem structure or function. In other cases, damage to the ecosystem may be so pervasive that the system may be degraded for decades or effectively lost forever. We'll look at the potential for long-term impacts at tar sands mining sites in western Canada.

5. Terrestrial ecosystems provide an array of services to humans, including erosion and flood control in mountainous areas, insect pest control, water purification and, of course, sequestration of atmospheric carbon dioxide. If human activities interfere with such services, there can be widespread ramifications. It's important to note that you don't have to destroy an entire ecosystem to lose these services. An important concept is habitat

FIGURE 4.1.6
Habitat fragmentation in the Indiana Dunes National Seashore. *Source: US Geological Service.*

fragmentation, which occurs when structures like highways and power lines (Fig. 4.1.6) divide natural areas into smaller and smaller sections, restricting and eventually eliminating some species that require large areas to survive.

6. As was the case with aquatic ecosystems, terrestrial ecosystems are often exposed to multiple stressors. One case study will focus on the tropical rainforest of the Amazon Basin. Threats facing this biodiversity hotspot include industrial cattle ranching, construction of dams and roads, gold mining, unsustainable logging, small-scale farming, and climate change. As you might expect with a system this large, understanding the combined impact of all these stressors is challenging, as is enforcing effective policies designed to protect this ecosystem.

7. Remember those systems connections! As human developments spread across the landscape, system-wide changes may occur. The presence of more and more impervious surfaces like pavement and sidewalks that typically accompany urbanization leads to changes in the hydrological cycle, and these land-based changes can have serious ramifications for the hydrological cycle. If enough of an urban/suburban area is paved over, there can even be atmospheric impacts. The heat-island effect (Fig. 4.1.7), in which elevated temperatures occur in major cities, is well documented.

8. Terrestrial ecosystems are resilient. In fact, some ecosystems rely on occasional disturbance to reset growth and maintain ecosystem health. Unfortunately, in cases where disruption of the ecosystem occurs on a massive scale, ecosystem recovery is unlikely without substantial human intervention.

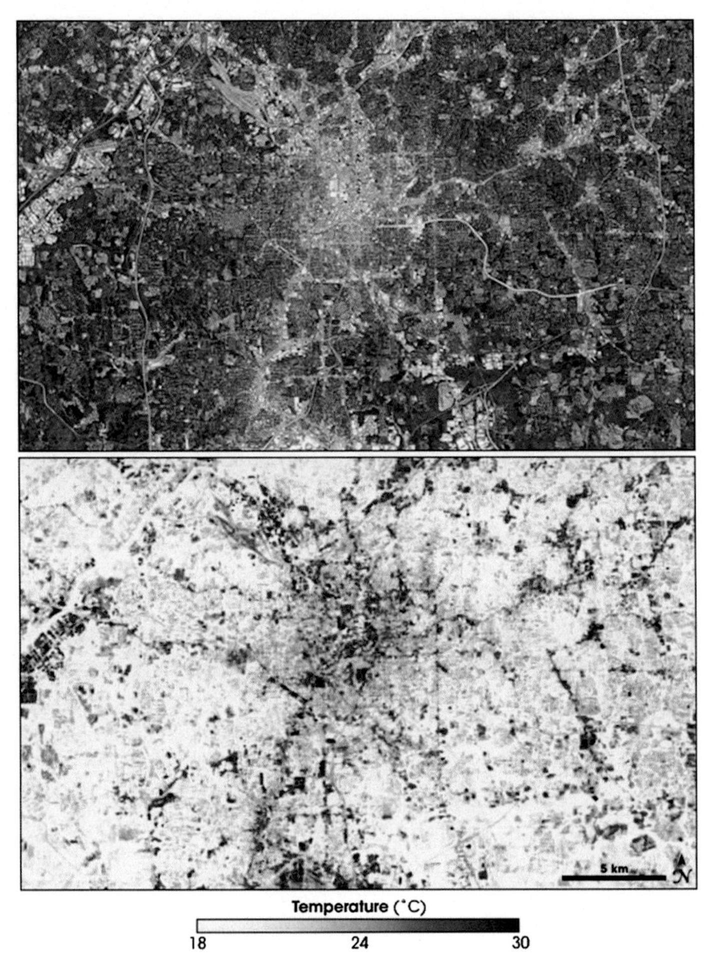

FIGURE 4.1.7

Surface temperatures show the "urban heat island," with temperatures in developed areas approximately 10°F warmer than adjacent vegetated areas. *Source: NASA.*

9. There is little doubt that the classic "western" lifestyle has led to substantial changes in the global landscape. As we think about the rest of the 21st century, a major question arises: can we support the degree of economic development necessary to provide all humankind with the basic necessities of life without causing irreparable harm to the planet that supports us?

Over the past 1,000 years, the world's population has exploded, with concurrent increases in the planetary footprint required to support our increasingly affluent lifestyles. Estimates of Earth's carrying capacity reported by various experts range widely, from fewer than 2 billion to more than 100 billion (UNEP Global Environmental Alert Service, 2012.).

In these Landscape case studies, we'll consider how ecosystems have been directly affected by our ever-increasing demand for the resources necessary to support a burgeoning global population. We'll investigate the following:

a. ***Energy***: The quest for carbon-based fossil fuels drives projects like the tar sands of Alberta, Canada. We'll consider both the environmental impacts of this massive project and the net energy balance from these extraction-intensive activities.

b. ***Food***: Agricultural issues will be the focus of two case studies. One explores the global phenomenon of desertification, which results, in part, from poor land management practices associated primarily with agriculture and forestry. A second case study will focus on the presence of genetically modified organisms (GMOs) in food products. The heated debate over the use of this technology continues in many parts of the globe. You'll have a chance to formulate and defend your stance on this issue.

c. ***Timber***: The tropical rainforest of the Amazon houses a remarkable portion of the Earth's biodiversity. Yet, poorly regulated timber harvesting and land clearing activities continue to threaten this remarkable resource. We'll also consider other stressors, including agriculture, both large- and small-scale, as part of this case study.

   Some ecosystem impacts don't result directly from human demands for resources. They may be the result of human carelessness, as in the case of waste disposal, or the indirect effects of climate change. These case studies include:

d. ***Natural pests***: A tiny insect, the mountain pine beetle (MPB), has laid waste to large tracts of western pine forests in North America beginning around 1999. We'll look at why this pest is having such devastating impacts and consider the implications for organisms at the top of the food web.

e. ***Solid waste disposal***: Our society for too long has used a "throwaway" approach to the products we consume (Fig. 4.1.8). Even the smallest, seemingly unimportant items like cigarette butts can harm the environment when disposed of by the billions. Recent years have seen an explosion of electronic technology. Lured by clever marketing, consumers often stand in long lines to buy the latest e-gadget. But what happens to

FIGURE 4.1.8
Solid waste littering a wetland. *Source: U.S. Fish and Wildlife Service Headquarters (CC BY 2.0) via Wikimedia Commons.*

your old gadget? Too often your e-waste finds its way to distant parts of the globe where human health and the environment suffer as the device is improperly recycled.

REFERENCES

National Geographic, 1996. EarthPulse: A Visual Guide to Global Threats: The Human Condition. http://www.nationalgeographic.com/earthpulse/food-and-water.html.

UNEP Global Environmental Alert Service, 2012. One Planet, How Many People? A Review of Earth's Carrying Capacity. www.unep.org/geas.

Chapter 4.2

Bark Beetle Infestation

FIGURE 4.2.1
Drought and bark beetle–induced mortality in high-elevation whitebark pine (*Pinus albicaulis*) forests, northern Warner Mountains (Drake Peak), Oregon, turn a once-green landscape into forests dominated by standing dead timber. *U.S. Forest Service photo by Constance Millar.*

OVERVIEW

Some of the worst environmental damage can be done by some of the least likely suspects. That's certainly the case with the mountain pine beetle (MPB) (*Dendroctonus ponderosae*), a small, weevil-like insect with a taste for pines, including lodgepole, ponderosa, jack, and Scotch varieties. Only about 5 mm long, about the size of a grain of rice, MPB has affected tens of millions of acres of forest from the US West Coast through the Rocky Mountains (Fig. 4.2.1). While bark beetles are native to US forests and play important ecological roles, they can cause extensive tree mortality. Climate change has led to an increase in these damaging effects. This increase is correlated with changes in temperature and increased water stress, which create conditions within trees that are favorable for beetle survival and growth.

PUTTING SCIENCE TO WORK

Exercise 1: While it is difficult to establish "cause and effect" relationships between changes in climate patterns and ecosystem responses, the US Forest Service (USFS) has developed a compelling synthesis of climate change's effects on native bark beetles.

- Explore the information presented by the USFS scientist Barbara Bentz linked on your Resource Page.
- Outline the "confluence of evidence" that lends credibility to climate change as a primary driver of increased bark beetle-induced mortality in the western US.

Environmental Sciences in the Field: Researchers at the US Department of Agriculture/USFS Rocky Mountain Research Station conducted an experiment to determine the role fungal invaders played in tree transpiration (the movement of water from the roots upward). They compared water dynamics in control trees to that in mechanically girdled trees and to trees that they intentionally infested with MPB.

They discovered that the movement of water through the trees started to decrease within 10 days of infestation, while control trees and girdled trees showed no change in transpiration. Because phloem girdling (as replicated by the mechanical girdling treatment) caused by the insect itself took much longer to cause impacts on pine health, the scientists concluded that fungi introduced by MPB, not the beetles themselves, are the main cause of tree mortality (RMRS, 2014).

SYSTEMS CONNECTIONS

Exercise 2: In addition to obvious damage caused to the trees they infest, MPB may have a number of additional impacts on the environment in general and the ecosystems they inhabit.

- Put on your "systems connections" thinking cap, and list as many possible indirect effects of the MPB infestation as you can. Consider not only impacts on other species, but also how widespread pine mortality might impact ecosystem processes such as carbon storage, successional processes, and fire dynamics. Do this before proceeding in order to lay the foundation for the rest of this case study.

BACKGROUND

The MPB (Fig. 4.2.2) is one of a number of beetles known to infest pine trees. Adult MPB bore into the bark of pines, laying egg masses as they burrow vertically and also shedding fungal spores present in their bodies. In an apparent effort to rid themselves of the invaders, trees exude pitch through the holes made by the beetles. The eggs hatch into white grubs, which dig their way laterally around the tree, reducing the flow of sap within the tree. The fungus, with a characteristic blue stain (Fig. 4.2.3), also clogs pore spaces, further slowing sap flow.

FIGURE 4.2.2
Adult mountain pine beetle. *Source: USFS Research and Development.*

FIGURE 4.2.3
The sapwood of an infested tree will be discolored by a blue staining fungi, but the heartwood will not be stained. *Source: (RMRS, 2014).* http://www.fs.fed.us/rm/boise/AWAE/briefing/AWAE_Science_Briefings-MountainPineBeetleEffectsOnLodgepoleTranspiration.pdf.

THE PROBLEM

MPB-caused tree mortality in western Canada and the US has been stunning, perhaps the largest insect outbreak ever seen in the US. Since 1996, an estimated 42 million acres of various types of pines have been infested, with Colorado leading the way in the US at 6.6 million acres (Handy, 2013). As of May 2013, pine trees in all 19 western US states had been affected, with an estimated 70,000 square miles of forest in the Rocky Mountains of the US and Canada dying between 2000 and 2010 (Robbins, 2010).

CRUNCHING THE NUMBERS

Exercise 3: Assume that each pine tree in the 70,000 sq. miles of forest that died in the Rocky Mountains between 2000 and 2010 was 100 feet tall and in good health before infestation.

- Do this back-of-the-envelope calculation: what would have been the total commercial value of the pine trees if they had been harvested before the infestation? (Hints: you'll need to do some digging to estimate the number of pine trees per square mile and the commercial value of each tree).
- As you recall from your review of cost–benefit analyses in the Tools and Skills section, the true costs of the deaths of those pines also include the loss of the many ecological services provided by the trees.
  - What are some examples of such benefits that could also be included in your calculation?
  - How might you estimate the value of such benefits?

Why has this outbreak been so severe? MPB outbreaks aren't a new problem; there have been many in the past. Why has this one been so bad? A drought across much of western North America in 2002 contributed to the MPB outbreak, as drought-stressed trees are less able to defend themselves against pests like MPB. By the time the drought ended, MPB populations had begun to increase substantially in many areas (Henson, 2013).

A second factor contributing to the outbreak was related to fire management. Many western US forest ecosystems are fire adapted, meaning that these ecosystems evolved to depend on regular, smaller-scale fires to maintain a stable structure. The importance of fire to natural systems is often overlooked. In the pine-dominated forests of the western US, fire initiated by lightning and rock slides was a natural part of the successional process. Many of the species in these ecosystems have serotinous cones that have a waxy coating, which opens in response to the heat, scattering seeds onto soil newly fertilized by nutrients in the ash (Fig. 4.2.4). They also have thick bark and limbs far enough above the ground to help prevent mortality from low-intensity surface fires. With intentional suppression of fires in some forests, there has been an increase in older trees particularly susceptible to MPB infestation.

A third factor was also at work: the gradual warming of the climate. Cold weather can kill MPB, with mortality occurring at −4 °F. However, recent warmer winters, particularly at higher elevations where low temperatures have held MPB numbers in check in the past, have allowed more beetles to survive. With the warming climate and longer growing seasons, MPB is able to complete their life cycle in one season instead of two, thus reducing the chances of winter kill.

Environmental Science in the Field: The scientific method starts with observations, which lead to the formation of a hypothesis, which is then tested by experimentation. The MPB provides an excellent example of the scientific method in action. During the Colorado MPB outbreak, a University of Colorado graduate student was walking through an infested stand of limber pines. S. Ferrenberg noted that sticky resin, which is expelled as trees try to repel the burrowing beetles, appeared only on trees with rough bark. Similar to natural variability in human hair color, trees of the same species exhibit a range of bark roughness (Fig. 4.2.5). Ferrenberg theorized that the beetles might be avoiding pines with smooth surfaces, resulting in increased resistance to MPB in certain trees.

FIGURE 4.2.4

This serotinous cone from a lodgepole pine tree was opened by fire, allowing it to release its seeds. *Source: US National Park Service.*

FIGURE 4.2.5

Two side-by-side lodgepole pines demonstrate the natural variability in bark roughness. *Photo by Scott Ferrenberg.*

He performed a simple experiment to test his hypothesis (Ferrenberg and Mitton, 2014). In the laboratory, Ferrenberg placed adult MPB on smooth and rough bark samples. While the beetles placed on rough bark remained there, beetles placed on smooth bark quickly fell off. The study also revealed that smaller limber pines had a higher proportion of smooth bark than older trees and thus were more resistant to infestation. In a second study, Ferrenberg and his colleagues also found that pine trees with more resin ducts were better able to deter the beetles (Ferrenberg et al., 2014).

This type of information is incredibly helpful to land managers who desire to slow the spread of beetle outbreak by removing infested trees. They may have better success if they target the trees most at risk, the rough-barked pines that produce less pitch.

PUTTING SCIENCE TO WORK

Exercise 4: Ferrenberg conducted his MPB bark experiments under the controlled conditions of a laboratory. But conditions in a lab are only approximations of how intact trees might respond in their natural environment.

- Design a study to test Ferrenberg's hypothesis in the field. Be sure to include all of the information required to describe your experimental design. Indicate how you might control for confounding factors while still capturing the natural variability inherent in nature.

Systems Connections: When you estimated the commercial losses from the destruction of millions of pine trees, you saw the most obvious cost of the infestation. Some other impacts are also quite apparent, such as loss of habitat for species living in the forest, an increased risk of erosion of forest soils as living trees are no longer available to hold soils in place during storms, and the aesthetic loss when a living forest dies. But there are even more subtle, and perhaps surprising, effects.

Take a look at that list of indirect effects you came up with at the beginning of this case study. How many of the following did you get? You'll note as we go through the list that the science isn't settled on some of these; that's the nature of the process of scientific inquiry.

a. ***Air Quality***: Volatile organic compounds (VOCs) are organic chemicals that trees emit to attract pollinators and repel harmful insects and animals. Trees also produce VOCs in response to stress. Amin et al. (2012) measured VOCs in the air in lodgepole pine forests with and without MPB infestations. The research team found a five- to 20-fold increase in VOC emissions in the forests with infestations. Atmospheric aerosols like VOCs can impact the climate, degrade air quality and visibility, and have detrimental effects on human health. Evidence also indicates that biogenic VOCs interact with nitrogen oxides and sunlight to produce ozone and other secondary pollutants.

The predominant VOC identified in the Amin et al. (2012) MPB study was a monoterpene, a compound contributing to smog and haze formation. While this has obvious negative impacts on the environment, there is also a potential positive outcome of this discovery. If monoterpenes in the air are significantly higher in MPB-infested stands, their presence could be used as an indicator of new infestations. This could speed the management response to help limit the spread and severity of MPB damage.

b. ***Hydrology/Water Cycle***: R. Maxwell and his colleagues at the Colorado School of Mines have discovered that pine mortality has measureable effects on the hydrology of impacted forests. Since the trees are no longer transpiring, more water enters the soils and filters down into groundwater reserves. The scientists found that this process leads to elevated water tables and a 30% increase in summer groundwater flow in these forests. This enhanced groundwater flow can increase the amount of water entering

DRILLING DEEPER

Exercise 5: Pine mortality can impact the hydrology of mountain ecosystems in many ways beyond reduced evapotranspiration in the summer (Fig. 4.2.6). In the mountainous

FIGURE 4.2.6
Looking up into pine trees infested by pine beetles at Fraser Experimental Forest, one can see how the loss of once-dense canopies can affect solar radiation, snowpack density, and snow melt. *Photo: David Moore/UA School of Natural Resources and the Environment.*

Continued

DRILLING DEEPER—Cont'd

western US, snow and snowmelt contribute significantly to annual water budgets for the surrounding landscape.

- Investigate how MPB-induced mortality might impact the quantity of snow and timing of snowmelt.
- What are the physical mechanisms that cause these changes in snow?
- What might this mean for ecosystems that depend on snowmelt as a substantial source of water?

nearby headwater streams, helping to maintain surface water flows during dry conditions. Sounds like good news? Not necessarily, as we'll see below.

c. ***Water Quality***: The increased groundwater flow entering headwater streams in decimated high-elevation Colorado forests is able to maintain streamflow during dry weather, but there is a downside. Groundwater also transports pollutants into streams. Bearup et al. (2014a,b) measured increased levels of the mobile forms of the heavy metals cadmium and zinc in soils underneath forests suffering from mortality. Subsequent movement of these metals into nearby streams could threaten water quality.

   Mikkelson et al. (2013) even suggested the possible existence of a human health threat. Naturally occurring organic substances like humic and fulvic acids released by vegetation and carried by groundwater beneath infested forests into streams may enter downstream water treatment plants. Here they may combine with chlorine, which is used to disinfect water supplies, and form harmful by-products like trihalomethanes, some of which are carcinogenic. The scientists analyzed water samples from treatment facilities located in watersheds containing areas of tree die-off and confirmed the presence of elevated contaminant levels.

d. ***Greenhouse Gases***: Some scientists believe that beetle-driven pine forest die-off may upset the global CO_2 balance. Forests affect the carbon budget through photosynthesis, during which plants take CO_2 out of the atmosphere and lock it up in organic compounds, and respiration, during which plants and soil microbes release CO_2 back into the atmosphere. The balance of these processes determines whether a particular forest is a carbon source or a carbon sink.

   After a massive tree die-off, conventional wisdom has it that forests change from carbon sink to carbon source as trees no longer photosynthesize but continue to decompose and release large amounts of CO_2 to the atmosphere. A team of Canadian Forest Service scientists (Kurz et al., 2008) estimated that by 2020, the amount of CO_2 release attributable to MPB just from pine forests in British Columbia could reach 270 megatons, turning the forests from a carbon sink to a net carbon source.

Other scientists disagree. D. Moore and his team of scientists measured CO_2 released from two Colorado forests, one heavily impacted by MPB and the other not. They found little difference in the amount of CO_2 released, suggesting that both photosynthesis and respiration slowed in infested forests. These scientists believe that any contribution of the beetle-impacted forests would be very gradual (Moore et al., 2013).

CRITICAL THINKING

Exercise 6: Forest carbon dynamics are, well, dynamic. Read through the Kurz and Moore papers (links available on your Resource Page).

- Make a list of the arguments supporting each paper's contrasting results about the impact of MPB mortality on carbon storage in these forests.
- Consider the time frame of each study. How does this affect your evaluation of dead forests as carbon sources or sinks?
- What other ecosystem processes could alter these results (consider what natural processes following die-off could push forests to sequester more, or emit more, CO_2 in the long-term)?

DRILLING DEEPER

Exercise 7: We've looked at some fairly unusual indirect effects of the MPB invasion. Remember that we're talking about an insect the size of a rice grain. Here's one additional indirect effect for you to figure out. The outbreak of the MPB has had negative effects at the highest levels of food webs in parts of the western US. Do some research to find out what top predator is suffering because of the MPB infestation. (Hint: the tree being affected is the whitebark pine.)

Answer the following:

- What animal species is being affected and where?
- Describe the role climate change is playing in this case
- What is an unpleasant outcome for humans in the region in question?
- Suggest a solution to the problem

Making lemonade out of lemons: You have probably heard the old adage about making something good out of something bad. That's happening to an extent with the MPB-infested forests.

1. **Biofuels**: The huge number of dead pines in western North America may represent an opportunity to use the deadwood as biofuels. The dead trees can be used as feedstock to form n-butanol, which can be blended into gasoline, converted into jet fuel, or used in paints and other products. While converting these dead pine trees into biofuels is a complex process, it does have the potential to make a positive use of the trees.

2. **Furniture**: The blue stain left in the dead pines by fungi introduced by the MPB has created a new market. Furniture and flooring made from the blue wood are now in high demand in Colorado. While dead trees that have stood for long periods lose their integrity and aren't useful for building houses, items ranging from high-end furniture to skis are now being made from this beautiful blue–gray wood (Whelan, 2012).

Alternate Scenarios: We've focused on the outbreak of MPB in the western US. Obviously, forests around the world face a broad array of other biotic threats. Pests and pathogens have evolved with forested ecosystems, resulting in a complex balance. But humans are altering that balance in several ways:

1. As global trade has become the norm, many insects and pathogens are transported across oceans and continents to ecosystems that have not evolved with feedbacks to control their spread or impact. These invasive species can wreak havoc on forested ecosystems, leading to widespread mortality and, in some cases, loss of keystone tree species.

2. Changes in climate conditions and patterns impact pests and pathogens directly. A consortium of scientists in the northeastern US has summarized

EXPLORE IT

Exercise 8: The USFS has an online map viewer that allows you to visualize both the spatial extent and severity of various forest stress agents. Visit the forest pest conditions viewer at http://foresthealth.fs.usda.gov/portal/Flex/FPC and linked from your Resource Page to explore county-level maps of major forest insect and disease conditions throughout the US.

- The bar graph at the bottom of the web viewer provides a side-by-side comparison of the most damaging insect pests in the US, with acres of mapped damage as squares and the number of counties with reported infestations in circles (Fig. 4.2.7). Based on these data, what is the most widespread insect pest?
- Considering that counties are included if an insect is detected and acres are mapped during aerial overflights to capture severe canopy stress and mortality, which of these insect pests is the most damaging to forest health (not just the most widespread)? Why?

EXPLORE IT—Cont'd

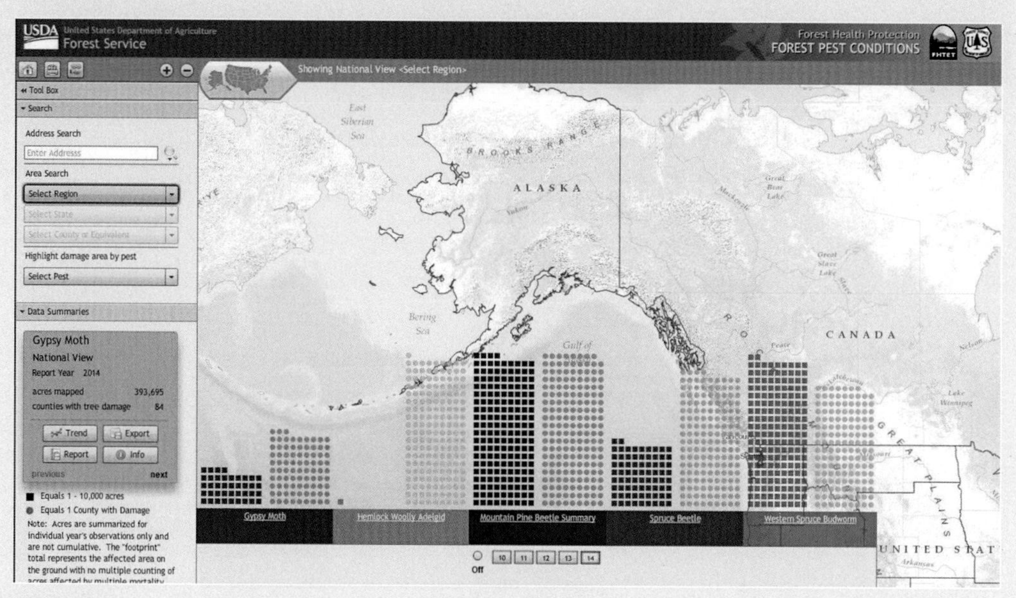

FIGURE 4.2.7

The US Forest Service's forest insect and disease reporting portal. The Forest Pest Conditions mapper includes a myriad of state-, county-, and local-level forest insect and disease conditions data. *Source: USFS.*

- Using the left-hand bar, you can access data summaries for each of these pests, including data on the trends in area damaged over the past 5 years. Examine the trends for counties. Which pests continue to spread into new counties, and which appear to have stabilized or reduced their impact range?
- Another insect pest of great concern is the Emerald Ash Borer (EAB). Using the "highlight damage area by pest" dropdown menu in the left tool bar, compare the range of EAB to MPB, along with their current trends. Which pest do you believe represents the greater risk of spread?

expected changes in invasive species with continued climate change (Dukes et al., 2009). Their conclusion was that the majority of pests and pathogens will increase in frequency, intensity, and virulence as our climate continues to warm.

While invasions are a part of ecosystem evolution, changes in global trade and climate patterns have increased the rate of insect and pathogen invasion over the past century, their rate of spread upon invasion, and their impact. Many organizations are working together to detect and track the movement of invasive forest pests, with many efforts led by the USFS.

CRUNCHING THE NUMBERS

Exercise 9: The USFS has developed a stand susceptibility index to guide land managers in identifying forest stands most susceptible to MPB attack and direct their management activities.

Fig. 4.2.8 outlines this general formula. Use this to compare two different forest stands.

Calculating the Stand Susceptibility Index

To determine a stand's ***risk index***, determine first its ***susceptibility index*** using stand data in the following relationship: The ***susceptibility index*** will range from 0 to 100. Highest values indicate the most susceptible stands.

Susceptibility Index (S) = P x A x D x L where:
P = Percent of susceptible pine basal area
A = Age factor
D = Stand density factor
L = Location factor

P is calculated: Average basal area per hectare of LPP > 15 cm d.b.h. / Average basal area per hectare all species ≥ 7.5 cm. d.b.h. x 100
(Note: Metric units must be used in formulas to assure appropriate table values are obtained.)

Susceptibility index is a measure of stand characteristics which describe their attractiveness to beetles and is based on four variables:

1. Susceptible host basal area (as percent of stand basal area),
2. Age of dominate and codominant host,
3. Stand density, and
4. Location (latitude, longitude and elevation).

Age factor from table:

Age of dominant or co-dominant LPP	Age Factor
< 60 years	0.1
61-80 years	0.6
> 80 years	1.0

Density factor from table:

Stems per Hectare (all species ≥ 7.5 cm d.b.h.)	Density Factor
≤ 250	0.1
251-750	0.5
751-1,500	1.0
1,501-2.000	0.8
2,001-2,500	0.5
> 2,500	0.1

There are three possible location factors, based on a combination of longitude, latitude and elevation. Location factor is determined from the formula Y = [24.4 longitude] -[121.9 latitude] -[elevation (in meters)]+ 4545.1. For most areas on a District, and perhaps over a Forest, the location factor, once determined, will remain constant.

Y	Location Factor
≥ 0	1.0
between 0 and -500	0.7
< -500	0.3

FIGURE 4.2.8
USFS MPB hazard rating identifies the likelihood of an outbreak within a specific time period as a function of stand conditions. *Source: USFS.*

Stand A is a pure lodgepole pine stand with approximately 500 stems per hectare and a mean stand age of 75years. The location factor is 1.29.

CRUNCHING THE NUMBERS—Cont'd

Stand B is a mix of lodgepole pine (50% basal area) and quaking aspen (50% basal area) with approximately 2200 stems per hectare and a mean stand age of 45 years. The location factor is 1.59.

- Calculate the hazard rating for both stands.
- Which stand is more susceptible to MPB infestation?

SOLUTIONS

Exercise 10: Review some of the management options presented in the USFS Management Guide for Mountain Pine Beetles linked on your Resource Page.

- For which of the two stands examined in Exercise 9 would you recommend taking management actions?
- What might those management actions entail? (Be sure to consider all possible options, from silvicultural to biological and chemical treatments).
- Assume that you are the forest manager for this large public forest. Considering the costs and expected benefits of various management strategies, what is your recommendation for these two stands?

Solutions: There is no "silver bullet" solution to the MPB outbreak. Once the pests are established in a pine forest, there is relatively little that can be done to stop the infestation. However, steps can be taken to limit the spread. The strategy of preventive management is to keep beetle populations below injurious levels by limiting the beetles' food supply. Because outbreaks usually develop in mature to overmature forests and trees with rough bark, forest management techniques to remove favorable hosts could prevent MPB populations from reaching critical levels.

It appears that the outbreak in the western US has peaked. Experts believe that the beetle numbers in Colorado hit a maximum in 2012 and are now declining. But it also appears that MPB is spreading east. Scientists have detected MPB infestations in jack pine in Alberta, Canada, from beetles apparently blown in by the winds from forests in British Columbia. A concern is that jack pines stretch across Canada and down into the eastern US. While the cold climate of the boreal forest has limited MPB spread in the past, with a warming climate, there is less confidence that the current spread will be limited to western areas.

REFERENCES

Amin, H., Atkins, P.T., Russo, R.S., Brown, A.W., Sive, B., Hallar, A.G., Hartz, K.E.H., 2012. Effect of bark beetle infestation on secondary organic aerosol precursor emissions. Environ. Sci. Technol. 46 (11), 5696–5703.

Bearup, L.A., Maxwell, R.M., Clow, D.W., McCray, J.E., 2014a. Hydrological effects of forest transpiration loss in bark beetle-impacted watersheds. Nat. Clim. Change 4, 481–486.

Bearup, L.A., Mikkelson, K.M., Wiley, J.F., Navarre-Sitchler, A.K., Maxwell, R.M., Sharp, J.O., McCray, J.E., 2014b. Metal fate and partitioning in soils under bark beetle-killed trees. Sci. Total Environ. 496, 348–357.

Dukes, J.S., Pontius, J., Orwig, D.A., Garnas, J.R., Rodgers, V.L., Brazee, N.J., Cooke, B.J., Theoharides, K.A., Stange, E.E., Harrington, R.A., Ehrenfeld, J.G., Gurevitch, J., Lerdau, M., Stinson, K., Wick, R., Ayres, M.P., 2009. Responses of insect pests, pathogens and invasive species to climate change in the forests of northeastern North America: what can we predict? Can. J. For. Res. 39, 231–248.

Ferrenberg, S., Mitton, J.B., 2014. Smooth bark surfaces can defend trees against insect attack: resurrecting a "slippery" hypothesis. Funct. Ecol. 28, 837–845.

Ferrenberg, S., Kane, J.M., Mitton, J.B., 2014. Resin duct characteristics associated with tree resistance to bark beetles across lodgepole and limber pines. Oecologia 174 (4), 1283–1292.

Handy, R.M., 2013. Beetle Kill Forests' Uncertain Future After the Epidemic. The Coloradoan. December 21 Edition.

Henson, B., 2013. The Bark Beetle Blues. National Center for Atmospheric Research/University Corporation for Atmospheric Research. www2.ucar.edu/atmosnews/perspective/9366/bark-beetle-blues.

Kurz, W.A., Dymond, C.C., Stinson, G., Rampley, G.J., Neilson, E.T., Carroll, A.L., Ebata, T., Safranyik, L., 2008. Mountain pine beetle and forest carbon feedback to climate change. Nature 452, 987–990.

Mikkelson, K.M., Dickenson, E.R.V., Maxwell, R.M., McCray, J.E., Sharp, J.O., 2013. Water-quality impacts from climate-induced forest die-off. Nat. Clim. Change 3 (3), 218–222.

Moore, D.J.P., Trahan, N.A., Wilkes, P., Quaife, T., Stephens, B.B., Elder, K., Desai, A.R., Negron, J., Monson, R.K., 2013. Persistent reduced ecosystem respiration after insect disturbance in high elevation forests. Ecol. Lett. 16 (6), 731–737.

Robbins, J., 2010. What's Killing the Great Forests of the American West? Environment 360. Yale University. 15 March Report.

Rocky Mountain Research Station, February 24, 2014. Mountain Pine Beetle Impacts to Lodgepole Pine Forests. Science Briefing. http://www.fs.fed.us/rm/boise/AWAE/briefing/AWAE_Science_Briefings-MountainPineBeetleEffectsOnLodgepoleTranspiration.pdf.

Whelan, R., 2012. Insect chic. In: Colorado, Beetles Create Décor Trend. The Wall Street Journal. November 30, 2012 Edition.

Chapter 4.3

Tar Sands

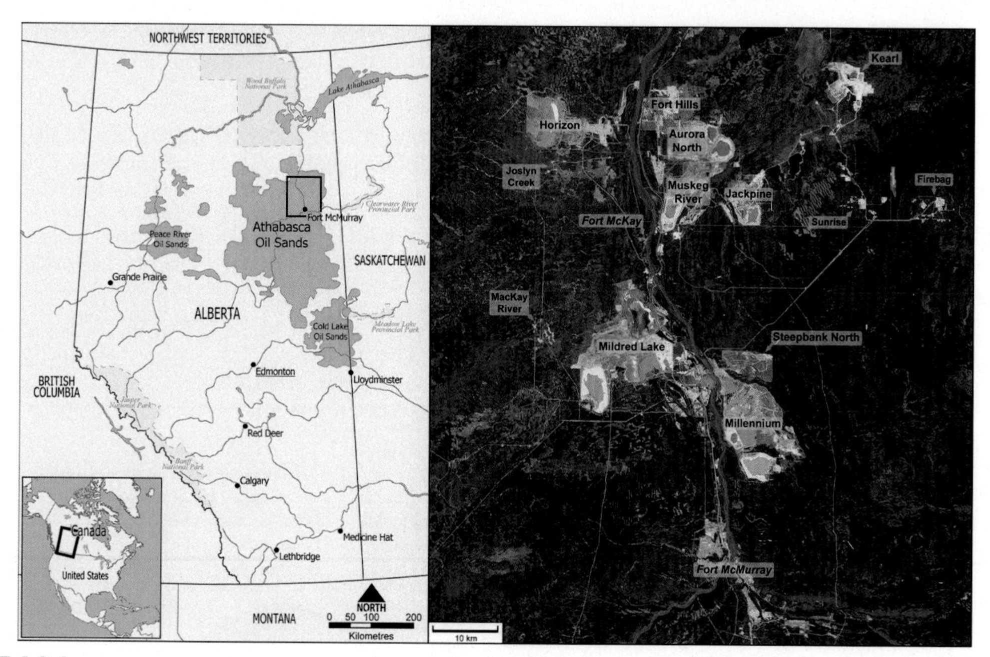

FIGURE 4.3.1
Map of Alberta, Canada, showing extent of oil sand occurrences (left) and labeled satellite image (right) with oil sand open-cast mines and *in situ* production sites along the Athabasca River. *Source: Gretarsson (CC) via Wikimedia Commons.*

OVERVIEW

While the use of alternative sources of clean energy is on the rise, the current world economy and the modern societies it supports continue to depend on fossil fuels. But fossil fuels are a finite resource. What took millions of years to create is being extracted at rates that could consume the remaining estimated ultimately recoverable global oil during your lifetime. The majority of studies suggest that world oil production is likely to peak sometime between 2010 and 2030. But many of these predictions fail to consider the development of new technologies that can extract harder-to-reach fossil fuels.

As obtaining oil from more traditional locations like oil fields has become more limited or difficult due to geopolitical challenges, more attention has been focused on alternatives like oil shale and tar sands (Fig. 4.3.1). Called "synfuels" in earlier decades, these fossil fuels are held in subsurface deposits that must be extracted for use.

CRUNCHING THE NUMBERS

Exercise 1: Earlier studies estimated that oil production would peak somewhere between 2010 and 2030. But these studies didn't take into account the creation of new technologies that could extract harder-to-reach sources of fossil fuels. More recent models (Fig. 4.3.2) show slightly longer time ranges for global oil production.

Open the spreadsheet titled "Total Oil Production.xls" from your Resource Page. These data from the US Energy Information Administration quantify the global production of crude oil over the past 35 years.

- Which countries produced the most oil in 2014? Do any of these countries surprise you? Are there other countries that you are surprised aren't at the top of the list? Where does your country rank?
- Use these data to graph oil production both globally and for each of the top five producers you identified above. How does production vary over time, both globally and for the top producers, compared to the estimates shown in Fig. 4.3.2?

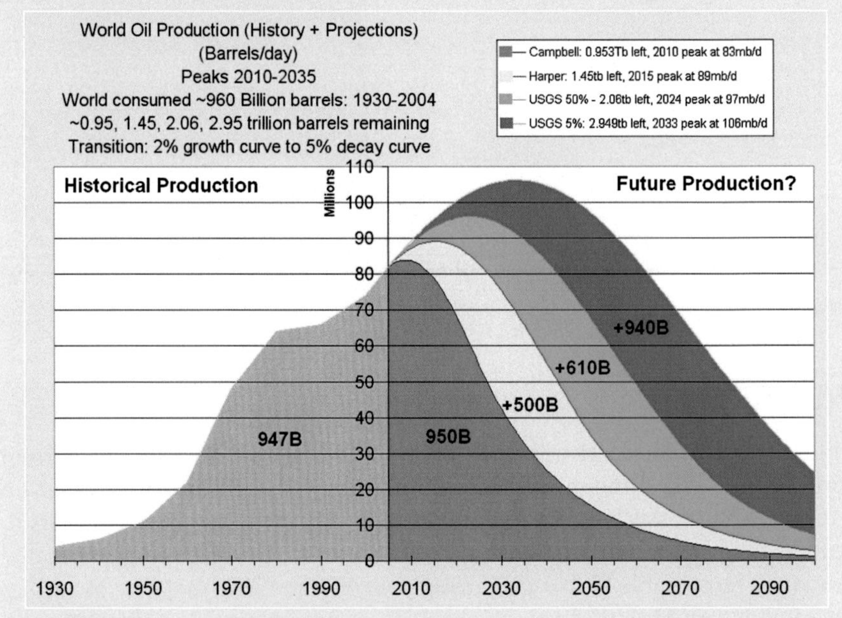

FIGURE 4.3.2
Projections of oil production based on different publicly offered world ultimate reserve estimates. This graph does not consider political and technical limitations on oil production, but simply assumes the less oil that remains, the more slowly it will be extracted. *By Tomruen (public domain) via Wikimedia Commons.*

In this case study, we'll take a look at the Alberta oil (or tar) sands of Canada. We'll start by considering what exactly tar sands are, how the oil is extracted from the sand, and some of the environmental and human health costs associated with the exploitation of this energy source. Then we'll take a broader look at the life cycle of fossil fuels and ask you to consider the environmental impacts at every step, from removal of the crude oil from the ground to the production of the gas you use to fill up your car. We'll also take a look at the bottom line for tar sands: how much energy do we get out compared to what we use to extract it, and how do tar sands compare to other sources in terms of CO_2 emissions?

Background: The Athabasca Oil Sands Region (AOSR) in northern Alberta, Canada (Fig. 4.3.1) has the third largest reservoir of crude oil in the world after Venezuela and Saudi Arabia (Bari et al., 2014), with about 12% of global oil reserves in 2010. The tar sands have an interesting history. Typically, extracting oil from sand was not considered profitable because of the high cost of extraction, but, during the oil embargo of the 1970s, tar sands began to be considered as an alternate source.

The amount of oil in the ground in AOSR is immense, an estimated 170 billion barrels that can be recovered with today's technology and another 1.63 trillion barrels if all the embedded product could be extracted from the sands. With this much reserve, the fate of AOSR will play a key role in our energy future. It is projected that output from AOSR will likely increase to 3.7 million barrels/day by 2025 (Bari et al., 2014). But because of the high cost of extracting and processing these fuels, fluctuations in oil prices will ultimately determine how much is extracted. While demand and prices were on the rise through much of the 21st century to date, the global drop in prices for crude oil that started in 2015 may have an impact on the future growth of Canadian tar sand crude.

CRUNCHING THE NUMBERS

Exercise 2: Let's put the amount of crude available in this region into perspective.

- Recall that each gallon of gas burned by your car releases about 8.9 kg of CO_2. Assume that all of the oil that can be extracted from the AOSR with current technologies (170 billion barrels) is converted to gasoline for automobiles. How many kg of CO_2 would be released by burning all of this AOSR product?
- Typically, CO_2 emissions are presented in units of million metric tons. Convert your kilogram estimate above to million metric tons.
- Consider the historical global emissions from transportation (Fig. 4.3.3). How do the potential emissions from the AOSR compare to annual global transportation emissions?

Continued

CRUNCHING THE NUMBERS—Cont'd

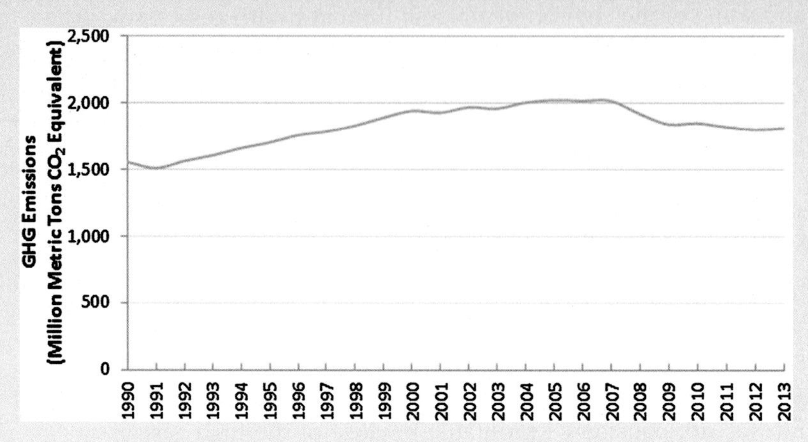

FIGURE 4.3.3

Global greenhouse gas emissions from transportation. *Source: US EPA, 2014.* http://www.state.gov/e/oes/rls/rpts/car6/index.htm.

EXPLORE IT

Exercise 3: To get a better feeling for the AOSR area and the magnitude of the tar sands operation, you can visit some of the websites linked on your Resource Page. These sites provide more detailed information and photographs documenting tar sands extraction and its impact on the landscape.

However, to truly gain perspective on the geographic extent of these operations, we need to employ geospatial tools. The province of Alberta has created an online interactive map (Fig. 4.3.4) that allows us to explore the extent and nature of both the mining and monitoring operations.

- Start by exploring this interactive map linked on your Resource Page and found at http://osip.alberta.ca/map/. Notice how you can choose which data layers to visualize, zoom in or out on the map, and change your base layers.
- Change the base layer to "Google satellite" imagery so that you can visualize the landscape impact of mining activities.
- From the layer options on the left, select from the OIL SANDS PROJECTS tab the **Operating Oil Sands Mines** box. Zoom in to one of these active mines and answer the following questions:
  - What is the predominant land cover type in this region (i.e., what ecosystem was stripped away to get at the tar sands)?
  - What is the proximity of these strip mines to natural surface waters (i.e., rivers, lakes, or streams)?
- Now select the **Oil Sands Tailings Ponds Locations** box under the TAILINGS tab.
  - Zoom in to some of these ponds, and describe what you can tell about them from the satellite imagery.
  - How does the water quality in these tailing ponds differ (or not) from surrounding natural water bodies?
- When you open the LAND DISTURBANCE AND RECLAMATION tab, a new layer that shows you many categories of both disturbed area and reclaimed area opens up. Zoom around and explore the mapped areas that are disturbed. Consider the overall extent of these areas.

EXPLORE IT—Cont'd

- Now select the layer boxes such that only the **Land Reclaimed** is showing in the viewer. What proportion of the total disturbed area has been reclaimed? What proportion has been certified as reclaimed?
- One of the classes of disturbance mapped is **"In situ well sites."** These locations highlight where mining activities are not possible because of the depth of the deposits. How does the spread and density of these well sites compare to the open mines?

FIGURE 4.3.4
The Alberta Department of Environment and Parks interactive tar sands map interface found at http://osip.alberta.ca/map/.

Terms to Learn

...**bitumen**: viscous petroleum that saturates sands that are either loose or partially consolidated sandstone (Fig. 4.3.5).

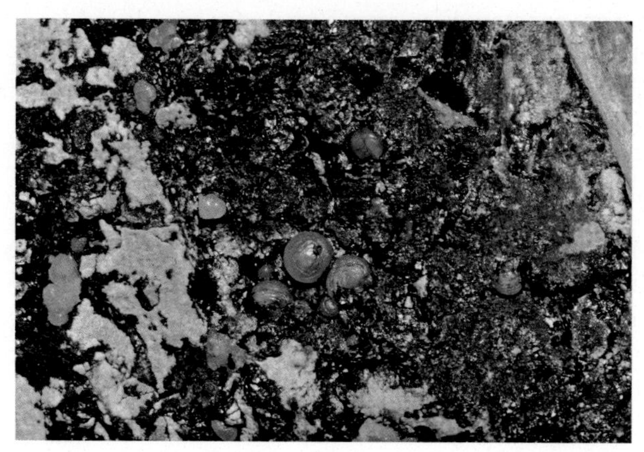

FIGURE 4.3.5
Bitumen. *Source: Parent Géry (CC BY-SA 3.0) via Wikimedia Commons.*

FIGURE 4.3.6
Petroleum coke. *Source: romanm (CC BY-SA 2.5 si) via Wikimedia Commons.*

...**upgrading facilities**: once the bitumen has been removed from the ground, it must be processed to convert it to a form of synthetic crude oil that can be handled by refineries.
...**dilbit**: bitumen after it's been processed and diluted to facilitate transport by pipeline.
...**pet coke**: petroleum coke, or pet coke (Fig. 4.3.6), a solid waste product very similar to coal, is produced as a by-product of the oil refining process. Emitting at least 30% more CO_2 per ton than the lowest quality coal when burned, pet coke is accumulating at US refineries that process AOSR bitumen.
...**tailing ponds**: ponds where the wastewater and tailings from bitumen processing are stored (Fig. 4.3.7).
...**OSPW**: oil sands process-affected water. Two to four barrels of this wastewater are generated for every barrel of oil produced.
...**naphthenic acids (NAs)**: carboxylic acids contained in crude oil and commonly found in OSPW. They have exhibited both acute and chronic toxicity to aquatic life.
...**EROI**: energy returned on investment; in other words, how much energy does it take to get a unit of energy from a particular source?

The Process: About 20% of the tar sands in AOSR are close enough to the surface to be strip-mined. This typically includes formations within 75 m of the surface where the soil layer can be removed. Once the sands are mined, the oil is extracted using hot water and caustic soda that separate the bitumen from clay, sand, and other soil constituents. The wastewater from the process, containing dissolved metals and organic compounds like polycyclic aromatic compounds (PACs) and NAs, is stored in onsite tailings ponds. About four cubic meters of OSPW are produced from each cubic meter of tar sand processed.

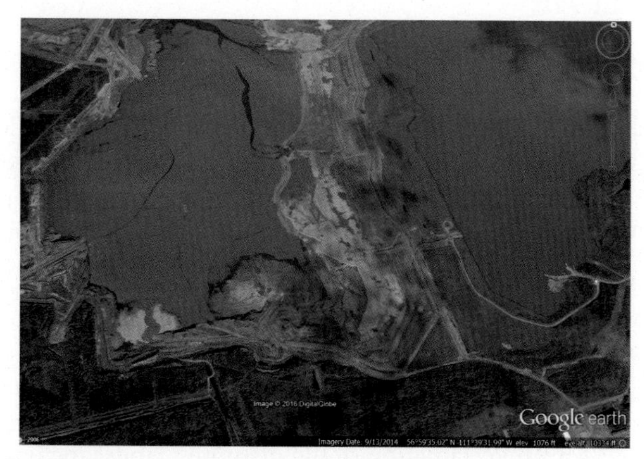

FIGURE 4.3.7

Alberta's massive tar sands operations use large amounts of water every year to process the bitumen found in sand formations. The wastewater is customarily placed into tailings ponds, which now cover 66 square miles of northern Alberta. Here are tailings ponds south of Fort MacKay, Alberta. *Image from Google Earth.*

The remainder of the tar sands lies deep underground. Removal of this deep material requires *in situ* extraction techniques, such as injecting steam into the bitumen to liquefy it so that it can then be pumped to the surface. One pipe carries steam deep underground, while a second pipe collects the flowing bitumen and carries it to the surface.

The very heavy bitumen must then be converted to a synthetic crude oil prior to transportation and processing. The upgrading process may involve vacuum distillation and a number of other steps to yield a product that can be sent via pipelines to refineries for further processing. To help make movement by pipeline possible, a diluting agent such as natural gas condensate, particularly the naphtha component, is added. This lowers viscosity and prevents the bitumen from precipitating out in the pipeline.

DRILLING DEEPER

Exercise 4: Accidents happen, and pipelines are no exception. Occasionally, pipelines carrying oil rupture and leak. This can be particularly problematic with underground pipelines where leaks may not be immediately apparent. Often, these leaks result in contamination of both ground and surface waters.

This is further complicated for tar sands because there's an important difference between the behavior of synthetic crude from tar sands and regular light crude when a leak occurs.

- Do some digging to understand how the properties of these two forms of oil differ and how these differences affect pollutant movement and impacts on water quality.
- Find and discuss one example of a leak of synthetic crude and its environmental effects.

THE PROBLEMS

Habitat Destruction: One obvious initial impact is the damage to the natural environment done during exploration and digging of surface mines. By the end of 2013, 813 km^2 of forested land had been cleared or disturbed (Government of Alberta, 2014). Figure 4.3.8 compares the area of AOSR in 1984 to 2011. Note the extensive areas cleared for surface mining over the 25-year period. The native vegetation must be completely stripped away to get at the soils beneath.

Global Forest Watch has listed this region as one of the five most overlooked deforestation hotspots worldwide. Their forest cover loss animation linked on your Resource Page shows the annual change in forest cover as new pipelines are laid and the ground is cleared for open-pit mining. Between 2001 and 2014, there was a dramatic reduction in forest cover associated with mining operations (Fig. 4.3.9). Reclamation of the disturbed landscape is ongoing.

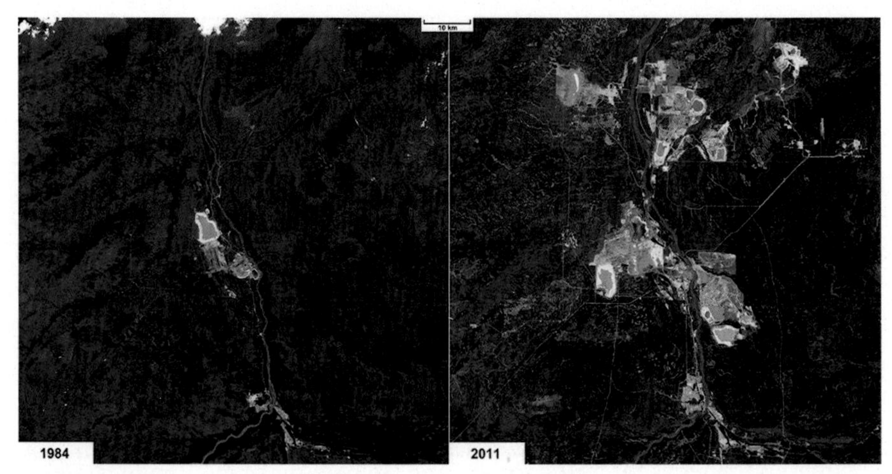

FIGURE 4.3.8

Athabasca tar sands open-cast mining area north of Fort McMurray, Alberta, Canada; comparison of satellite images taken in 1984 and 2011, respectively. *Source: NASA, edited by Gretarsson, via Wikimedia Commons.*

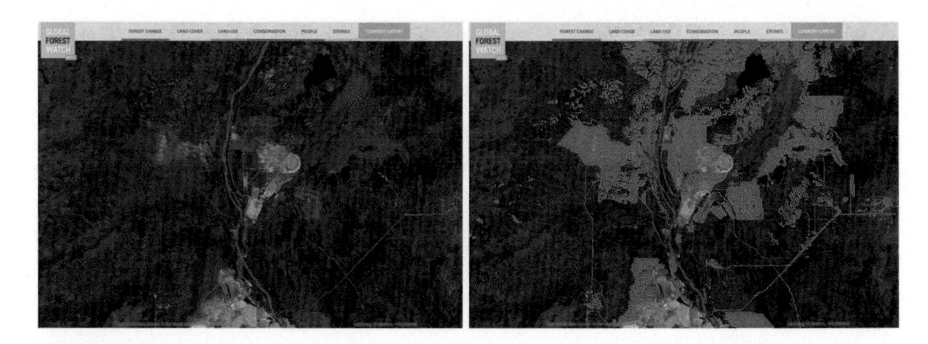

FIGURE 4.3.9

Global Forest Watch's remote sensing assessment of forest cover loss in the Fort McMurray region resulting from mining activities. Pink areas indicate forest loss. *Source: Global Forest Watch Interactive Forest Change Map (CC 4.0).*

Air Pollution: The mining and upgrading of bitumen in the AOSR lead to a number of environmental effects. We'll focus on one type of impact in particular, the movement of air pollutants from the site into surrounding areas. Evidence that areas surrounding AOSR are being affected by the development of the tar sands has been increasing. Kurek et al. (2013) collected core samples of sediments from five lakes within a 35-km radius of AOSR and from one remote lake for comparison. They found increased levels of PACs, organic pollutants associated with tar sands processing. Overall, PAC enrichment ranged between 2.5 and 23 times levels found in the control lake.

CRITICAL THINKING

Exercise 5: Review the Kurek et al. (2013) paper linked on your Resource Page. Their goal was to compare lakes located near AOSR operations to pristine lakes far from AOSR.

- Consider their experimental design. Do you feel that their selection of treatment (near AOSR) and control (pristine) lakes was sufficient to correlate proximity to AOSR activities to water quality?
- How did the researchers make the link between air pollution and water quality?

Some additional evidence of the impact of AOSR on the region's air quality has been provided by satellite data. McLinden et al. (2012) analyzed satellite remote sensing observations of the gases NO_2 and SO_2. High-resolution maps showed increased concentrations of the gases over a 30 by 50 km area surrounding AOSR, with the total mass of pollution released comparable to the largest ever seen from an individual Canadian source.

Bari et al. (2014) directly measured wintertime deposition of contaminants in the AOSR using field monitoring equipment. The investigators collected samples at two sites within 20 km of the tar sands development and at two additional sites more than 45 km distant from AOSR from January to March 2012. Both inorganic and organic contaminants were measured.

For each pollutant measured, a clear trend of decreasing deposition with increasing distance from AOSR was noted (Fig. 4.3.10). Of interest were deposition rates for PACs, compounds associated with fossil fuel sources. Deposition of PACs known or suspected to cause cancer, such as benzo(a) pyrene and chrysene, was up to 75 times greater at the two sites near tar sand development than at more distant sites. Values were, however, lower than or similar to those at other sites lying in urban or industrial environments elsewhere.

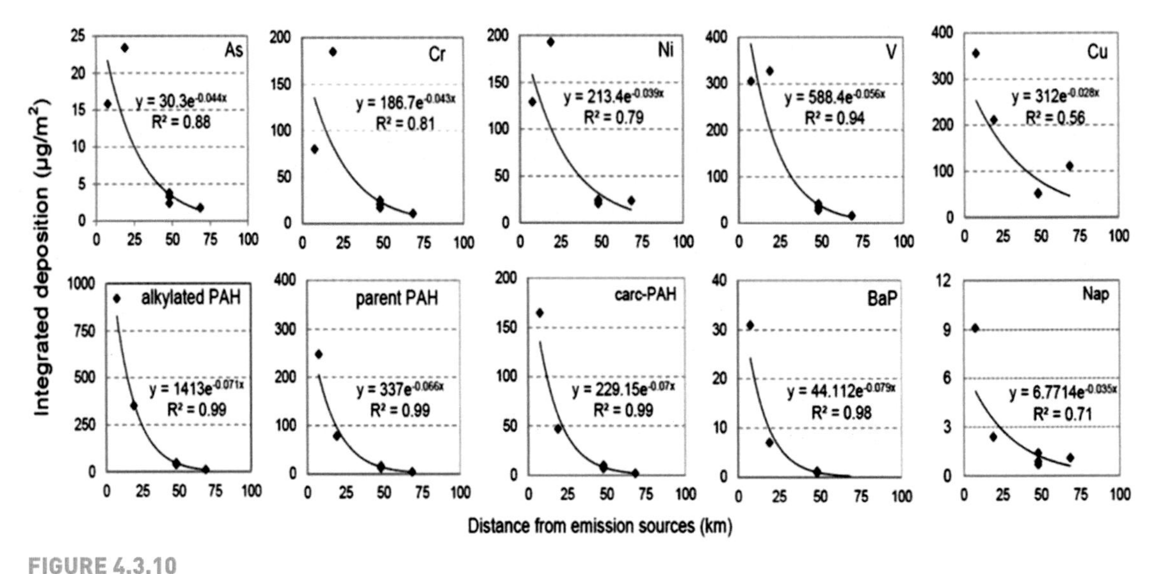

FIGURE 4.3.10

Integrated 3-month winter deposition of selected metals and alkylated and parent polycyclic aromatic hydrocarbons at the sampling sites versus distance from an arbitrary reference location between the Syncrude Canada Ltd. and Suncor Energy Inc. tar sands facilities. *Source: Bari et al. 2014.*

In addition to concerns about habitat disruption and air pollution, there are a number of additional environmental issues related to AOSR.

Groundwater Contamination: Using sophisticated analytical approaches, a team of scientists led by Frank et al. (2014) measured levels of organic pollutants, including NAs, in OSPW tailing ponds and in the groundwater below these ponds. The objective was to identify specific chemical components that could directly link OSPW to contamination of the groundwater. The scientists found a similar assemblage of pollutants in both sets of samples, suggesting that wastewater from the tailing ponds was directly contaminating groundwater beneath. Several samples taken immediately beneath the Athabasca River contained the same pollutant markers, suggesting that the substances were also making their way into the river.

Impacts on the Athabasca River: A team of scientists (Kelly et al. 2009, 2010) measured levels of both PACs and trace elements in water from both the Athabasca River and its tributaries upstream and downstream of the tar sands development and in melted snow collected from each site. Analyses identified elevated levels of the contaminants in samples closest to the industrial activity (Fig. 4.3.11). Applicable guidelines for the protection of aquatic life were exceeded for contaminants in melted snow and/or water collected near tar sands development.

PUTTING SCIENCE TO WORK

Exercise 6: Examine the results from the Kelly et al. study shown in Fig. 4.3.12. Given the differences between sites, seasons, and elements, describe the potential implications of these findings:

- Considering these locations [upstream (UP), near development (DEV), downstream (D&D), and the lake (LA)], which contaminants measured here can be linked to tar sands development?
- Which elements are transported the greatest distance? What are the implications of this pollutant movement?
- Which locations are at greatest risk of contamination (and by what elements) considering the magnitude of increases in concentration compared to the upstream baseline? What does this suggest about focusing future monitoring or mitigation activities?
- What do differences between summer and winter concentrations tell you about the process of pollutant movement? How might these data help inform mitigation or clean up efforts?

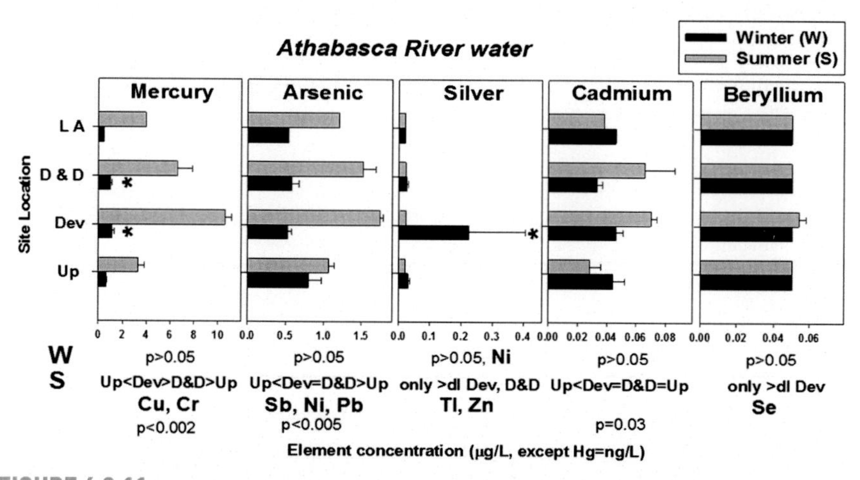

FIGURE 4.3.11

Element concentrations in water collected in winter (*W, black bars*) and summer (*S, gray bars*). *D&D*, downstream and Athabasca delta; *Dev*, downstream/near development; *LA*, Lake Athabasca; *Up*, upstream. *Source: Kelly et al. 2010.*

Environmental Health Risks: The environmental effects of Alberta's tar sands have long been controversial, but now residents of northern Alberta say that emissions from oil production in the AOSR have been having severe effects on their health as well. Air pollutants released by oil sand development activities in AOSR have been linked to a variety of human health issues among residents living close to the area (Edwards, 2014). Symptoms including headaches, sinus infections, throat congestion, and unexplained weight loss have been reported. The long-term effects caused by exposure to PACs and other hazardous compounds from AOSR remain an unknown threat.

Wildlife Impacts: The use of tailings ponds during tar sands operations creates large water bodies filled with contaminated water. Even when contained, these waters can have a variety of adverse effects on organisms exposed to the OSPW they contain. Many instances have been documented, the most recent in 2010, when flocks of ducks mistook a tailings pond for a safe place to land. An estimated 550 ducks died in the 2010 incident, while an estimated 1606 birds, mostly mallards, perished in a similar 2008 episode.

One suggestion to reduce the toxic nature of OSPW is to use reclamation ponds (Fig. 4.3.12) to hold the wastewater until levels of toxic components can be reduced by natural processes. However, Anderson et al. (2012) measured the toxicity of both untreated OSPW and reclaimed wastewater to a common invertebrate species. Even OSPW aged in reclamation ponds maintained its toxicity to the test organism. NAs were suggested as the source of toxicity.

SYSTEMS CONNECTIONS

Exercise 7: The Anderson et al. (2012) paper described the impacts of OSPW on the larvae of *Chironomus dilutus*, a common midge. Consider the role of this invertebrate in the larger ecosystem.

- How might reduced survivorship and fecundity of this species have repercussions elsewhere in the food web?
- How might the loss of this species have repercussions beyond its immediate food web (i.e., cascading impacts through the larger ecosystem)?

FIGURE 4.3.12
Syncrude's base mine tailings pond. *Source: TastyCakes and Jamitzky (public domain), via Wikimedia Commons.*

CRITICAL THINKING

Exercise 8: Above we have briefly reviewed various environmental issues that arise from AOSR activities. On your Resource Page, you will find links to several scientific studies examining some of these issues. Scan through each of these and consider the following questions:

- Which of these environmental issues do you feel poses the greatest threat to local ecosystems?
- Do you think air or water pollution poses the greatest risk to human health in the vicinity of AOSR? Defend your choice.
- As you've learned in some of your other case studies, establishing cause and effect is challenging. Which of these papers presents the strongest evidence that AOSR activities are directly linked to the environmental problems presented? How to they make this connection?
- Which of these studies presents the most tenuous direct link to AOSR activities? How might additional research strengthen this connection?

Implications for Climate Change

Environmental concerns emerge at many different steps in the life cycle of fossil fuels like tar sand oil, from the mining process to extraction and refining to the ultimate consumption (e.g., fueling your car). Some refer to this accumulation of environmental impacts as "well to wheels" impacts. We've considered several examples of effects that have occurred in the vicinity of AOSR related to the extraction and processing of bitumen mined at the site and the disposal of waste produced during production. But to understand the full scope of impacts and compare them to the benefits of exploiting this energy source, we need to dig deeper. Let's look at the bigger picture.

How Does Athabasca Oil Sands Region Rank as a Net Energy Source?

Recall our definition of EROI above. For oil extracted from the ground using conventional technology, the ratio is about 25 to 1 (Fig. 4.3.13). In other words, it takes about 1 unit of energy to obtain 25 units of oil-based energy. However, according to Nuwer (2013), surface-mined tar sands energy has an EROI of about 5 to 1, and for tar sands retrieved from deep underground, the value decreases to about 2.9 to 1, with the drop due mostly to the natural gas that must be burned to heat the deep bitumen. If the full life cycle of tar sands production is considered, including such components as transportation, refining at distant plants, and end use efficiency, the actual value may be closer to 1 to 1. Since the remaining reserves of tar sands, another 143 billion barrels

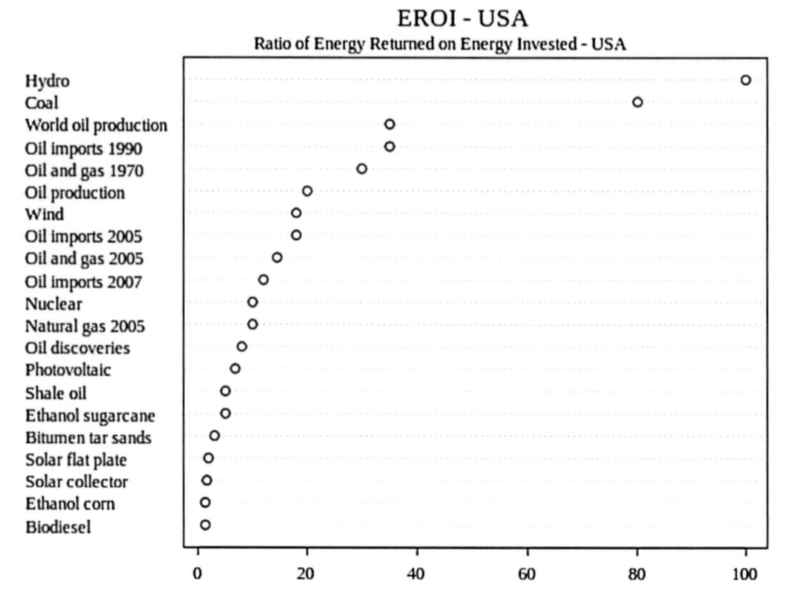

FIGURE 4.3.13

US EROI, ratio of energy returned on energy invested, for the most common fuel sources. *Source: Murphy, and Hall 2010.*

worth, lie mostly under Alberta's boreal forest, with only about 8% accessible by surface mining, the EROI for the tar sands doesn't seem likely to improve.

Representatives of the petroleum producers argue that the tar sands EROI is closer to 6 to 10 to 1. They also counter that there are other fuel sources with even lower EROIs; corn ethanol has an estimated EROI of 1.3 to 1, although ethanol has the advantage of being considered a renewable source of energy.

WHAT ABOUT THE ADDITIONAL RELEASE OF GREENHOUSE GASES FROM ATHABASCA OIL SANDS REGION?

Of course, another important part of the energy equation is the amount of greenhouse gases released from the tar sands operation: how much will the extraction of tar sands contribute to climate change? According to Biello (2013), the greenhouse gas emissions released during the mining and upgrading of tar sands add up to about 116 kg/barrel of oil produced. This does not include the actual burning of the oil extracted and processed. Biello estimates that if all the oil from AOSR were burned, the added CO_2 from this one region would increase mean global temperatures by about 0.4 °C.

But there's more to this part of the story. Once dilbit makes it to a refinery in the US, its long hydrocarbon chains are broken into shorter, lighter hydrocarbons, forming gasoline and other products. A by-product of the refining process is the pet coke we defined in your Terms to Learn. Canadian upgrader plants produce an estimated 10 million metric tons of pet coke annually, and US refineries, another 61 million tons. Although used as fuel for cement manufacture and power plants, pet coke is one of the dirtiest fuels available, releasing at least 30% more CO_2 than the lowest-quality coal. Thus, pet coke used as a fuel results in even higher total carbon emission estimates from tar sands activities.

CRUNCHING THE NUMBERS

Exercise 9: When considering the production of an item, the term "externalities" refers to costs that aren't included in the actual price of the item. This can include costs associated with environmental damage or reclamation and human health effects. The Canadian Energy Research Institute published a cost-benefit analysis of the bitumen refining process. On your Resource Page, we've also included a link to the Victoria Transportation Policy Institute's chapter on the External Costs of Petroleum Consumption. Start by reviewing this report to get an idea of the types of considerations that go into such an analysis.

Now, let's perform a similar (although more rudimentary) analysis for the entire "well to wheels" impact of tar sands extraction.

- Start by listing as many externalities associated with the production of petroleum products from tar sands as you can.
- Based on the range of cost estimates for the various externalities presented in the documents provided, estimate what the true cost of a gallon of gas made from tar sands bitumen would be if all the environmental and human health costs were included in the price. Feel free to adjust these numbers based on your own online research of externality values associated with carbon.
- Consider the current cost of a barrel of oil. How different is the "true" cost from the market price? How might this "true" cost be recovered or reclaimed to pay for some of the external damages associated with oil? How might the adoption of these "payback" activities impact the demand for tar sands oil?

Some Good News: In response to significant pushback from environmental and community leaders, the oil industry in AOSR is attempting to reduce greenhouse gas emissions that result from their activities. For instance, adding hydrogen to bitumen reduces the amount of carbon released. In addition, efforts are underway to add carbon capture technology to operations in AOSR. The province of Alberta is considering levying a carbon tax, an additional incentive for oil producers to reduce carbon emissions.

SOLUTIONS

Exercise 10: In 2011, the Pembina Institute of Alberta, Canada, released a report entitled "Solving the puzzle: environmental responsibility in oilsands development." Read this report, linked on your Resource Page at www.pembina.org/reports/solving-puzzle-oilsands.pdf and answer the following:

- The report makes 19 recommendations designed to reduce the environmental impacts of tar sands development in five areas: land, water, air, greenhouse gases, and monitoring. Identify and discuss one specific recommendation from each of the five areas covered.
- Which of your five choices do you believe would, if accomplished, lead to the greatest overall improvement in the environment?
- Which one of the 19 recommendations do you believe would be the easiest to accomplish? Why? The hardest? Why?

CONSIDER THIS

Exercise 11: While it is easy to think of the tar sands extraction as simply an environmental problem, there are a number of social/environmental justice issues related to this case study.

- Think about all the things we've covered in this case study, and discuss two instances that you think are examples of environmental injustice that arise from the "well to wheels" activities associated with tar sands extraction.

REFERENCES

Anderson, J., Wiseman, S.B., Moustafa, A., El-Din, M.G., Liber, K., Giesy, J.P., 2012. Effects of exposure to oil sands process-affected water from experimental reclamation ponds on *Chironomus dilutus*. Water Res. 46, 1662–1672.

Bari, M.A., Kindzierski, W.B., Cho, S., 2014. A wintertime investigation of atmospheric deposition of metals and polycyclic aromatic hydrocarbons in the Athabasca Oil Sands. Sci. Total Environ. 485–486, 180–192.

Biello, D., January 23, 2013. How Much Will Tar Sands Oil Add to Global Warming? Scientific American.

Edwards, J., 2014. Canada's oil sands residents complain of health effects. Lancet 383, 1450–1451.

Frank, R.A., Roy, J.W., Bickerton, G., Rowland, S.J., Headley, J.V., Scarlett, A.G., West, C.E., Peru, K.M., Parrott, J.L., Conty, F.M., Hewitt, L.M., 2014. Profiling oil sands mixtures from industrial developments and natural groundwater from source identification. Environ. Sci. Technol. 48 (5), 2660–2670.

Government of Alberta, 2014. Alberta's Oil Sands: Reclamation. http://oilsands.alberta.ca/reclamation.html.

Kelly, E.N., Short, J.W., Schindler, D.W., Hodson, P.V., Ma, M., Kwan, A.K., Fortin, B.L., 2009. Oil sands development contributes polycyclic aromatic compounds to the Athabasca River and its tributaries. Proc. Natl. Acad. Sci. U.S.A. 106 (52), 22346–22351.

Kelly, E.N., Schindler, D.W., Hodson, P.V., Short, J.W., Radmanovich, R., Nielsen, C.C., 2010. Oil sands development contributes elements toxic at low concentrations to the Athabasca River and its tributaries. Proc. Natl. Acad. Sci. U.S.A. 107 (37), 16178–16183.

Kurek, J., Kirk, J.L., Muir, D.C.G., Wang, X., Evans, M.S., Smol, J.P., 2013. Legacy of a half century of Athabasca oil sands development recorded by lake ecosystems. Proc. Natl. Acad. Sci. U.S.A. 110 (5), 1761–1766.

McLinden, C.A., Fioletov, V., Boersma, K.F., Krotkov, N., Sioris, C.E., Veefkind, J.P., Yang, K., 2012. Air quality over the Canadian oil sands: a first assessment using satellite observations. Geophys. Res. Lett. 39 (4), L04804.

Murphy, D.J., Hall, C.A.S., 2010. Year in review–EROI or energy return on (energy) invested. Ann. N. Y. Acad. Sci. 1185, 102–118.

Nuwer, R., February 19, 2013. Oil Sands Mining Uses Up Almost as Much Energy as it Produces. Inside Climate News.

US EPA, 2014. Climate Action Report. http://www.state.gov/e/oes/rls/rpts/car6/index.htm.

Chapter 4.4

Desertification

FIGURE 4.4.1
Drought and land use practices take a toll on both land and people in Brazil. *By LeoNunes (CC BY-SA 3.0) via Wikimedia Commons.*

OVERVIEW

In spite of its broad geographic scale and catastrophic environmental effects, desertification receives little attention in much of the developed world. Desertification is the process in which fertile land gradually transitions to desert (Fig. 4.4.1), typically because of human activities that disturb fragile ecosystems delicately balanced in marginal climates. Common where water is scarce and human populations alter natural environments, this phenomenon threatens approximately 40% of the Earth's land surface. Many of the areas affected by or vulnerable to desertification lie in developing nations (Fig. 4.4.2), where the loss of fertile land can be a severe blow to poor populations least able to cope with such changes.

In this case study, after a brief review of the causes and consequences of desertification, we'll look at a series of investigations focused on measuring its extent,

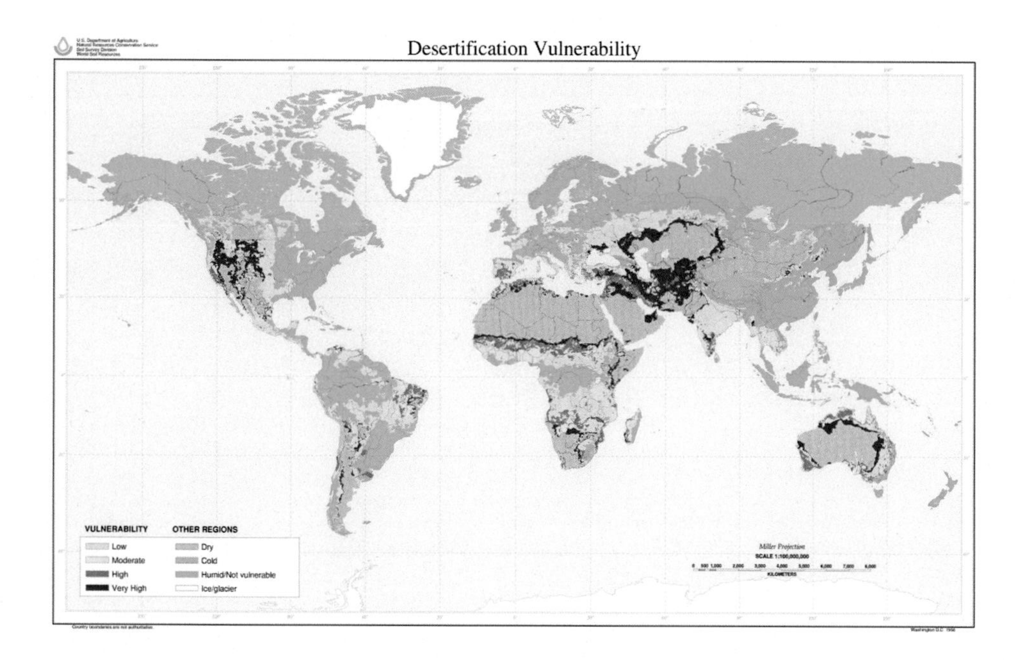

FIGURE 4.4.2

The Desertification Vulnerability map is based on a reclassification of the global soil climate map and global soil map into four vulnerability classes. USDA-NRCS, Soil Science Division, World Soil Resources.

assessing its impacts, and slowing or reversing the process. We'll conclude by considering some of the social implications of desertification, particularly those linked to the changing climate.

BACKGROUND

Despite its name, desertification doesn't refer to the advance of existing deserts. Instead, it's the transformation of normally dry but useable lands into desert-like conditions (Fig. 4.4.3), with some of the worst examples occurring in poorer areas of Africa and Central Asia, where it threatens the livelihoods of some of the most vulnerable human populations. Some of the causes of desertification include unsustainable farming practices that result in poor soil stability, overgrazing, deforestation, and climate change.

Land degradation can occur anywhere, but desertification, by definition, occurs only in areas characterized by low precipitation, large variations in temperature, and soils low in organic matter. Typically, these marginal drylands are populated by plants adapted to a high degree of climatic variability. These hearty species are able to stabilize and maintain the structure of the dry soils.

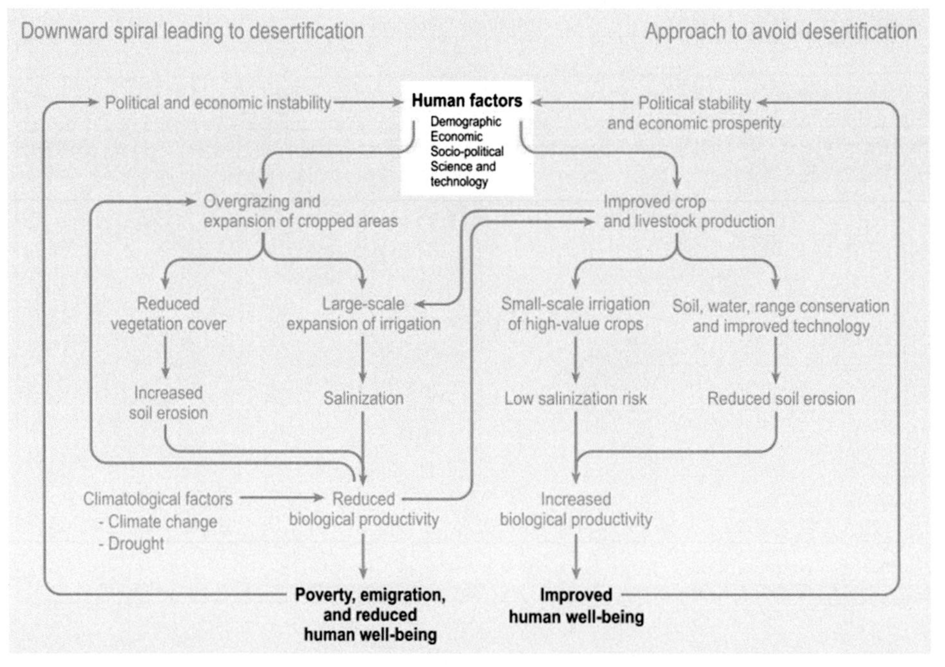

FIGURE 4.4.3
A graphic showing how drylands can be developed in response to changes in key human factors. The left side shows developments that lead to a downward spiral of desertification. The right side shows developments that can help avoid or reduce desertification. Both development pathways occur today in various dryland areas. *Source: Millennium Ecosystem Assessment, 2005.*

Typically, desertification starts as vegetation that helps bind soils is removed for firewood or to clear land for crops.

Overgrazing by livestock may also destroy grasses, making soils vulnerable to erosion. Intensive farming done unsustainably can deplete soils of nutrients and organic matter. Wind and water erosion can then remove exposed soils, and the landscape is gradually converted into a desert-like environment (Fig. 4.4.4). Projections for the future are grim, as increased drought severity, duration, and frequency associated with climate change will likely expand the extent of vulnerable drylands.

Terms to Learn

Drylands: arid or semiarid land defined by a scarcity of water
Arable: used or suitable for growing crops
Desertification: conversion of drylands to desert by human-linked or natural causes
Deflation: the erosion of soil due to the loss of stabilizing vegetation

FIGURE 4.4.4
Sand dunes advancing on Nouakchott, the capital of Mauritania. *Source: NASA.*

Land Degradation: any change in the landscape considered to be undesirable. Desertification is one form of land degradation; others include chemical pollution and mining activities
Albedo: the fraction of solar energy reflected from the Earth's surface back into space
Desertification reversion: the process of restoring lands previously desertified

DIGGING DEEPER

Exercise 1: Before continuing with this case study, do some online research to learn more about drylands and desertification. A good place to start might be the Millennium Ecosystem Analysis Desertification Synthesis linked on your Resource Page.

a. Where are the major regions of the globe where drylands are found? What do populations living in these regions often have in common?
b. Desertification is not a new phenomenon; it has contributed to the downfall of many empires. Do some research on historical aspects of the issue and describe the role that desertification played in the collapse of an empire.
c. Identify an area that is undergoing or currently threatened by desertification outside of Asia or Africa. Describe the causes and consequences of desertification in this region. Keep this information in mind, as we'll come back to this choice in a later exercise.

THE PROBLEM

The process of desertification has a host of environmental and social consequences.

Environmental

1. *Soil quality*: Desertification has a profound impact on soils. While it often takes centuries for soils to develop, they can be eroded away quickly once

vegetative cover is removed. Other observed impacts include a reduction in overall soil quality, decreased storage of elements like carbon and nitrogen, loss of soil fertility, and compression and compaction of the soil surface. Irrigation of these soils can lead to an increase in mineral salts at the soil surface as water evaporates.

2. *Air quality and visibility*: Loss of vegetative cover can lead to increased wind erosion, sometimes with far-reaching impacts. Visibility in Beijing, China, can be affected by dust storms originating in distant areas undergoing desertification. Some of this dust can even reach Korea, Japan, and the western US.

3. *Sedimentation and water quality*: Loss of vegetative cover in areas undergoing desertification can lead to soil loading and sedimentation in rivers and reservoirs, often at great distances from the source of the problem.

4. *Hydrology*: Natural vegetation and healthy soils in drylands serve to both store and transport infrequent precipitation. The hard soil crust that typically forms as a result of desertification restricts water infiltration, increasing surface water flow and erosion. Periods of high rainfall can result in devastating floods and landslides.

5. *Biodiversity*: Land degradation severely impacts the ability of vegetation to maintain itself in these fragile environments. Animal species dependent on this vegetation are also affected, forced to either migrate to a more suitable habitat or disappear from the region entirely.

FIGURE 4.4.5
Darfur refugee camp in Chad. *By Mark Knobil (CC BY 2.0) via Wikimedia Commons.*

Societal

1. *Food security*: Loss of productive land as a result of desertification can lead to famine and mass migrations into refugee camps (Fig. 4.4.5) and cities, often contributing to urban sprawl and increased demands on natural resources.

2. *Poverty*: Desertification limits people's ability to provide for their basic needs such as food but also their ability to earn a livelihood. Poverty then drives populations to overexploit the remaining natural resources to survive, triggering a vicious cycle of continued land degradation and poverty. Poverty is thus both a cause and a consequence of desertification.

3. *Political instability*: Mass migration from areas of desertification can contribute to political instability and global security concerns. Harsh conditions for many migrating populations threaten social stability and cultural identity, sometimes contributing to ethnic or religious conflicts.

4. *Human health*: Migrating populations are often forced to reside in appalling conditions in refugee camps or temporary shelters. Other impacts far from the desertification itself have been linked to increased wind-blown dust (Fig. 4.4.6). These include an increase in respiratory conditions like asthma on the Caribbean islands of Trinidad and Barbados linked to dust blowing in from Africa (Schmidt, 2001).

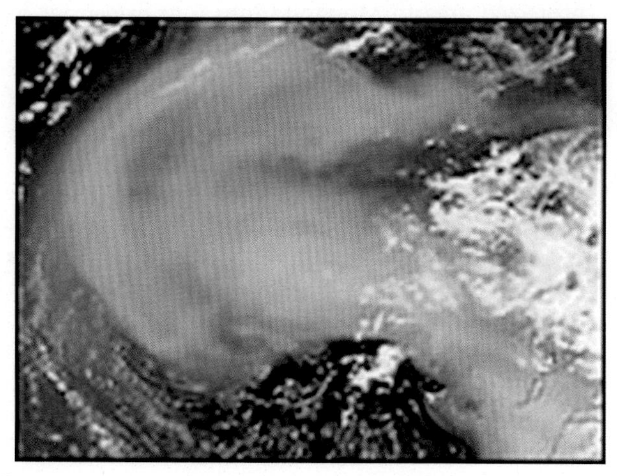

FIGURE 4.4.6
A dust storm sweeps from Africa into the Atlantic. *Source: NASA.*

FAST FACTS ABOUT DESERTIFICATION

...Between 10% and 20% of drylands are already degraded.*
...Between 6 and 12 million km^2 of the globe are affected by desertification.*
...Current arable land loss is estimated to be 30 to 35 times the historical rate.**
...Seventy-four percent of the world's poor are directly affected by land degradation.**
...As many as 50 million people may be displaced by desertification during the next decade.**

Sources: *http://www.greenfacts.org/en/desertification/l-3/3-impacts-desertification.htm#2p0 and **http://www.un.org/en/events/desertificationday/background.shtml.

Environmental Sciences in Action: Now, we'll take a look at several investigations focused on (1) measuring the extent of desertification, (2) quantifying its overall impact, and (3) managing and reversing the process.

Measuring the Extent of Desertification

The best way to understand the true magnitude of desertification is to view it from space. Satellite remote sensing not only allows for an accurate estimation of the true extent of land degradation across vast landscapes, but it also provides the means to continuously monitor change over time. Lamchin et al. (2015) examined rates of desertification over a 20-year period in a local region in Mongolia using the Landsat Enhanced Thematic Mapper (ETM) sensor. Utilizing vegetation indices related to biomass, soil grain size, and surface albedo, the investigators were able to classify each $30\,m^2$ pixel across the Hogno Khaan protected area in Mongolia into four categories based on the degree of desertification, from none to severe.

The overall desertification of the study area was shown to be increasing each year, but the nature of those changes differed among time periods. The desertification map (Fig. 4.4.7) for 1990–2002 shows a decrease in areas of no and low desertification and an increase in areas of high and severe desertification. From 2002 to 2011, the areas of no and severe desertification both increased significantly, while areas of medium and high desertification saw little change.

As you examine Fig. 4.4.7, you'll notice that the extent of severe deterioration (the red color in the figure) in 2002 actually decreased compared to 1990, but it increased again by 2011. The authors suggested that this later increase in severe desertification resulted from wind erosion and sand dune

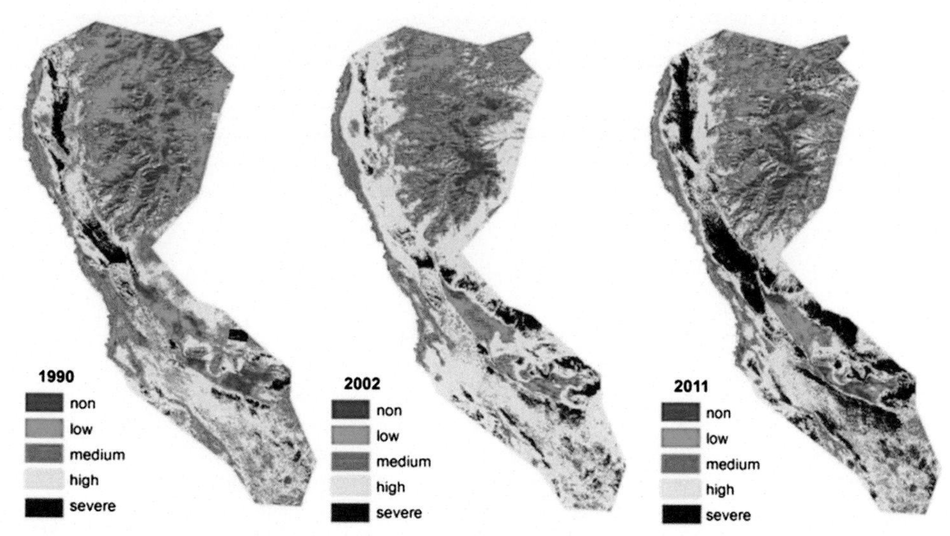

FIGURE 4.4.7
Desertification maps for Hogno Khaan for 1990, 2002, and 2011. *Source: Lamchin et al., 2015.*

movement that led to reduced grassland vegetation and crop growth. A second article discussed below explored in greater depth why such variability may be commonplace.

Xu et al. (2014) used a similar approach based on satellite measurements of surface albedo, a normalized difference vegetation index (NDVI), and an infrared reflectance band to assess the progress of desertification in a farming–pastoral region of North China. Satellite images showed that between 2000 and 2010, the amount of land recovering from desertification (desertification reversion) was only slightly less than the amount of land undergoing desertification (Fig. 4.4.8).

The authors found a large degree of variability in desertification rates. They concluded that climate change and government regulation were the dominant factors leading to desertification reversion. Areas where desertification reversion occurred had experienced increased humidity levels, possibly as a result of warming temperatures, favoring revegetation. In addition, government policies restricting activities that disturb vegetation, such as grazing by livestock, also contributed to the recovery of some areas. In contrast, human activity was the dominant factor that controlled the desertification expansion process between 2000 and 2010.

CRUNCHING THE NUMBERS

Exercise 2: Download the Landsat Lamchin.csv file from your Resource Page. This file contains the mean Landsat index values for each year in the time series considered in the Lamchin et al. (2015) paper. We'll examine these data to better understand the nature of each index and how sensitive each might be as a monitoring tool moving forward.

Start by graphing an overlay time series for each index:

- Which index is the most highly variable over time? Which is the least variable?
- Do any show a clear, consistent trend over time? What is the nature of that trend?
- Seeing BIG changes isn't hard, but often, it is the small, incremental changes over time that can be important to catch early on. Run a *t*-test for differences in each of these indices between 2010 and 2011. You are given the mean and standard deviation. The sample size for this analysis is 928 observations. Based on the significance of the change between 2010 and 2011, which is the most sensitive index?
- Do some research to find out what NDVI, Topsoil Grain Size Index (TGSI), and Albedo indices tell you about the Earth's surface. Based on this, which year in the time series had the most widespread desertification?
- Based on what you know about dryland ecosystems and their management, what might account for the high degree of variability in these data?

SYSTEMS CONNECTIONS

Exercise 3: The take-away message from the Xu et al. study is that a complex combination of factors determines the degree of desertification over relatively small areas and that these drivers can vary from region to region. These connections between driving forces are essential to understand if desertification rates are to be slowed or reversed.

a. Xu et al. concluded that, in their study area, increasing humidity led to an increase in vegetation cover in areas that showed desertification reversion. They attributed this to higher dew points and water-holding capacity of warmer air masses. What are some other anticipated climatic changes that might slow down or speed up the rates of desertification expansion or reversion?
b. Conversely, how might the expansion of desert globally impact the Earth's climate?
c. While we can control land use and management approaches, the climate drivers cannot be altered as easily. What land/atmosphere characteristics might explain the high degree of spatial variability in humidity?

SYSTEMS CONNECTIONS—Cont'd

d. How might this influence where restoration or prevention efforts should be focused?
e. How might this spatial variability present opportunities or obstacles for government agencies hoping to limit desertification?

PUTTING SCIENCE TO WORK

Exercise 4: While many of the potential solutions to desertification lie in good government regulation and land management, how and where to implement these actions need to be informed by science to be most effective.

Assume that you're an official concerned about the spread of desertification in the Sahel region of Africa. You've just read the article by Xu et al. and are wondering if climate change is also affecting conditions in your region and if you might have regions where particular management strategies might be most effective.

- Design a field experiment that might help you examine the spatial variability in desertification rates in your region. This should also include an assessment of potential drivers of desertification at each of your sites.
- Are there social, political, or economic constraints that might affect your chances of implementing the necessary steps in some areas? How might these be incorporated into your field study?

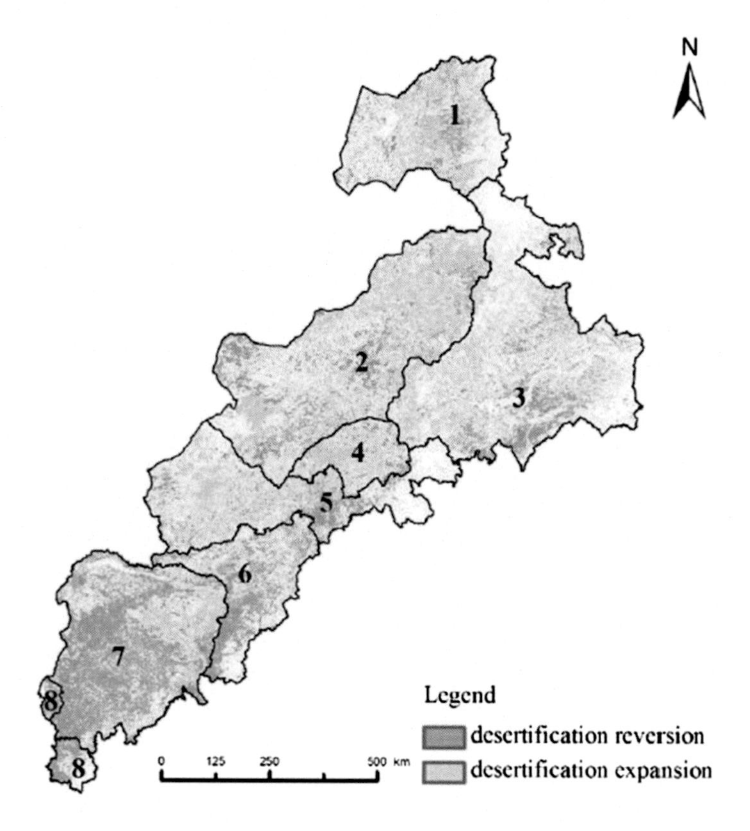

FIGURE 4.4.8

The spatial distribution of lands that experienced desertification reversion and expansion between 2000 and 2010 in the farming–pastoral region of North China. *Source: Xu et al. 2014.*

EXPLORE IT

Exercise 5: So far, we have focused on how poor land management has contributed to desertification. The Aral Sea presents an interesting contrast in that water management was the primary instigator of desertification for the region. We'll use Google Earth to explore how the loss of water has contributed to desertification in the Aral Sea region (Fig. 4.4.9).

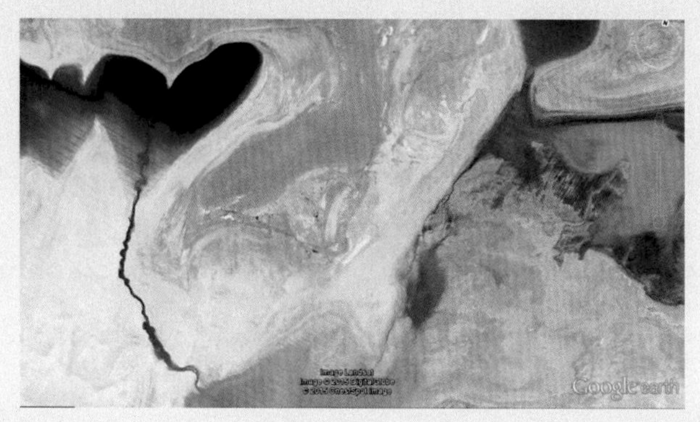

FIGURE 4.4.9
The narrow channel connecting what remains of the two halves of the Aral Sea. *Source: Google Earth.*

Start by opening Google Earth and navigating to the Aral Sea.

- Click on the historical imagery tab so that you can explore changes in the water level over time. Move back in time to the earliest available imagery in the Google Earth timeline. Using the Distance tool, measure the approximate width (east to west) of the extent of the water (in km).
- Examine each of the imagery dates over time. While small portions of the region update imagery sporadically, imagery over the full sea is available for December 1983, December 1986, December 1999, December 2004, September 2006, and, most recently, April 2013 (with the historical imagery tab off). At what point did the Aral Sea effectively become two much smaller bodies of water connected by a narrow channel?
- Only the western body of water is effectively deep enough to be navigable. What is the approximate width of this water body at its widest?
- Calculate the annual rate of width reduction between your earliest measurement and the 2013 measurement. Based on this rate of shrinkage, how many more years will it take for the Aral Sea to cease to exist?

Rocky Desertification in Southwest China

Similar to the Aral Sea, there are other examples of desertification that don't have poor agricultural land use as a leading cause. In this case, karst deposits, which are buried or exposed carbonate rock, are involved. Such deposits are characterized by frequent exposure of the underlying rock, a shallow soil cover, and poor water-holding capacity. The poor soils are susceptible to erosion and rapid water seepage, allowing only sporadic plant communities to develop. Any disturbance to these fragile areas can lead to rapid degradation.

Steep, rocky landscapes with thin soil cover are fragile and vulnerable to dramatic desertification (Fig. 4.4.10). This most often occurs as a result of deforestation (Jiang et al., 2014) in areas with high population density and low availability of alternate sources of fuel.

China has approximately 3.44 million km^2 of karst areas, about 36% of its total land, representing 15.6% of all the karst areas in the world. Of China's total, about 0.51 million km^2 are currently exposed/outcropped carbonate rock areas. Jiang et al. (2014) found that rocky desertification affected 35.6% of the exposed carbonate rock areas in Yunnan province in 2000 and averaged about 22% in the region (Fig. 4.4.11).

Quantifying the Impacts of Desertification

We've already summarized some of the most serious environmental and social effects of desertification. Let's look at another important impact. In arid and semiarid ecosystems, desertification is characterized by wind-driven soil erosion, which transports soils and the nutrients they hold far from their source. A study by Tang et al. (2015) focused on Yanchi County in Ningxia, China, an area characterized by land degradation and severe wind erosion during dry conditions common from April to mid-June.

The scientists measured carbon (C) and nitrogen (N) concentrations in plants and soils across five different desertification severity classes. They found that C and N storage in plants and soil decreased with increasing severity of desertification. Lands suffering from very severe desertification held 60.9% and 76.5% less C and N, respectively, than areas with no desertification (Fig. 4.4.12). This represents a dramatic decrease in the overall fertility and production capacity of these areas. Not only do such losses reduce the productivity of affected

FIGURE 4.4.10
An example of rocky desertification in Southwest China. *Source: Jiang et al., 2014.*

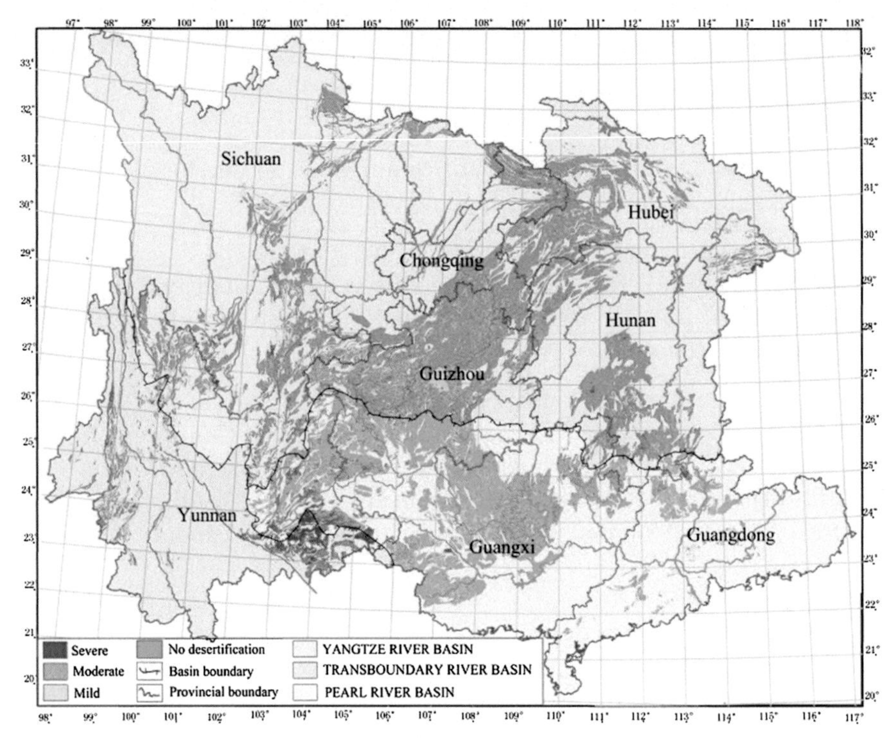

FIGURE 4.4.11

Distribution and classification of rocky desertification areas in Southwest China. *Source: Jiang et al., 2014.*

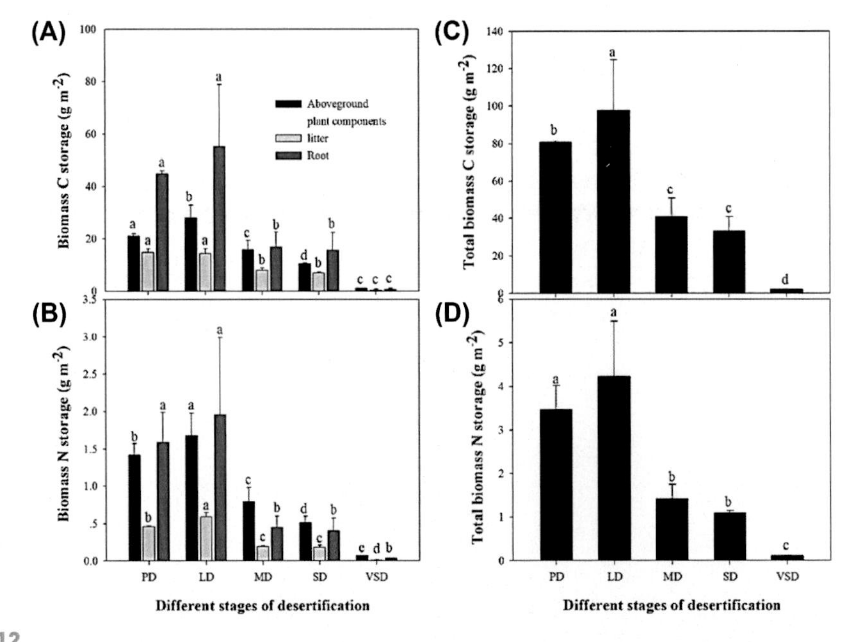

FIGURE 4.4.12

Changes in soil C and N storage in the desert steppe ecosystem. (A) C storage; (B) N storage; (C) C loss; (D) N loss. *LD*, light desertification; *MD*, moderate desertification; *PD*, potential desertification; *SD*, severe desertification; *VSD*, very severe desertification. Different lowercase letters indicate variations significant at the 0.05 level. *Source: Tang et al., 2015.*

lands, the release of CO_2 to the atmosphere resulting from desertification can be substantial. The loss of vegetation and reduction in soil organic matter in areas experiencing rocky desertification reduce the uptake and storage of atmospheric CO_2, adding to the greenhouse gas burden.

Systems Connections: Two important concepts in the environmental sciences are feedback loops and ecosystems services. Both come into play with desertification. An environmental feedback loop is a circular process that results when the output of one occurrence, climate change for instance, speeds up (a positive feedback) or slows down (a negative feedback) the original process. In the case of desertification, both positive and negative feedback loops can be at work.

For example, the research of Xu et al. suggested that in some areas, increased humidity resulting from a warming atmosphere could support the growth of more vegetation. This increased plant growth would, in turn, absorb more atmospheric CO_2, slowing climate change and representing a negative feedback loop.

The work of Tang et al. illustrates a positive feedback loop. As desertification proceeds and soil quality declines, a reduction in amounts of vegetative cover reduces the amount of greenhouse gases absorbed by plants, increasing concentrations in the atmosphere, speeding climate change, which, in turn, can hasten the process of desertification. The co-occurrence of both types of feedback mechanisms complicates efforts to effectively manage desertification.

A second important connection exists between ecosystems affected by desertification and the human populations that are part of these ecosystems. Dryland human populations rely on ecosystems for food production, water supply, and the growth of wild plants that can serve as alternate medicines. These are classic examples of ecosystem services. As desertification proceeds, these critically important services may be lost.

CRUNCHING THE NUMBERS

Exercise 6: While Tang et al. (2015) found significant variations in carbon and nitrogen storage, depending on the degree of desertification, the two elements are impacted differently by the same process. Review the Tang et al. article linked on your Resource Page.

- After comparing the nature of the differences between the two elements due to desertification, which element is impacted earlier in the desertification process? What about carbon and nitrogen cycles might explain this?
- Arid regions cover roughly a quarter of the global landmass (148.94 million km^2), and an estimated 70% of these dryland areas currently experience moderate to severe desertification (UNCCD, 2007). Using Tang et al.'s (2015) Fig. 4.4.12 to estimate the kg of carbon lost from those areas designated as having "severe desertification," calculate the total CO_2 released into the atmosphere from the degradation of these areas.

Continued

CRUNCHING THE NUMBERS—Cont'd

- Look online to find yearly CO_2 emissions for various countries. How does the impact of desertification compare to emissions from automobiles and industry? What are the implications of your findings for global climate change?

SOLUTIONS

While the widespread expansion of desertification seems daunting, there are many steps that can be taken to protect drylands from desertification and even reverse degradation when it is ongoing. The United Nations Convention to Combat Desertification (http://www.un.org/en/events/desertificationday/pastobs.shtml) lists a number of steps that can reduce desertification:

- integrating land and water management to protect soils from erosion and salinization;
- protecting the vegetative cover, including replanting and reforesting drylands to stabilize soil;
- stabilizing soil by using sand fences, shelter belts, etc.;
- managing water use, reusing treated water, and harvesting rainwater;
- farmer managed natural regeneration (FMNR): selective pruning of shrub shoots and using the residue to mulch fields;
- integrating the use of grazing and farming to enrich soil with nutrients within agricultural systems; and
- turning to alternative livelihoods that are less demanding on natural resources yet provide sustainable income.

While many of these seem like logical steps, putting them into practice can be challenging. Effective restoration and rehabilitation of desertified drylands require a combination of policies and technologies and the close involvement of local communities. Let's take an in-depth look at some activities at the local, regional, and global level to combat desertification.

Local Initiatives: Reusing Wastewater

Barbosa et al. (2015) reviewed studies in which wastewater was used to irrigate fiber crops to help restore water-starved areas that have been affected by desertification.

Certain grass fiber crops, including maiden grass, bamboo, soft rush, and papyrus, have the potential to remove pollutants from wastewater, produce high biomass yields, stabilize soils, store carbon, and restore degraded lands.

Using wastewater to irrigate fiber crops carries considerable risks, however. The presence of a variety of potentially harmful pollutants, such as heavy metals and persistent organic substances, depending upon the type of wastewater, might pose a risk to the plants themselves, or they might migrate through the soil into the groundwater. Any pathogens present in the wastewater could also pose a human health risk. While there is a great potential to use wastewater to help reverse desertification, care must be taken to avoid the negative impacts or negative perception of its use.

Regional Initiatives: Great Green Walls

There have been large-scale regional attempts to slow the progress of desertification. In 1978, in response to growing desertification resulting primarily from overgrazing and logging, China initiated the world's largest tree planting project. Trees have been planted manually as well as by air dropping millions of seeds in capsules. By 2014, 66 billion trees had been planted by citizens. By 2050, this "green wall" is expected to stretch 4500 km along the edge of China's northern deserts. To see more about this massive tree planning initiative, view the videos linked on your Resource Page, including the BBC report at https://www.youtube.com/watch?v=VS-v0b8GkFs.

DIGGING DEEPER

Exercise 7: In addition to the massive nature of the project, scientists and conservation groups are concerned about the long-term viability of the program. Do some digging into China's green wall project. While some of the results have been very impressive, there have been a number of problems with the approach, implementation, and unintended consequences of the green wall construction.

- What are some of the challenges that have occurred while creating this green wall?
- How are recent efforts attempting to correct past problems?

A second "great green wall" project is underway in Sub-Saharan Africa. Originally envisioning a line of trees from east to west through the Sub-Saharan desert in 2007, African heads of state agreed to support a "Great Green Wall for the Sahara and the Sahel Initiative." Now backed by many regional and international organizations, including several United Nations programs and the World Bank, this initiative spans different countries and cultures, each requiring its own host of management steps. In southern Niger

(Fig. 4.4.13), for example, farmers have rehabilitated more 5 million ha of land using specific pruning and mulching techniques outlined by FMNR (FAO, 2013). In Senegal, 11 million trees have been planted on 27,000 ha of degraded land. With 83% of the rural population in Sub-Saharan Africa depending on the land for their livelihoods, the various programs supported by the Green Wall initiative are also playing a critical role in the stability of the region.

SOLUTIONS

Exercise 8: Planting trees in dryland regions to combat desertification is not a new idea. Some of the most famous large-scale projects include the Algerian Green Dam project, which, as indicated in published reports and findings, met with some success, but also quite a few failures, resulting in major project changes. The best approach to any new project, particularly for those as big as the transnational Great Green Wall in Africa, is to begin by examining lessons learned from previous efforts.

From your Resource Page, download and review the African Great Green Wall project pdf fact sheet (http://www.csf-desertification.eu/combating-desertification/item/the-african-great-green-wall-project) (Bellenfontaine et al., 2011).

Considering what you've learned about the Great Green Wall project, go back to Exercise 1 to your choice of another part of the world other than China or Africa that is suffering from desertification. Assume that you've been given an unlimited budget to stem the tide of desertification in your region. Develop a two-page proposal detailing:

a. the size of area threatened by desertification;
b. the causes of the problem;
c. specific solutions appropriate for your region designed to reverse the process;
d. the role of various groups in achieving your program; and
e. any problems you'd likely encounter.

Global Initiatives: Grazing Livestock

One of the activities that is often cited as a primary cause of desertification is the overgrazing of livestock on grasslands. Not everyone agrees, however. Dr. Allan Savory, an ecologist by training, believes that using properly managed livestock grazing can reverse desertification. He cites several examples of areas where his holistic management has led to dramatic reversals of desertification.

CRITICAL THINKING

Exercise 9: The jury is still out on livestock grazing as either a catastrophic cause of, or potential solution to, desertification. Let's examine the case both for and against grazing.

Start by watching Dr. Savory's TED talk on grazing and desertification linked on your Resource Page and answer the following:

- Exactly what does Savory advocate for in his "holistic management and planned grazing" approach?
- How does he believe that his recommendations would reverse desertification?
- What evidence does he cite to support these claims?

Now read some of the viewpoints of those who disagree with Savory's approach. Links to a 2013 Wildlife News blog by Ralph Maughan and a 2015 review article in the International Journal of Biodiversity by Carter et al. are available on your Resource Page to get you started.

- What are the major rebuttals to Savory's approach to grazing?
- What evidence do they cite to support their claims?
- What other studies can you find that show the detrimental impacts of any grazing in areas at risk of desertification?

Who do you think makes the stronger case: Savory or his opponents? Why?

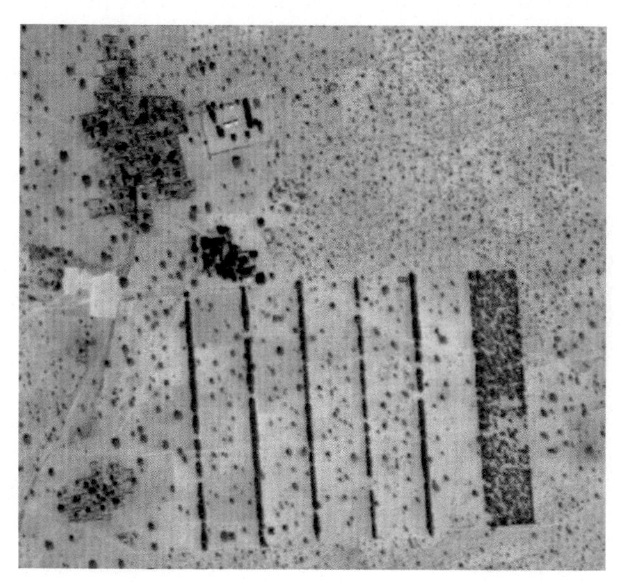

FIGURE 4.4.13

Reforestation outside Maradi, Niger. *Source: Google maps.*

LINKS BETWEEN DESERTIFICATION, CLIMATE CHANGE, AND GLOBAL SECURITY

Globally, drylands are home to one-third of the human population. Some 10–20% of drylands are already degraded, with desertification threatening much of the remainder. Therefore, desertification is one of the greatest environmental challenges today and a major barrier to meeting basic human needs in drylands.

Desertification and land degradation also carry with them implications for global security. As once-useful lands lose their ability to support human populations, entire societies may be forced to uproot and migrate, potentially displacing those living on still productive lands and adding to the estimated 25 million or more people already identified as climate refugees (Fig. 4.4.14).

Climate change also plays a role. A recent study by Kelley et al. (2015) suggests that the drought in Syria from 2007 to 2010 is consistent with climate change predictions for the region. The severe drought sped the process of desertification, decreasing agricultural output and contributing to the political unrest in the country.

IT'S DEVELOPED COUNTRIES TOO!

While much of this case study has focused on desertification in some of the poorest regions of the world, developed nations are not immune. The 1930s Dust Bowl in the US Great Plains remains as a stark reminder of what the combination of drylands, drought, and unsustainable farming practices can lead to.

FIGURE 4.4.14
Bread flour being distributed to mothers and children in Dakhla refugee camp in southwestern Algeria. *By US Mission/Rome Humanitarian Attaché's visit to Algeria's Saharawi refugee camps via Wikimedia Commons.*

Because of low population densities and the availability of alternate food sources in vulnerable parts of the western US, we've avoided some of the catastrophic effects of desertification seen elsewhere in the world. However, even today, parts of 17 western US states remain at risk of desertification (Fig. 4.4.15).

With prolonged droughts expected to become commonplace as the climate changes, we must pay attention to vulnerable areas in developed nations and follow sustainable land management practices in these areas.

FIGURE 4.4.15
Desertification of arid grassland near Tucson, Arizona, 1902 to 2003. *By Rob Wu (CCSP SAP 4.3).*

CONSIDER THIS

Exercise 10: Natural deserts are actually healthy, functioning ecosystems that contribute to the Earth's diversity and wealth. How can you reconcile the concept of deserts as a valuable natural ecosystem with the current expansion of desertification and global efforts to combat its spread?

REFERENCES

Barbosa, B., Costa, J., Fernando, A.L., Papazoglou, E.G., 2015. Wastewater reuse for fiber crops cultivation as a strategy to mitigate desertification. Ind. Crops Prod. 68, 17–23.

Bellefontaine, R., Bernoux, M., Bonnet, B., Cornet, A., Cudennec, C., D'Aquino, P., Droy, I., Jauffret, S., Leroy, M., Malagnoux, M., Réquier-Desjardins, M., 2011. The African Great Green Wall Project: What Advice Can Scientists Provide? Comité Scientifi que Français de la Désertification (CSFD). www.csf-desertification.org/great-green-wall.

Food and Agriculture Organization of the United Nations, 2013. Africa's Great Green Wall Reaches Out to New Partners. http://www.fao.org/news/story/en/item/210852/icode/.

Jiang, Z., Lian, Y., Qin, X., 2014. Rocky desertification in Southwest China: impacts, causes, and restoration. Earth-Sci. Rev. 132, 1–12.

Kelley, C.P., Mohtadi, S., Cane, M.A., Seager, R., Kushnir, Y., 2015. Climate change in the Fertile Crescent and implications of the recent Syrian drought. Proc. Natl. Acad. Sci. U.S.A. 112 (11), 3241–3246.

Lamchin, M., Lee, J.-Y., Lee, W.-K., Lee, E.-J., Kim, M., Lim, C.-H., Choi, H.-A., Kim, S.-R., 2015. Assessment of land cover change and desertification using remote sensing technology in a local region of Mongolia. Adv. Space Res. http://dx.doi.org/10.1016/j.asr.2015.10.006.

Millennium Ecosystem Assessment, 2005. Ecosystems and Human Well-being: Desertification Synthesis. World Resources Institute, Washington, DC.

Schmidt, L.J., 2001. When the Dust Settles. NASA Earth Observatory. http://earthobservatory.nasa.gov/Features/Dust/.

Tang, Z.-S., An, H., Shangguan, Z.-P., 2015. The impact of desertification on carbon and nitrogen storage in the desert steppe ecosystem. Ecol. Eng. 84, 92–99.

United Nations Convention to Combat Desertification, June 17, 2007. World Day to Combat Desertification. http://www.un.org/en/events/desertificationday/pastobs.shtml.

Xu, D., Li, C., Song, X., Ren, H., 2014. The dynamics of desertification in the farming-pastoral region of North China over the past 10 years and their relationship to climate change and human activity. Catena 123, 11–22.

Chapter 4.5

Amazon Rainforest

FIGURE 4.5.1
Aerial view of the Amazon rainforest near Manaus, the capital of the Brazilian state of Amazonas. *By Neil Palmer/CIAT (CC BY-SA 2.0), via Wikimedia Commons.*

OVERVIEW

The Amazon's tropical rainforest (Fig. 4.5.1) has been called the "lungs of the world." In addition to providing habitat for an amazing array of plants and animals, the rainforest is so immense that it has an important influence on the region's weather as well as the global climate. Because of this influence, the health of the Amazon is directly linked to the health of the planet. Unfortunately, this rainforest has been disappearing, and it continues to be threatened by a wide variety of human activities.

DIGGING DEEPER

Exercise 1: Environmental issues come and go from the media spotlight. Deforestation of the rainforest was on the world's radar screen in the 1980s when development in the nine countries it spans was booming at the expense of the region's forests. Since then, deforestation has remained a significant threat, even if overshadowed by other environmental issues.

Before you get started, visit one of the web sites linked on your Resource Page to get some additional background on the ecology of the Amazon basin and the history of deforestation in the region.

DIGGING DEEPER—Cont'd

- Describe some of the unique ecological characteristics of the Amazon rainforest.
- What are some of the primary threats to efforts to protect the Amazon rainforest?
- How have these threats changed over the past 30 years?
- What are some of the potential consequences (local, regional, and global) of the loss of the Amazon rainforest?

In this case study, we'll review some of the basics of the Amazon rainforest and identify some of the major stressors affecting this ecosystem. We'll also consider some of the approaches designed to help protect and manage this vital global resource. An important feature of this case study will be an assignment tasking you with analyzing the forest–climate–fire feedback loops currently at play in the Amazon and drafting a policy statement detailing the best approach for protecting the forest, given the realities of the complex relationships and feedbacks among these stress agents.

BACKGROUND

The Amazon River basin is home to the largest rainforest on Earth. Fifteen million years ago, this river actually flowed west, but with the rise of the Andes Mountains, the Amazon become a vast inland sea that gradually gave way to a massive swamp and lake. Marine inhabitants of that time adapted to life in freshwater. For example, more than 20 species of stingrays, most closely related to those found in the Pacific Ocean, can be found today in the fresh waters of the Amazon. But this is not the only reason for the rich species diversity that exists in the Amazon.

Tropical rainforests across the globe are characterized by relatively stable warm climates and diverse vegetation structure (Fig. 4.5.2) that have allowed for the evolution of many highly specialized species. In addition to its vast biodiversity, the Amazon rainforest is also an important sink for carbon. Saatchi et al. (2013) estimated that the total biomass of the Amazon basin contained about 86 billion metric tons of carbon in 2007. At the time of the study, the scientists estimated that this represented at least 11 years' worth of global CO_2 emissions.

Fast facts about the Amazon basin: One need look at only a few of the following facts about the Amazon basin and its tropical rainforest to appreciate its unique position among the globe's ecosystems.

...The basin accounts for about one-third of the tropical forest acreage on the globe, covering more than 5.5 million km^2 (Fig. 4.5.3). This is roughly the same size as the continental US.

...It covers parts of nine South American countries: Brazil, Venezuela, Columbia, Ecuador, Bolivia, French Guiana, Guyana, Suriname, and Peru.

FIGURE 4.5.2

Rainforests are characterized by a complex mosaic of ground vegetation, shrubs, understory, canopy, and emergent trees. *CSIRO (CC BY 3.0) via Wikimedia Commons.*

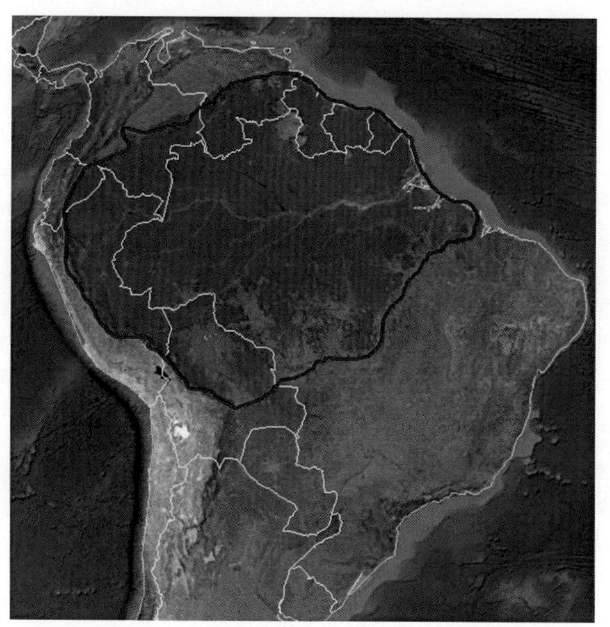

FIGURE 4.5.3

The Amazon River basin (red) and national boundaries it encompasses (*yellow*). *Source: Google Earth.*

...The Amazon River contains about 20% of the volume of the world's rivers. For reference, the Amazon's daily freshwater discharge into the Atlantic is enough to supply New York City's freshwater needs for 9 years.
...Vegetation in the Amazon rainforest, a moist, broadleaf forest, releases about 20% of the globe's oxygen.
...About one-half of the world's species of plants and animals are native to the Amazon rainforest, with more than 60,000 species of plants alone.

In addition to providing habitat for a broad array of plants and animals, the Amazon rainforest provides humans with a variety of ecosystem services, just a few of which include:

...aesthetics and ecotourism
...tropical fruits, with as many as 200 used in the Western world, and nuts
...ingredients for about one quarter of all pharmaceuticals
...stabilization of the global climate by absorbing vast quantities of CO_2 from the atmosphere

THE PROBLEM

About one-fifth of the Amazon rainforest has already been deforested. But this isn't simply driven by global demand for wood or illegal logging. The threats to the Amazon rainforest are much more complex (Fig. 4.5.4). We'll examine some of the biggest problems as well as some possible solutions.

a. *Cattle ranching*: Clearing forest land to raise beef is the leading cause of deforestation in the Amazon. About 60% of deforested land in the Amazon basin ends up as cattle pasture, with Brazil the world's top beef exporter. Pasturing in Amazonia isn't very efficient, as each cow requires about 1 ha of land. Once a pasture is "grazed off," a new area must be deforested to support the cattle (Marull, 2013).

   Solutions: Some sustainable cattle grazing is now being practiced in the Amazon rainforest. By coupling field rotation with better pasture quality, farmers can avoid deforestation. One farmer practicing a more sustainable approach has been able to graze 2.3 head of cattle per hectare (Marull, 2013).

CRUNCHING THE NUMBERS

Exercise 2: Let's put this into perspective with some "back-of-the-envelope" calculations. Consider the following:

- Assume each cow requires approximately 1 ha of pasture to grow to maturity in the Amazon and that rainforest was cleared to provide all pasture land.

CRUNCHING THE NUMBERS—Cont'd

- In Brazil, there were 49.9 million head of cattle in 2014 (source: indexmundi.com).
- A standard football playing field in the US is approximately 53 yards by 100 yards.

Based on this information, how many American football fields worth of rainforest were cleared to accommodate Brazilian cattle in 2014?

How many American football fields would have to be cleared per day to accommodate this demand for pasture? Per hour? Per second?

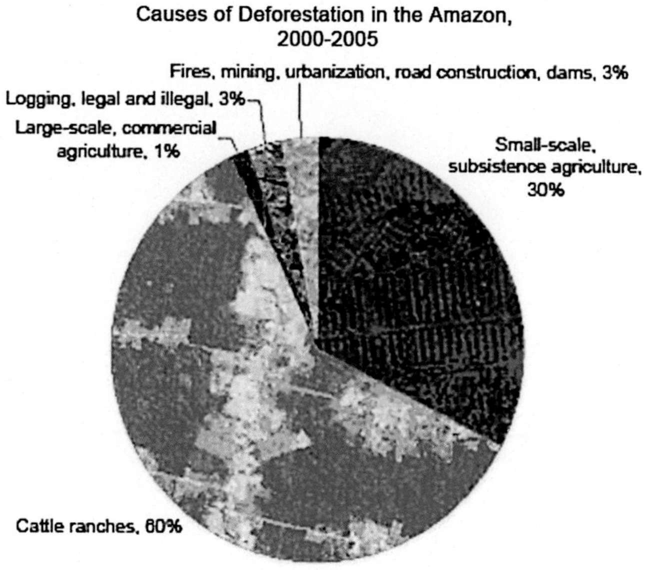

FIGURE 4.5.4
The primary drivers of deforestation in the Amazon between 2000 and 2005. *Source: Rhett A. Butler, Mongabay.com.*

b. ***Industrial logging***: Timber, pulp, and wood fiber are used to create building materials and consumer products, and rainforest species have previously been in high demand because of their large size, wood strength, appearance, and resistance to rot. Logging tropical hardwoods, done legally or illegally, does more damage than simply removing the trees.

   Logged-over forests (Fig. 4.5.5) are vulnerable to fire, loss of biodiversity, and conversion to pasture or crop land. Soils may be compacted and eroded on steep terrain and overall soil nutrient content depleted. In addition, illegal dirt roads built to access logging areas serve as conduits for others to penetrate even more deeply into once undisturbed areas

FIGURE 4.5.5
Industrial logging in the Amazon. *By Wilson Dias/Agência Brasil (CC BY 3.0) via Wikimedia Commons.*

of forest. Deforestation also has the net effect of adding CO_2 to the atmosphere in two ways: less uptake of CO_2 from the atmosphere and greater release into the atmosphere (rapidly if trees are burned, slowly if they decompose naturally).

Solutions: Brazil has begun to introduce electronic logging certificates and is using satellite technology and remote sensing to monitor illegal logging operations. However, the challenge of tracking illegal activities in an area as large as the Amazon rainforest across nine different national borders cannot not be overstated. Conservation groups such as the Rainforest Alliance have also created a Forest Product Certification program to verify that wood products purchased from the Amazon are compliant with sustainable forest management practices.

c. ***Subsistence farming***: Unlike large-scale industrial activities, subsistence farmers typically use a smaller-scale slash-and-burn approach to clear

FIGURE 4.5.6
Typically, land is cleared and then burned to promote the growth of grasses for pasture. *Source: USFS via Wikimedia Commons.*

land to grow crops to support themselves and their families. Because the soils in the Amazon rainforest are so low in nutrients, the burned vegetation (Fig. 4.5.6) initially helps fertilize the soils to support crop growth, but is quickly depleted. Once the nutrient levels drop, the farmers move on to clear another plot.

Solutions: Helping small farmers raise and harvest items such as fruits and nuts that don't require deforestation has been successful in slowing the amount of forest falling to slash-and-burn agriculture. Also, educating farmers about how to maintain soil health and reduce water use for irrigation can increase yields and reduce the need for clearing additional lands.

DIGGING DEEPER

Exercise 3: Another major agricultural product in the Amazon is palm oil. Do some exploring to find out what impact growing oil palms is having on the rainforest and describe one approach being used in Brazil to minimize the impact of oil palm on the forest.

d. ***Agribusiness***: Brazil is the world's second leading grower of soybeans after the US. The bean's use as a feedstock for biofuel production has helped stimulate the surge in production. Soybean growers take over land that has been deforested for cattle ranches. Having exhausted the landscape, ranchers move on to clear new forest for additional ranches.

   Solutions: In 2006, a moratorium on buying soybeans grown on deforested land helped slow the deforestation associated with this crop. While not all countries have signed the moratorium, it has slowed the rate of deforestation in the Amazon.

e. ***Mineral mining***: The Amazon basin has substantial mineral assets. Deposits are scattered throughout river channels and floodplains of the Amazon rainforest. To get at the gravel deposits where minerals may lie, rainforest is cleared (Fig. 4.5.7). Additional forest is removed to construct roads to the mining sites and to provide housing for miners. A secondary, well-publicized threat associated with Amazon mining is the release of mercury and other harmful pollutants into the Amazon River. More than a kilogram of mercury is required to extract a similar amount of gold.

SOLUTIONS

Exercise 4: Mining represents a complex problem with no easy solution. For example, tax incentives are often provided by governments for large-scale projects to boost development. Even where mining is restricted, many deposits are located in "low-governance" regions, where minimal monitoring or enforcement is possible to limit illegal mining. In this exercise, you will lay out a host of possible actions to help mitigate the extent and impact of mining activities in the Amazon.

- Make a list of concrete actions that could help address specific aspects of the mining problem. For example, this might include activities designed to locate illegal mines, alter harmful practices in legal mines, strengthen economies of local populations, and support reclamation efforts.
- Do some research to see what ongoing activities are in place to reduce the impact of mining in the Amazon or in other rainforest biomes.

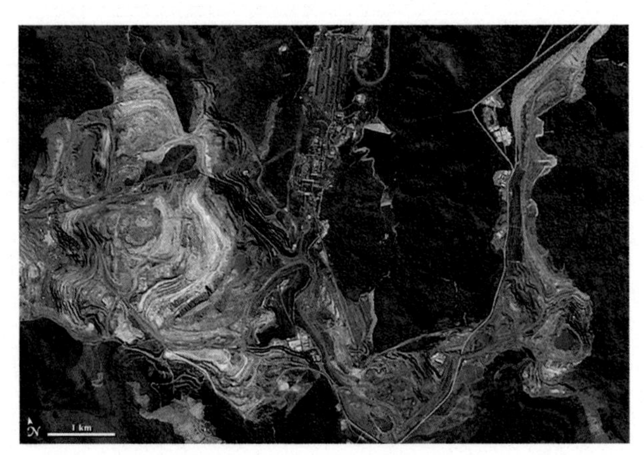

FIGURE 4.5.7
Areas which were once covered with lush rainforest have been turned into barren and toxic wastelands as a result of gold mining. *Source: NASA.*

Feedback Loops and the Future of the Forest: Here we've shown how many different activities represent ongoing threats to the health of the Amazon rainforest, with nearly 20% of the forest cut down over the past 40 years (Wallace, 2007). There are a number of other stressors threatening the Amazon rainforest that we have not covered, including construction of reservoirs which flood portions of the forest, construction of oil and gas pipelines and power lines, and development of various industrial projects. But the Amazon is a complicated ecosystem, with a complex series of ecological feedback loops that may represent an even greater challenge currently facing the Amazon rainforest: climate change, drought, and fires.

IMPORTANT CONSIDERATIONS

1. A healthy Amazon rainforest has impressive effects on the local and regional climate. Unlike the case for most ecosystems, fully one-half of the rain falling on the forest originates from water vapor released by the rainforest itself via evapotranspiration. Rainfall in adjacent areas to the south and to the east of the Andes also comes from the forest. As forests are cleared, precipitation can be significantly reduced.

2. Climate change is affecting the Amazon rainforest. Recent years (2005, 2010) have seen major droughts in the Amazon basin. Some scientists believe that warming ocean temperatures associated with climate change played an important role in the 2005 drought. Droughts deplete soil moisture, eventually leading to increased tree mortality and a greater risk of fire (Saatchi et al., 2007). Climate models suggest that extreme droughts in the Amazon Basin may increase as the climate continues to change.

3. Fires are an important issue in the Amazon rainforest, and the Amazon basin is a hotspot for human-caused biomass burning, accounting for 15% of total global fire-related emissions (Mishra et al., 2015).

CRUNCHING THE NUMBERS

Exercise 5: Mishra et al. (2015) used satellite imagery to map smoke (aerosol optical depth, AOD) and two fire products: fire radiative power and fire counts. Tracking the spatial extent of these three indices over 12 years shows interesting temporal patterns (Fig. 4.5.8).

Examine the temporal trends in the fire counts and AOD to answer the following questions:

- When are fires most common during the year?
- How did fire frequency change over the 10-year study period?
- Compare the fire counts to AOD levels at the top of Fig. 4.5.8. What does this tell you about the impact of fire on air quality in the region?

CRUNCHING THE NUMBERS—Cont'd

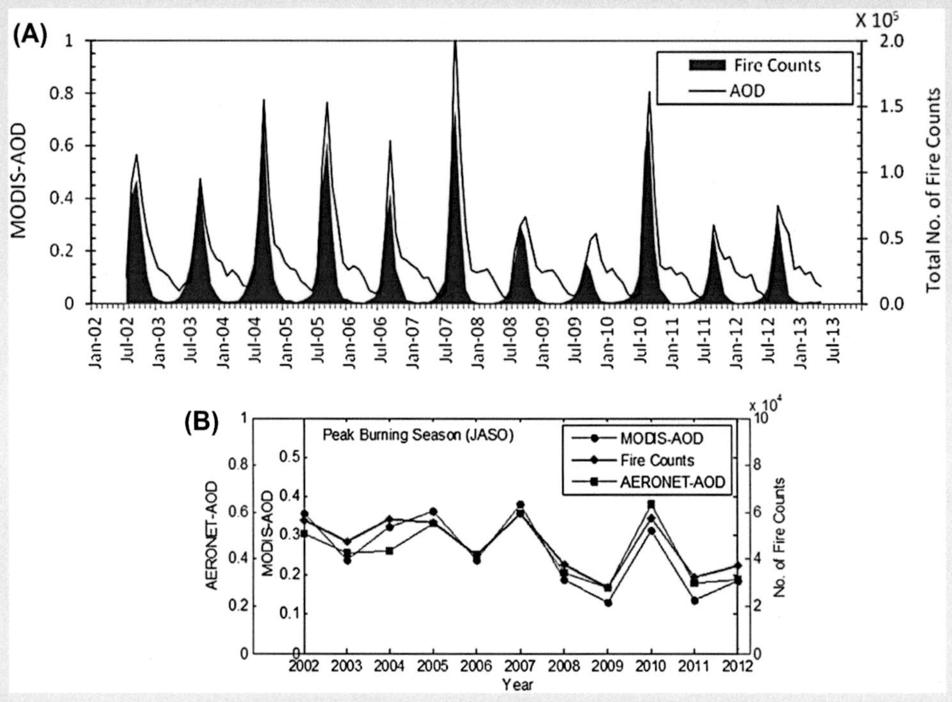

FIGURE 4.5.8

(a) Monthly time series of mean of the moderate resolution imaging spectroradiometer (MODIS)-aerosol optical depth (AOD)550 (*left axis, black*) and total number of MODIS-fire counts (FC) (*right axis, red*) over the Amazon; and (b) Mean MODIS AOD550 (*left axis, black*), mean AERONET-AOD500 (*left axis, blue*), and total MODIS-FC (*right axis, red*) for the ROI and peak biomass burning season (July–October) plotted as function of year. *Source: Mishra et al. 2015.*

EXPLORE IT

Exercise 6: We don't often think of fire as a problem in tropical rainforests compared to drier regions. Let's explore how the frequency of fires differs across the globe to get a sense of how common fires are in the Amazon.

NASA has created a nearly real-time web-mapping product to locate active fires across the globe using moderate resolution imaging spectroradiometer (MODIS) satellite imagery. Using satellites to identify fires is not complicated,

EXPLORE IT—Cont'd

but what is impressive is the turnaround. The NASA Fire Information for Resource Management System (FIRMS) makes active fire data available within 3 h of the satellite overpass.

- Navigate to the FIRMS web mapper linked on your Resource Page (https://firms.modaps.eosdis.nasa.gov/firemap/) (Fig. 4.5.9).
- By default, the map viewer shows you all active fires globally within the past 48 h. Change the settings so that you can see the extent of fires over the past 7 days.
- How does the number of fires in the Amazon compare to drier regions such as the US Southwest or southern India?
- Zoom in to the Amazon region and then go to the LAYERS tab on the right, and select PROTECTED AREAS. Continue to zoom and pan around the high fire density areas. Are any fires located within protected areas? What does fire activity in protected areas indicate?
- Are there any regions where illegal land clearing and fires are more common across the Amazon?
- How might this type of monitoring fit into management strategies to protect the rainforests in the Amazon?

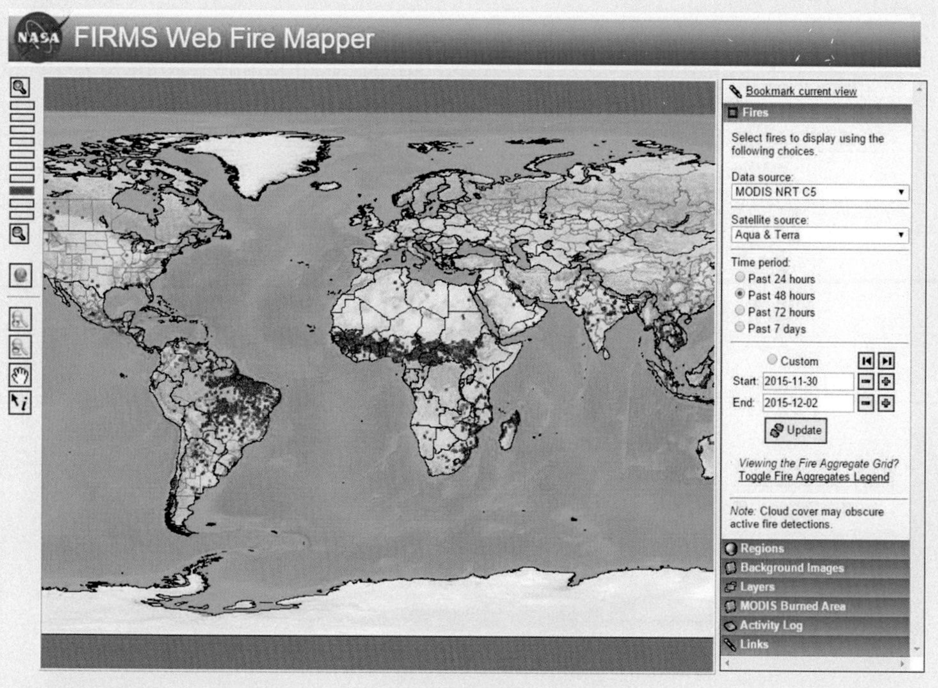

FIGURE 4.5.9
The Fire Information for Resource Management System (FIRMS) web mapper portal. *Source: NASA.*

Environmental Sciences in the Field: Fires in the Amazon (Fig. 4.5.10) differ from those in drier areas like the western US where wildfires are common and ecosystems adapted to fire. In the Amazon, the long-used practice of slash-and-burn for subsistence farming results in a much different pattern of fire across the landscape. Slash-and-burn fires tend to be small, leaving lots of dead wood standing (Fig. 4.5.11). When those dead trees fall, it creates gaps in the tree canopy. Entering sunlight dries out the understory, setting the stage for a new round of larger, more destructive fires (Lindsey, 2004).

A team of scientists from the Woods Hole Research Center in the US looked at 10 half-hectare forest plots in the eastern Amazon basin before and after a drought in 1998. Eight plots had experienced fire previously and two had not. While 23% of the previously unburned forest burned during the drought, the rate increased to 39% for forests burned once before and 69% for forests burned three times previously, indicating that fire predisposed the forests to additional fires (Lindsey, 2004).

Another study led by D. Morton of NASA's Goddard Space Flight Center (Morton et al., 2013) focused on less visible but very important understory fires. These long-burning, slow-spreading fires under tree canopies have been difficult to detect, but Morton's team used the MODIS satellite to track understory fires in the Amazon. Between 1999 and 2010, these typically unreported small-scale fires burned an estimated 85,500 km^2 of forest. Surprisingly, deforestation had little to do with the understory fires. Instead the research team believes that low nighttime humidity, associated with climate change, leads to conditions favoring understory fires.

FIGURE 4.5.10

On an unusually cloud-free day at the height of the dry season, several fires were burning in Amazonia, giving rise to a broad smoke pall easily seen from the International Space Station. *Source: NASA via Wikimedia Commons.*

SYSTEMS CONNECTIONS

Exercise 7: As we have seen, cutting down the rainforest creates feedback loops that further affect the health and fate of the forest ecosystem.

For each of the following scenarios, brainstorm any possible feedback loops that might come into play. Use lists or create flowcharts to show how the activities listed below might impact the broader Amazon ecosystem over the long term by creating feedbacks that either increase or reduce degradation of the Amazon. Consider the following in your discussion of potential impacts: freshwater systems, including the Amazon River, agricultural production, wildlife, human health, fire patterns, soils, weather, and climate.

Scenarios to consider:

1. Continued deforestation of the Amazon rainforest leading to the widespread conversion for forests to grasslands;
2. Dramatic reduction in deforestation as a result of successful management actions maintaining current forest cover;
3. Stable forest cover but a significant acceleration of climate change through 2050; and
4. Significant conversion of forest to agriculture and continued acceleration of climate change.

FIGURE 4.5.11
Amazon rainforest before and after slash-and-burn. *Source: NASA.*

DEVELOPING A MANAGEMENT STRATEGY FOR THE AMAZON RAINFOREST

A number of management approaches are being used in an attempt to reduce the rate of deforestation of the Amazon. Below, we identify and briefly describe four of these. In the exercises that follow, you will have an opportunity to do some research on other potential management approaches for the Amazon and to develop a proposal for effectively controlling deforestation in the Amazon.

1. ***Sustainable Use Conservation Reserves***: Designed to be much more than simply setting aside forest for preservation and minimal human use, sustainable use conservation reserves set aside existing forest areas where natives can earn a living by growing/harvesting forest products in a sustainable way. Rubber tapping, growing Acai palm, harvesting Brazil nuts, and supporting ecotourism are ways to generate income for those who live in the rainforest without engaging in deforestation. Shade grown coffee is another example of preserving forests while providing income for local communities (Fig. 4.5.12).

2. ***Clean Development Mechanism***: These United Nations–supported projects help developed countries reduce greenhouse gas emissions by implementing projects in developing countries to counter their emissions. In the case of the Amazon rainforest, clean development mechanism (CDM) projects include afforestation, the establishment of plantation projects for which developed countries can claim carbon-capture credits. For example, The

FIGURE 4.5.12
A shade-grown coffee plot helps protect water quality, provides higher yields, reduces irrigation, and even provides wildlife habitat. *Photo credit: Edwin Mas. USDA via Flickr.*

Vale Florestar CDM project aims to plant 150,000 ha with eucalyptus in the most degraded part of the Brazilian Amazon. Additionally, approximately 150,000 ha will be legalized as areas for the protection of native vegetation (legal reserve and permanent preservation areas).

3. ***Alternative Timber Harvesting Techniques***: Several alternative approaches to clear-cutting exist for harvesting timber in the Amazon. Selective harvesting removes only one or two target species, leaving the rest of the forest intact. This approach still allows for economic gain from the forest resource while minimizing impacts on a diverse forest. However, evidence from the Amazon suggests that this approach is still damaging, with one study indicating an average of 19 years for forest biomass to recover (Costa et al., 2015) to preharvest levels. Another approach is reduced-impact logging (RIL) in which standard harvesting techniques are applied to reduce the ecological impacts of silvicultural activities themselves. RIL focuses on reducing soil and canopy damage, protecting future crop trees and decreasing waste generated during logging by at least 50%.

4. ***Soy Moratorium***: Brazil's soy moratorium (SoyM) was the first zero-deforestation agreement implemented in the tropics. In response to pressure from retailers and nongovernmental organizations, major soybean traders signed the SoyM, agreeing not to purchase soy grown on lands deforested after July 2006 in the Brazilian Amazon. It is hoped that this pact, now permanent, will serve as a model for other commercial entities like palm oil and beef.

SOLUTIONS

Exercise 8: Do some research to identify additional potential management, policy, or outreach activities that could help minimize deforestation in the Amazon basin.

- Describe at least three additional steps that could be taken in the short term (next 3–5 years)
- Describe at least three additional steps that could help minimize deforestation in the long term (10–20 years).

Considering these options and the ones described in this case study:

- Prioritize your action steps, and justify your choices.
- Create a timeline for their implementation.
- Identify any major hurdles that would need to be overcome in your approach.
- List any major stakeholder groups that must be engaged.

In spite of decades of efforts by governmental, nonprofit, and local communities, deforestation of the Amazon rainforest continues. While deforestation rates decreased in the late 2000s, recent years have seen an uptick. In fact, an analysis by UN's Food and Agriculture Organization of Landsat images done in 1990, 2000, 2005, and 2010 found a global increase of 62% in the rate of tropical deforestation, with Brazil topping the list at 5957 km^2 per year.

CONSIDER THIS

Exercise 9: Most of our time in this case study has been focused on the environmental impacts of deforestation. But deforestation (and many potential solutions to deforestation) in the Amazon have a tremendous impact on local communities as well.

- Do some digging to identify some examples of environmental injustice that natives are experiencing, as well as efforts to correct those injustices.
- Are there any steps that you, as a consumer of wood products in the developed world, might take to help address the injustices you've identified?

REFERENCES

Costa, J., Fernanda, V., et al. 2015. Synthesis of the first 10 years of long-term ecological research in Amazonian Forest ecosystem–implications for conservation and management. Natureza & Conservação 13 (1): 3–14.

Lindsey, R., 2004. From Forest to Field: How Fire Is Transforming the Amazon. NASA Earth Observatory. June 8 Feature.

Marull, Y., 2013. Cattle Ranching Goes Green in the Brazilian Amazon. Phys.org: http://phys.org/news/2013-09-cattle-ranching-green-brazilian-amazon.html.

Mishra, A.K., Lehahn, Y., Rudich, Y., Koren, L., 2015. Co-variability of smoke and fire in the Amazon basin. Atmos. Sci. 109, 97–104.

Morton, D.C., Le Page, Y., DeFries, R. Collatz, G.J., Hurtt, G.C., 2013. Understorey fire frequency and the fate of burned forests in southern Amazonia. Phil. Trans. R. Soc. B. 368, 2012.0163.

Saatchi, S.S., Houghton, R.A., Dos Santos Alvala, R.C., Soares, J.V., Yu, Y., 2007. Distribution of aboveground live biomass in the Amazon basin. Glob. Change Biol. 13 (4), 816–837.

Saatchi, S.S., Asefi-Najafabady, S., Malhi, Y., Aragao, L.E.O.C., Anderson, L.O., Myneni, R.B., Nemani, R., 2013. Persistent effects of a severe drought on Amazonian forest canopy. Proc. Natl. Acad. Sci. U.S.A. 110 (2), 565–570.

Wallace, S., 2007. Last of the Amazon. National Geographic, pp. 40–72.

Chapter 4.6

Electronic Waste

FIGURE 4.6.1
Workers burning plastic off wires for copper recovery. *Photo by Jcaravanos (CC-SA-4.0) via Wikimedia Commons.*

OVERVIEW

A number of our case studies have demonstrated the broad scope of impacts that even seemingly minor human actions can have on global resources. Many people are aware of the impacts of CO_2 emissions from their cars or homes. Fewer realize the danger posed each time they upgrade their laptop or cell phone. We live in a culture that associates status with having the newest and best electronics.

But what happens to our old equipment? Most of us assume that these materials, often containing toxic materials, are neatly recycled with little impact on the environment or our communities. Disposal of solid waste is one human impact on the landscape that is too often overlooked, with the old adage "out of sight, out of mind" describing many individuals' attitudes about waste disposal.

FIGURE 4.6.2

A US Geological Survey (USGS) report in 2006 estimated that, even by 2005, as many as 500 million obsolete cell phones were being stored in drawers and closets, with most of these likely destined for landfills. *Photo: davepatten/Flickr.*

While some parts of the world have been very successful in recycling materials and composting various types of food waste, one type of waste, in particular, has outstripped others in recent years. The exploding numbers of devices, from cell phones to computers to big-screen televisions, have created a solid waste problem that is truly international in scope (Fig. 4.6.1). In our rush to get the latest electronic gadgets, too often we forget about "closing the loop" and considering what happens when we're done with the latest version and ready to buy the newest product (Fig. 4.6.2).

We'll consider two related issues in this case study. First, we'll look into a number of externalities, those real but mostly ignored costs to the environment and human health that result from the production, use, and disposal of everyday technological equipment like cell phones. We'll consider additional measures of environmental impact, such as ecological footprints and carbon budgets. Before proceeding with this case study, review each of these in the Tools and Skills section.

The second portion of this case study focuses on the international aspects of e-waste, considering the environmental and human health consequences that result from our failure to adequately consider how to deal with our electronic product flow. Not surprisingly, there is a strong element of environmental justice in this story. We'll conclude by looking at some steps that we can take to reduce the production of e-waste, both from the individual's perspective and from that of those responsible for protecting our environment.

BACKGROUND

Fast Facts About Cell Phones

- Of the world's seven plus billion inhabitants, more than 6 billion have cell phone access, while only 4.5 billion have working toilets.
- The life expectancy of cell phones in the US is about 12 months and 18 months globally.
- In 2014, global smart phone sales exceeded 1.2 billion units.
- The number of cell phones recycled remains low, with most estimates at less than 10%.

Terms to Know

(Review each of these in the Tools and Skills section of Chapter 01.)

Life Cycle Assessments look at all the impacts of a given product or process from the beginning of its production to its disposal. Impacts are all-inclusive, including resource consumption, release of greenhouse gases, and generation of solid waste.
Carbon footprints or budgets specifically focus on the amount of carbon released into the atmosphere by an activity or from the production of a good or service. Not surprisingly, different products or activities may have very different carbon budgets, depending on the amount of energy used to produce them and a host of other factors.
Ecological footprints assess the strain that humans are placing on the Earth. They look at the impact our lifestyles have on the globe's terrestrial and aquatic resources. Key elements include lifestyle choices, energy consumption, and waste generation.

THE PROBLEM

It's clear from the numbers above that the constant stream of new cell phones represents a significant environmental problem. We're buying millions of them each year, and far too many end up in landfills, where toxic constituents like lead, cadmium, beryllium, mercury, and flame retardants pose a long-term threat to the environment.

To understand the "big picture" impacts of products like cell phones, we need to consider what happens from the time the phone is made until it is discarded. To do this, we'll use a life cycle assessment (LCA) approach to dig deeper into the real costs associated with cell phones. Remember, although an individual cell phone is small, the problem is magnified by the numbers involved. In 2014, the estimated number of active mobile devices globally (7.22 billion) exceeded the world population of 7.2 billion.

DIGGING DEEPER

Exercise 1: From the outside, cell phones appear to be predominantly composed of plastic. But much more goes into your typical mobile device. Do some digging, and then complete the following (note that we will use this information as we work through the LCA later in the case study):

- List all the materials you think go into making your phone (glass, electronic components, raw materials, etc.)
- List the environmental impacts that you think could result from the production, use, and disposal of your phone

LIFE CYCLE ASSESSMENTS

Assessing the environmental impacts of a product throughout its lifespan (Fig. 4.6.3) may have several benefits. Suppose you're a company committed to the best environmental practices possible. By completing an LCA for the products you produce, you can determine the steps in your product's life cycle where the greatest environmental impact is occurring and where changes might result in the most improvement. For example, let's say your company makes glass bottles and that your LCA indicates that the greatest environmental impact results from the amount of water used during manufacture. Water conservation steps that you could employ might not only make your glass bottle a "greener" product, but they might also save you money as well.

LCAs can benefit consumers as well. Let's say you need to buy a trash bin. You find two that are the same price; one is plastic and the other is metal. You'd like to make the best choice for the environment. By reviewing a thorough LCA of the two options, you could make an informed choice and select the trash bin that has the least environmental impact.

The Life Cycle of a Cell Phone

Manufacture

Cell phones (Fig. 4.6.4), while small in size, contain a surprising number of parts, many of which are made of materials with substantial environmental negatives. The handset alone contains metals, plastics, and

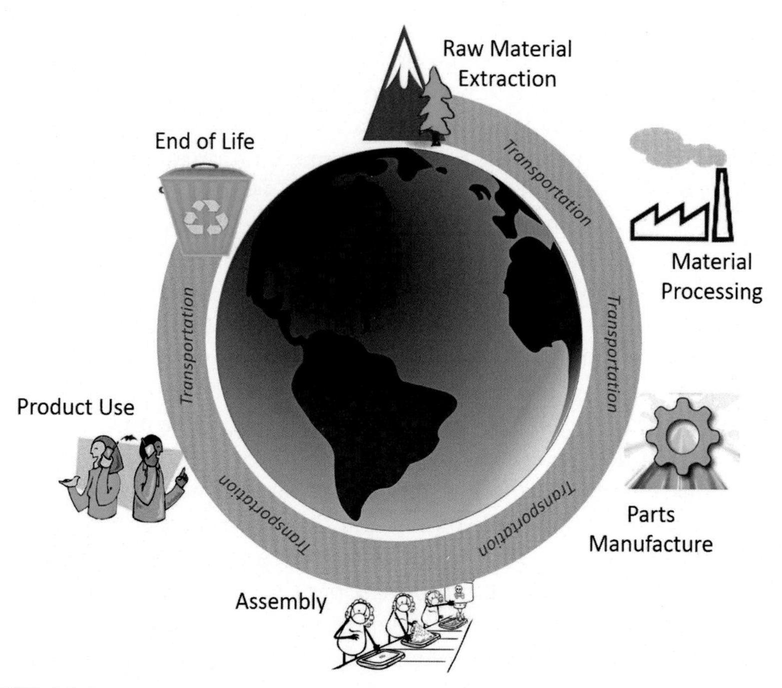

FIGURE 4.6.3

The typical life cycle assessment incorporates impacts of a given product or process from the beginning of its production to its disposal.

FIGURE 4.6.4

A cell phone's life cycle is surprisingly complex. *Source: US EPA.*

glass. The copper circuits on the circuit board are coated with a layer of tin–lead to prevent oxidation, while contact fingers are plated with tin–lead, then nickel, and finally gold to maximize conductivity. The liquid crystal display, which displays the cell phone's information, is made of glass or plastic and contains mercury. The rechargeable battery may contain nickel, cobalt, zinc, cadmium, and copper.

Materials extraction: To produce the necessary metals, metal ores must be mined, with a variety of environmental impacts resulting from the extraction process (Fig. 4.6.5). Crude oil is used for the plastics, and sand and limestone in the manufacture of any fiberglass components.

Materials processing: Raw materials must be processed as a next step. For instance, extracted metal ores are smelted to separate the pure metal from the nonmetallic components (Fig. 4.6.6). This requires the addition of chemicals and often the use of significant quantities of both electricity and water. Converting crude oil into plastic also requires energy and additional chemicals. Once materials have been produced, they still may need to be shaped into the appropriate parts by the manufacturer. Each of these steps may have a variety of environmental consequences.

Manufacturing: Once the basic components have been assembled at the plant, more energy and additional substances, like glue, will be required to assemble the cell phone.

Packaging and Transportation

Packaging: While some packaging is necessary before the consumer purchases a cell phone, very often, excessive packaging is used (Fig. 4.6.7), sometimes as a marketing ploy to help convince the consumer to make

FIGURE 4.6.5

Bingham Canyon open pit copper mine, Oquirrh Mountains near Salt Lake City, Utah. *Photo: Ken Lund/Flickr.*

FIGURE 4.6.6
Western Mining Corporation (WMC) nickel smelter south of Kalgoorlie, Western Australia. *Photo: Jeff Krisdale/Flickr.*

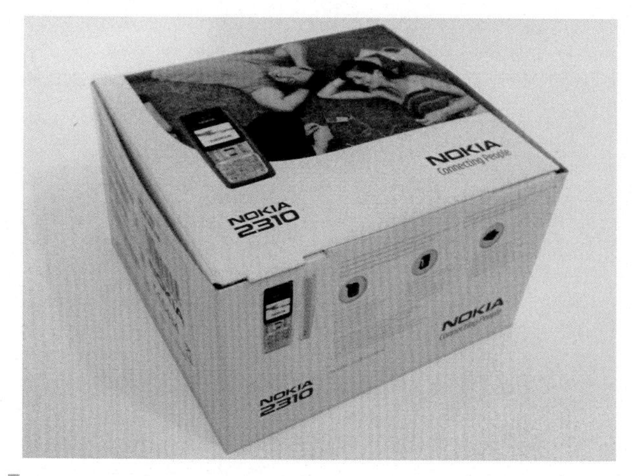

FIGURE 4.6.7
Packaging like this is not uncommon across the industry. *By Tomhannen (public domain) via Wikimedia Commons.*

the purchase. Packaging may include paper (from trees), plastics (from petroleum), and other materials. Energy is also required to manufacture, assemble, and transport packaging.

Transportation: The finished, packaged product must still make it to the marketplace. Transportation by air, truck, ship, or rail will have its own environmental impacts, with some means of movement more energy efficient and less polluting than others.

Use

Even routine use of a cell phone has an environmental cost. Charging the phone requires electrical power, which must be generated, most commonly by burning fossil fuels. There's additional energy required to transmit calls across the network. We'll take a closer look at these costs when we look at the cell phone's carbon footprint below. One could even consider the environmental impact of the cell towers themselves (Fig. 4.6.8). With towers often positioned on ridgetops, their construction and the opening of access roads can impact otherwise undeveloped environments.

Disposal

Disposing of cell phones in landfills has additional environmental costs. The metals in a cell phone include a number that are considered toxic, and, if they are released over time in the landfill, they could impair the quality of groundwater flowing beneath the landfill. We'll talk about how to reduce and avoid this problem later in the case study.

FIGURE 4.6.8
Sunset over an array of mobile phone towers. *Photo by Catherine (CC by 2.0) Flickr.*

The Bottom Line: At each step of the cell phone's life cycle, there will be a variety of environmental impacts. Just consider the number of potential impacts on air, water, and land from the mining activities necessary to extract metal ores alone. The tough job is to evaluate the relative severity of each impact identified in the life cycle. Typically, impacts are lumped into broad categories (e.g., water consumption, greenhouse gases released, nonrenewable energy or minerals consumed) and assigned a value. In this manner, the most damaging steps in the life cycle can be identified for corrective actions. As you ponder the list of environmental impacts that might occur in the four steps described above, how does this list compare with the one you started out with in Exercise 1?

CRUNCHING THE NUMBERS

Exercise 2: Let's do some back-of-the-envelope calculations about those metal components of cell phones to determine just how much we might be throwing away when we upgrade our phones.

- A December 2014 estimate from the Annual Wireless Industry Survey (Wireless Association) reported that there were 355.4 million active mobile devices in the US, more mobile devices than residents.
- A 2006 US Geological Survey (USGS) analysis reported that a typical cell phone contains 16 g of copper, 0.35 g of silver, 0.034 g of gold, 0.015 g of palladium, and 0.00034 g of platinum.
- Assume that all of the mobile devices in use today in the US have that same mass of metals.
- Find today's monetary value for each of those metals.
- Determine the total value of each of the metals contained in the active mobile devices in the US in 2014.
- If the annual life expectancy of a cell phone is 18 months, what is the value of the metals being discarded each year?

Given your calculations, it's probably no surprise that there is an international market for retrieving metals from cell phones and other electronics. One expert claims that today's global landfills contain 32 metric tons of gold!

CRITICAL THINKING

Exercise 3: Completing an LCA on even a fairly simple product can be surprisingly complicated. Read the article "Analyzing the environmental impacts of laptop enclosures using screening-level life cycle assessment to support sustainable consumer electronics" by Meyer and Katz (http://www.sciencedirect.com/science/article/pii/S095965261500801X) (linked on your Resource Page) and answer the following questions:

a. What were the four different types of laptop computer enclosures that were assessed?
b. Examine Figures 2 through 5 and identify one unique step in the life cycle of each of the four types of enclosures.

CRITICAL THINKING—Cont'd

c. According to Section 3.5.1 of the article, which type of enclosure had the least impact considering global warming, smog, acidification, noncarcinogenic human health, and fossil fuel depletion metrics?
d. What feature of the bamboo enclosure made it less sustainable than it might have otherwise been?
e. What feature of all the laptop enclosures contributed substantially to the overall life cycle impact, despite constituting only a small portion of the final product?

DIGGING DEEPER

Exercise 4: Cell phones and many other devices contain rare earth elements, with many found in the lanthanide series, those elements with atomic numbers 57 through 71. An example is dysprosium (atomic number 66). Not only is the supply of this important rare earth element running low, its unique magnetic powers make it difficult to replace. To make matters even more interesting, China has cornered the market in rare earth elements, controlling 97% of the world's production. Any time one nation corners the market for a valuable resource like these elements, there is a possibility of economic manipulation and global instability. For example, Jones (2013) reported that the price of rare earth elements increased by 750% in 2013.

- Dig into the literature to unearth another example of a rare earth element that might limit the production of mobile devices in the future.
- Describe its sources, uses, and alternative plans if it vanishes or becomes too expensive to use.

CARBON FOOTPRINTS

Given the important place climate change occupies on the list of global environmental concerns, it is not surprising that many LCAs focus heavily on the amount of CO_2 and other greenhouse gases emitted during the lifespan of a product. For instance, burning a liter of gas releases about 2.3 kg of CO_2, while a liter of diesel releases about 2.7 kg, and a liter of home heating oil, about 3 kg. Carbon footprints (Fig. 4.6.9) estimate the total amount of greenhouse gasses emitted by the manufacture, transport, and use of the object in question. You, and everything you own, has a carbon footprint. Because there are several greenhouse gases to be concerned about, we'll express values as CO_2-e. Review this concept in your Carbon Footprint entry in the Tools and Skills section.

Does my cell phone have a carbon footprint? You bet! While there are greenhouse gases emitted during the production and delivery of a phone into your hands, you might suspect that the carbon footprint of an individual cell phone

FIGURE 4.6.9
Simple show explains the carbon footprint. *Link to the video available on your Resource Page (CC 3.0 via Wikimedia Commons).*

would be quite small. While manufacturing the average cell phone doesn't have a huge carbon footprint, about 16 kg CO_2-e, the problem lies in the number of devices. With billions of cell phones in operation globally, hundreds of millions of tons of CO_2-e can be traced back to cell phones. Berners-Lee (2010) estimated that in 2009, the 2.7 billion mobile phones in use accounted for about 125 million tons of CO_2-e, about 0.25% of global emissions.

We also tend to overlook the fact that even once we have our phones, we continue to add to the carbon footprint of that phone just by using it. Berners-Lee estimated that using a cell phone for a minute has about the same carbon footprint, 57 g of CO_2-e, as producing an apple or most of a banana. However, chatting for an hour daily adds about 1250 kg CO_2-e annually, about the same as flying from London to New York.

Solutions: How can you reduce the carbon footprint of your cell phone?

- Texting is a much lower carbon option.
- Unplug your charger when not connected to the phone. It continues to draw energy even when not in use.
- Turn off your phone at night to conserve the battery.
- The battery will also last longer if you reduce screen brightness to 70% or 80%.
- Keep your phone for as long as possible. Consider repairs or refurbishing your phone before replacing it.

CRUNCHING THE NUMBERS

Exercise 5: Now it's your turn to figure out the carbon footprint of an object. To see an example before you start, visit http://www.openthefuture.com/cheeseburger_CF.html to see how they calculate the carbon footprint of a cheeseburger. After you've read this article, answer the following:

- What are the carbon-generating activities included in the cheeseburger's life cycle?
- What are some of the uncertainties in the numbers used in this exercise?
- The article compares the footprint of all burgers consumed in the US to all the SUVs on the road. What are the findings? Are you surprised by the outcome?

Now, work through these same steps to compute the carbon footprint of cell phones, assuming the following:

- There are approximately 7.2 billion active cell phones in use.
- Manufacturing a cell phone has a carbon footprint of about 16 kg of CO_2-e.
- The global lifespan of these phones is 18 months.
- Talking on the phone for an hour a day has an additional carbon footprint of about 3.4 kg of CO_2-e.
- The average person spends about 1.5 h per day talking on their cell phone.

What is your estimate for the carbon footprint of the world's cell phones over the next 3 years? Assume that all 7.2 billion phones were purchased on day 1 of the 3 year period.

If you're concerned about climate change, want to do something about it, and are good with numbers and tracking down information, you might want to consider a career in carbon counting. Major corporations around the globe are now looking for ways to reduce their outputs of carbon. According to Gunther (2014), Wells Fargo plans to achieve a 35% reduction in greenhouse gas emissions from its buildings, and General Motors' manufacturing plants are shooting for a 20% reduction in carbon intensity. Environmental scientists able to perform an LCA and identify the best places to reduce carbon emissions should be in demand as more and more industries and utilities look for ways to cut back on carbon emissions. If a carbon tax is adopted, the need for expertise in carbon analysis will only increase.

ECOLOGICAL FOOTPRINTS

Unlike carbon footprints, which are focused specifically on carbon emission equivalents, ecological footprints give us a measure of the overall strain we are placing on the planet. Humans depend on global resources for their existence. Our homes, cars, food, and clothing all come from natural resources. The more we have and the more we use, the more strain we put on these resources.

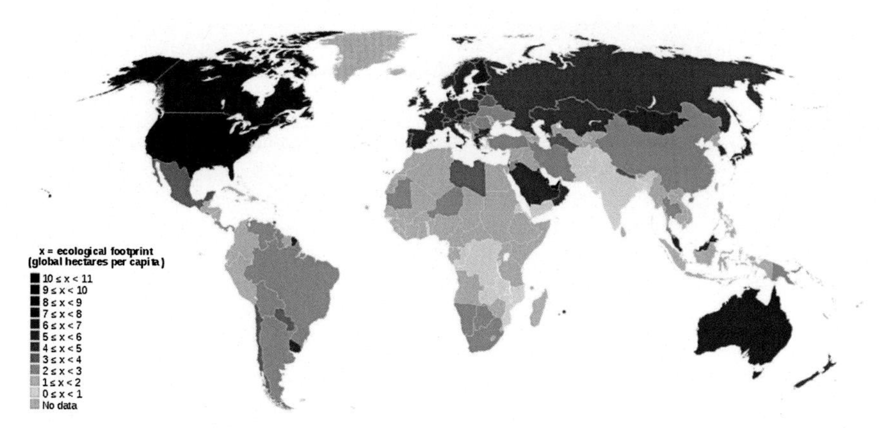

FIGURE 4.6.10

World map of countries shaded according to their ecological footprint in 2007 (published on October 13, 2010, by the Global Footprint Network) measured as the number of global hectares that are affected by humans per capita of the country. Lighter shades denote countries with a lower ecological footprint per capita, and darker shades, countries with a higher ecological footprint per capita. *Source: Jolly Janner (CC 3.0) via Wikimedia Commons.*

Experts have developed tools for estimating our ecological footprints. The calculations are expressed as the numbers of hectares of the Earth (terrestrial and aquatic) needed to support an individual, with the value depending on the lifestyle of the individual. For large scale assessments (global as opposed to individual) (Fig. 4.6.10), the total ecological footprint is typically measured as the sum of six factors: cropland footprint, grazing footprint, forest footprint, fishing ground footprint, carbon footprint, and built-up land.

CRUNCHING THE NUMBERS

Exercise 6: There are several online tools available to help you calculate your ecological footprint. Choose one of these from the list below (and linked on your Resource Page) to estimate your footprint. If you are away at school, use data from your home base.

...http://www.footprintnetwork.org/en/index.php/GFN/page/calculators/
...https://islandwood.org/footprint-calculator/#
...http://www.ecocamp.us/eco-footprint-calculator
...http://footprint.wwf.org.uk/

Now answer the following questions:

1. What is your ecological footprint?
2. How many Earths would be required if everyone on the planet had your ecological footprint?
3. What's the biggest contributor to your ecological footprint?

CRUNCHING THE NUMBERS—Cont'd

4. List three ways that you could easily reduce your ecological footprint.
5. Each of the calculators above is slightly different. Try several different ones and compare the results. If they gave different results, explain why.
6. Look at Fig. 4.6.10 that shows ecological footprints from around the globe. Does the footprint you've calculated match the number for your country of origin? If not, why might that be?
7. Many of the values you see represented in the figure are probably not surprising. Several might be, however. What do you think might account for the very high numbers for the United Arab Emirates? For Ireland?

ELECTRONIC OR E-WASTE

Background: Cell phones and other mobile devices are one component of electronic or e-waste. But e-waste also includes computers, televisions, VCRs, stereos, copiers, and fax machines. Many components of e-waste contain hazardous materials, including cathode ray tubes in televisions and trace metals like gold and silver in various parts of many electronics.

The amount of e-waste generated globally is staggering. A report by Arrow Value Recovery (http://www.arrowvaluerecovery.com/resources/reports-insights/e-waste-recycling/) estimates that in 2014, 37.2 million tons of e-waste, or about 5.3 kg per capita, were generated globally, with an expected growth rate of 4–5% annually.

Proper disposal and recycling of this new kind of waste are posing a serious challenge for both developed and developing countries. E-waste is commonly shipped to developing nations with limited regulatory oversight, and serious health and pollution problems can result. Even in developed countries, end-of-life management of e-waste may involve significant risks to workers and communities (Fig. 4.6.11).

Fast Facts on E-waste

- It takes about 227 kg of fossil fuels, 23 kg of chemicals, and 1363 L of water to produce a single computer and monitor.**
- The US discards 30 million computers each year.
- 81% of the energy associated with a computer is used during manufacture, not during its lifetime of use.*
- US EPA estimates that 2.44 million tons of electronics were disposed of in the US in 2010 and only 27% was recycled.**

FIGURE 4.6.11
E-waste accumulated in Birmingham, Alabama. *Photo: Curtis Palmer/Flickr.*

- In the US, an estimated 70% of heavy metals in landfills comes from discarded electronics.**
- Much of the e-waste from the US goes to nations like China, India, and Nigeria, where it is burned and disassembled with few, if any, human health or environmental considerations.*

Sources: *http://www.treehugger.com/lean-technology/crazy-e-waste-statistics-explored-in-infographic.html and **http://www.arrowvaluerecovery.com/resources/reports-insights/e-waste-recycling/

The Problem: We've established the basic problem with e-waste: there is too much generated globally, and we lack an effective system for keeping it and its hazardous components out of the environment. The average computer is 23% plastic, 32% ferrous metals, 18% nonferrous metals (e.g., lead, cadmium, mercury), 12% electronic boards (more nonferrous metals), and 15% glass (UNEP and Basel Convention, 2005). A report by McAllister (2013) estimated that of the roughly 40 million metric tons of e-waste produced globally each year, only about 13% of this mass is recycled.

In addition to the environmental issues associated with the production, use, and disposal of e-waste, and the dwindling supply of rare earth elements required for many of our electronic devices, there are other concerns about e-waste.

Data security: Carelessly discarded e-waste often contains sensitive information on hard drives that can fall into the wrong hands. Stealing passwords, banking, or personal identification information is a lucrative business that impacts individuals, institutions, and economies.

FIGURE 4.6.12
E-waste burning site, Agbogbloshie, Ghana. E-waste dumping sites concentrated in many poorer nations pose significant health threats to all who live in the vicinity. *Photo: Fairphone/Flickr.*

A 2009 PBS Frontline documentary focused on e-waste disposal centers like Ghana, where hard drives from discarded computers were being sold for about $35 US, and private financial data and credit card numbers from individuals and even private sector information about US government defense contracts were being accessed by cyber criminals.

"Conflict minerals": Some of the elements used in electronics, like gold and tungsten, are part of a violent economy in some African nations. The process to extract these raw materials typically comes with a tremendous human cost. A May 2015 story in the UK's Independent newspaper (Morrison, 2015) detailed the deadly violence in the Democratic Republic of Congo (DRC) centered around cassiterite, a tin ore used in phones and laptops. The story notes that armed groups have been carrying out mineral-related criminal activities in the eastern portions of DRC for years.

Human Health: Although a localized problem, the flourishing primitive recycling market in developing nations is perhaps the most disturbing. In villages where there are few ways to earn a living, many spend their days burning and sorting through this toxic waste (Fig. 4.6.12), hoping to collect enough raw materials to support their families. Often children participate in these efforts. Much of the e-waste generated in developed nations ends up in these e-waste villages. While we may be naively thinking that our e-waste is being recycled, the reality is that the international e-waste trade threatens both human lives and the environment.

According to McAllister (2013), nations like China and India handle between 50% and 80% of global e-waste by dismantling, shredding, and burning the materials under uncontrolled conditions. The global capital for recycling e-waste is Guiyu, a city of 150,000 in southeastern China (Figs. 4.6.13 and 4.6.14). Guiyu town, located in Guangdong Province, has been a center of

FIGURE 4.6.13
Sorting through e-waste in Guiyu, China. *Photo: basalactionnetwork/Flickr.*

FIGURE 4.6.14
E-waste dump in Guiyu, China. *Photo: By bleahbleahbleah (CC BY-SA 3.0) via Wikimedia Commons.*

unregulated e-waste disposal and recycling for about three decades. More than 6000 small-scale shops, employing 160,000 workers, dismantle and recycle e-waste (Xu et al., 2015) under mostly primitive conditions. Previous research has shown that both toxic trace metals and persistent organic pollutants are released into the environment in and around Guiyu. We'll take a look at several investigations that illustrate the extensive nature of this problem.

Environmental Sciences in Action

Xu et al. (2015) measured levels of lead and polycyclic aromatic hydrocarbons (PAHs) in the blood of 167 children from areas close to e-waste recycling sites in Guiyu and from a reference area distant from the recycling activities. In addition to blood contaminant levels, investigators measured children's growth parameters like height, weight, and chest circumference.

Blood values for total PAHs, including seven known or suspected carcinogens, were significantly higher in children living close to the e-waste recycling sites than in the reference group. PAH levels were negatively associated with height of children and chest circumference. E-waste recycling activities were assumed to be a strong contributing factor to PAH levels found in the exposed group.

CRUNCHING THE NUMBERS

Exercise 7: In addition to comparing contaminant levels in blood samples from children in Guiyu to those in a distant control village using *t*-tests, Xu and his colleagues also used continuous data to compare the blood levels to a host of exposure and impact factors (Fig. 4.6.15) using correlations.

	Σ16-PAH	Σ7 carcinogenic-PAH	Blood lead level
Location of child residence in Guiyu	0.688**	0.651**	0.486**
Residence adjacent to e-waste workshop	0.305**	0.279**	0.229**
Age	−0.267**	−0.290**	−0.205**
Father's engagement in work related to e-waste	0.321**	0.285**	0.317**
Time that father is engaged in work related to e-waste	0.324**	0.289**	0.347**
House as a family workshop	0.213*	0.211*	0.233**
Child milk consumption per month	−0.176*	−0.215*	0.129
Blood lead level	0.374**	0.327**	-
BMI	0.150	0.164*	0.080
Height	−0.105	−0.104	−0.158*

BMI: body mass index (kg/m^2); R_s: Spearman correlation coefficient.
$N = 167$; compared with the reference group.
* $P < 0.05$.
** $P < 0.01$.

FIGURE 4.6.15
Spearman correlations (Rs) of Σ16 polycyclic aromatic hydrocarbons (PAHs), Σ7 carcinogenic PAHs, blood lead levels and factors about children and potential sources of exposure. *Source: Xu et al., 2015.*

CRUNCHING THE NUMBERS—Cont'd

- Based on the results presented in the figure, describe the strength and nature of each significant ($p < .05$) correlate with blood levels of carcinogenic PAHs. Be sure to discuss how meaningful the results are, considering the sample size of 167 children in the exposed group.
- What story do these data tell about exposure to and impacts from PAHs emitted by e-waste facilities in Guiyu?

Other studies have examined the broader impact of e-waste dump sites on the natural environment. Luo et al. (2009) measured persistent halogenated compounds like polychlorinated biphenyls (PCBs) and the flame-retardant polybrominated diphenyl ethers (PBDEs) in the muscles of five waterfowl species collected from a delta near Guiyu. PCB levels were as high as 1.4 mg/g lipid, while levels of PBDEs were also elevated. A similar pattern was noted (Wu et al., 2008) in the same vicinity for a variety of aquatic biota, including the Chinese mystery snail, prawn, fish, and water snakes. The authors in both cases identified e-waste recycling as the likely source of contamination.

SYSTEMS CONNECTIONS

Exercise 8: Considering the "informal" recycling process used to extract the valuable components of electronics, as well as the fate of the remaining parts, it is easy to imagine a myriad of potential environmental impacts.

- Start this exercise by investigating the typical process employed to deconstruct e-waste at sites like Guiyu.
- Based on your findings, construct a flow chart of the myriad potential environmental impacts. Show which pollutants are involved for each impact.
- Which components of the ecosystem (e.g., surface water, air, soil, wildlife, flora, groundwater) in these locations are likely most at risk of degradation because of these activities? Why?

CONSIDER THIS

Exercise 9: The largely unregulated and uncontrolled global recycling of e-waste is an obvious case of environmental injustice. Given the extreme levels of e-waste contamination occurring in a number of developing countries around the globe and the documented threats to both human health and the environment, it is imperative that nations responsible for producing the waste take steps to eliminate these unacceptable risks occurring in other parts of the world.

View the two videos on e-waste found on your Resource Page, one from PBS' Dateline and the other from CBS' 60 Minutes. After viewing these, address the following questions:

CONSIDER THIS—Con'd

- What's your reaction to what you've seen? Are you surprised by the magnitude and extent of the problem?
- Who is most at risk from e-waste recycling?
- Who is predominantly responsible for creating e-waste?
- Consider the role that your family might play in what you've seen in the videos. Make a list of the new technology purchases made in your household along with a list of those disposed of. What steps might your family take to reduce this turnover in the future?

Solutions: There are many steps that could help to control the e-waste problem. A few include:

Government Regulation

...The EU requires electronics manufacturers to accept, free of charge, any used products brought in by their customers for recycling; other nations could follow suit.

...The Basel Convention is an international treaty signed in 1989 making it illegal to export or traffic in toxic materials such as e-waste. To date, the US has signed but not ratified the Basel Convention, which could increase regulation and control of the movement of e-waste to developing countries.

Corporate Initiatives

...Manufacturers could develop electronics that address end-of-life issues and are designed to make recycling easier.

...Manufacturers could sell their products with contracts that encourage customers to return and exchange their products for new models.

Individual Responsibility

...As an educated consumer, you could pledge to maintain and use your electronics for as long as possible, avoiding the marketing schemes trying to convince you to buy the "latest and greatest" device.

FIGURE 4.6.16
Volunteers help collect e-waste for recycling in San Diego, CA, where E-waste is a quarterly recycling event. *Photo by Seaman Stephen Votaw, US Navy.*

...Donate your old gadgets if they have some useful life left. There are many small companies that refurbish and reuse older electronics.

...When your gadget has reached the end of its useful life, take it to a certified e-waste recycler. There are lists of e-stewards and others who will recycle e-waste the correct way. Visit http://e-stewards.org/find-a-recycler/; click on the "Find a Recycler" link, and you can enter your zip code to locate the nearest facility.

- In some areas, public institutions hold e-waste recycling days for unwanted electronics (Fig. 4.6.16).
- Many stores that sell electronics have electronics recycling programs in their stores.
- More and more kiosks are popping up in malls and universities where you can "sell back" your cell phone.

...Take responsibility! If everyone took a little extra time to make certain that their old electronics were properly recycled, this problem would be greatly reduced.

SOLUTIONS

Exercise 10: While e-waste poses a daunting challenge, there are clearly many steps that could be taken to reduce the severity of the e-waste problem. Here, let's focus on your community. Your challenge is to reduce the impacts of e-waste in your local area.

- Begin with a review what's already being done in your town or school. Outline the various programs, policies, and initiatives that are currently in place to manage e-waste disposal.
- Work with your fellow students to design one program for your community and another for your school that would help reduce the problem of e-waste. Your program will likely incorporate steps that involve governments, businesses and other organizations, and individuals. Note that your proposed program should not duplicate what others are already doing.

BONUS: If either your town or school doesn't have an e-waste plan in effect, try to get them to consider the proposal you develop here.

REFERENCES

Berners-Lee, M., 2010. What's the Carbon Footprint of…Using a Mobile Phone? June 9 edition of The Guardian ecologicalfootprint.heroesoftheuae.ae.

Gunther, M., February 18, 2014. Counting Carbon: Why Emissions Targets Must Be Based on Science. The Guardian: Guardian sustainable business/sustainable business blog http://www.theguardian.com/sustainable-business/blog/counting-carbon-emissions-targets-scienc.

Jones, N., November 18, 2013. A Scarcity of Rare Metals Is Hindering Green Technologies. Yale Environment 360 Report.

Luo, X.-J., Zhang, X.-L., Liu, J., Wu, J.-P., Luo, Y., Chen, S.-J., Mai, B.-X., Yang, Z.-Y., 2009. Persistent halogenated compounds in waterbirds from an e-waste recycling region in South China. Environ. Sci. Technol. 43 (2), 306–311.

McAllister, L., 2013. The Human and Environmental Effects of E-waste. Report from Population Reference Bureau's Center for Public Information on Population Research http://www.prb.org/Publications/Articles/2013/e-waste.aspx.

Morrison, S., 2015. 'Conflict Minerals' Funding Deadly Violence in the Democratic Republic of Congo as EU Plans Laws to Clean up Trade. Independent: May 16 issue http://www.independent.co.uk/news/world/africa/conflict-minerals-bringing-death-to-the-democratic-republic-of-congo-as-eu-plans-laws-to-clean-up-10255483.html.

UNEP and Basel Convention, 2005. Viral Waste Graphics. Global Resource Information Database. www.grida.no/publications/vg/waste.

Wu, J.-P., Luo, X.-J., Zhang, Y., Luo, Y., Chen, S.-J., Mai, B.-X., Yang, Z.-Y., 2008. Bioaccumulation of polybrominated diphenyl ethers (PBDEs) and polychlorinated biphenyls (PCBs) in wild aquatic species from an electronic waste (e-waste) recycling site in South China. Environ. Int. 34 (8), 1109–1113.

Xu, X., Liu, J., Huang, C., Lu, F., Chiung, Y.M., Huo, X., 2015. Association of polycyclic aromatic hydrocarbons (PAHs) and lead co-exposure with child physical growth and development in an e-waste recycling town. Chemosphere. 139, 295–302. http://ac.els-cdn.com/S0045653515005615/1-s2.0-S0045653515005615-main.pdf?_tid=62fe92a2-7e5e-11e5-b273-00000aacb35d&acdnat=1446138135_ef8d0e7f4b601f0a8a0102321fa490af.

Chapter 4.7

Genetically Modified Organisms (GMO)

FIGURE 4.7.1
Protesters outside the offices of agriculture giant Monsanto in Washington, DC in 2011. *Photo by: Daniel Lobo (CC-BY-2.0) via Flickr.*

OVERVIEW

Agriculture has a variety of impacts on the landscape. Many of these concerns are long-standing and well-understood. For example, we know that soil erosion and subsequent runoff cannot only reduce agricultural productivity, but also degrade the quality of nearby streams and rivers. We also know that we can reduce the runoff of soils from fields by using proper soil and riparian management practices. However, some agricultural issues remain controversial. There probably is no better example of this than the growing use of genetically modified crops (Fig. 4.7.1).

For centuries, humankind has modified the genomes of organisms through traditional breeding programs. But advances in genetic engineering have allowed for precise control over the genetic changes introduced into an organism. Today, we can incorporate new genes from one species into a completely unrelated species, optimizing the desired traits. As such new technologies have entered the mainstream, governments have been grappling with policies and regulations surrounding their use (Fig. 4.7.2). One need only look at the very

FIGURE 4.7.2
Map showing nations that ban (red) or require labeling (green) of GMO containing food products. *By Co9man (CC0-BY-1.0) via Wikimedia Commons.*

different regulatory approaches to GMOs in Europe and the US to see that a consensus has yet to be reached.

GMOs have been in the headlines for years. Recent examples include "Chipotle bans GMO ingredients," "FDA approves genetically-modified salmon," and "Vermont becomes first state to require GMO labeling." While GMOs have been around for a long time, their use has become much more widespread than you might think. In 1982, the Food and Drug Administration (FDA) approved insulin produced by genetically engineered *Escherichia coli* bacteria. In 1994, the first GMO crop, a tomato, was approved for commercial sale. By 1999, more than 40 million hectares globally were planted with GMO crops (GMO Inside Blog, 2013).

Today, an estimated 75% of processed food in the US contains GMO ingredients, but because the FDA has declared GMOs "substantially equivalent" to foods produced through traditional breeding, many have no idea that they are eating foods with altered genes that have been engineered in a laboratory. As GMOs have increasingly appeared on grocery store shelves, controversy over their use has grown.

In this case study, we'll first give you some background on GMOs: what they are; how, where, and why they're used; and what some of the major questions are about their use. Then we'll ask you to decide where you stand on the GMO issue.

BACKGROUND

To create a GMO, certain genes are transferred from one plant or animal to another; the resulting organism is considered transgenic. By transferring genes, researchers are able to introduce one specific trait without also including dozens of unwanted traits, an approach sometimes called "gene splicing." The new trait might make a transgenic crop more nutritious, protect the crop from certain pests, or make it resistant to certain herbicides. The use of GMO organisms offers many benefits, including increased crop yields (Fig. 4.7.3), reduced costs for food or drug production, reduced need for pesticides, enhanced nutrient composition and food quality, resistance to pests and disease, ability of crops to grow in locations they might not otherwise, greater food security, and health benefits to the world's growing population.

Despite the various benefits resulting from their use, concerns about environmental and human health consequences of GMO use in crops and food products have plagued the technology throughout its history. Even the first commercial GMO crop, 1994's tomato, was withdrawn from stores after consumer concerns. One problem is that research around the impacts of GMOs on human health and the environment in which these modified organisms grow is still in its infancy. It is possible that there are unknown consequences to altering the natural state of an organism through foreign gene expression.

In addition to the specific trait selected for, such alterations can change other components of the organism, perhaps affecting not only the organism itself, but also the natural environment in which that organism grows. The transfer of resistance to pesticides, herbicides, or antibiotics from GMOs to their wild counterparts can result in reduction in the viability of their offspring, which may eventually threaten the viability of both the wild type and the GMOs. This can also lead to ecological imbalances, affecting species' composition, competition, and population dynamics.

FIGURE 4.7.3
Genetically modified organisms often produce higher yields, demonstrated by the large kernels in this new variety of corn. *Photo by Eneas De Troya (CC-BY-2.0) via Wikimedia Commons.*

Fast Facts on Genetically Modified Organisms

- Crops currently using GMO technology include corn, soy, canola, alfalfa, cotton, sugar beets, squash, and papaya.
- In the US, cotton, corn and soybeans products are the most likely to be modified, with about 90% of these crops genetically modified.
- As of 2013, GMO crops were growing on 170 million ha in at least 28 countries.*
- Of the 30 traits that are commonly engineered into commercially used plants, the most popular are those that confer herbicide or insecticide tolerance or both.*
- While a few countries ban the import or cultivation of GMOs, many others require labeling products that contain GMOs.

Source: *Nature, 2013: Tarnished Promise.

Some Examples of GMOs

Common Applications

1. ***Bt corn***: a toxic protein (Bt) made by the bacterium ***Bacillus thuringiensis*** ruptures the intestines of corn borers when it is ingested. The genes that code for the toxic protein are inserted directly into the genetic material of the corn. Any borers ingesting Bt corn die within 2 or 3 days. Butterfly larvae are susceptible to Bt toxin, and there are concerns that the corn or its pollen could affect populations of species like the monarch (Fig. 4.7.4).

2. ***Ht soybeans***: Several crops, including soybeans, have been modified to resist nonselective herbicides like Roundup. Genes inserted into the crops promote the breakdown of the active ingredient in the herbicide, rendering it harmless. Use of Ht crops may, however, lead to an increase in herbicide use and the development of herbicide-resistant weeds, which can threaten biological diversity in landscapes dominated by agriculture. Recent reports suggest that glyphosate may also be cancer-causing, raising additional concerns about its increased use on GMO fields.

Innovative Applications

1. ***Goats and spiders***: A gene can be transplanted from spiders like the golden orb-weaver into goat embryos. Adult goats maturing from these eggs produce milk containing an extra protein, which can be extracted and spun into spider silk threads. With their exceptional strength, these threads, light and resilient, might be used for a variety of products, including parachute cords and artificial ligaments (Moss, 2010).

SYSTEMS CONNECTIONS

Exercise 1: Every winter, monarch butterflies migrate from the Corn Belt in the US Midwest to fir forests in Mexico. According to a 2012 study, there was an 81% decline in monarch populations in the US Midwest from 1999 to 2010 (Pleasants and Oberhauser, 2012). We mentioned that genetic modifications to produce Bt corn can directly impact other species like the monarch. But there is also a very tight connection between declines in monarch populations and reductions in milkweed, its primary food source.

- Do some digging to find information about the use of other common genetic modifications on corn in the Midwest and how these might be indirectly impacting the monarchs by affecting the milkweed on which they depend.
- What other systems connections might exist in these agricultural landscapes? How might the decrease in monarch numbers impact other species? What is their role in these ecosystems?
- Research on the effects of Bt corn on monarch larvae has been controversial. Do some digging and describe some of the controversies in Bt-monarch research.

FIGURE 4.7.4
Monarchs are common in the midwestern US Corn Belt region, where milkweed serves as the primary food source. *Photo by: Kristi (CC by ND 2.0) via Flickr.*

FIGURE 4.7.5
Genetically altered fluorescent fish. *Photo by Matt (CC-BY-2.0) via Flickr.*

2. ***GloFish***: The first GMO pet, a zebrafish containing a gene from a bioluminescent jellyfish (Fig. 4.7.5), was initially developed as a pollution monitor but was introduced to US pet markets in 2003 (Frater, 2008).

Environmental Applications (Moss, 2010)

1. ***Enviropigs***: Pig waste contains high levels of phosphorus, which, if it enters rivers and lakes, can lead to water quality problems. By adding mouse DNA and genes from ***E. coli*** bacteria to pig embryos, scientists produced adult pigs that achieved as much as a 70% reduction in the amount of phosphorus released in their waste.

2. ***Low-methane cows***: Cows are significant sources of the greenhouse gas methane. The gas is released as bacteria in a cow's digestive tract act on their cellulose-rich diets. If we could breed cattle that grow faster and spend less time emitting methane over their lifespans, or produce cattle that are more efficient at converting feed into muscle, thus releasing less methane, scientists believe that the environment would benefit from the reduced levels of greenhouse gas emissions.

3. ***Pollutant-eating plants***: Scientists are working on GMO poplar trees designed to absorb groundwater pollutants into their roots. Once absorbed, pollutants like trichloroethylene (TCE), one of the most common groundwater pollutants, are broken down within the plant. In laboratory experiments, regular poplar seedlings removed only about 3% of the TCE present, while transgenic specimens were able to remove up to 91%.

SYSTEMS CONNECTIONS

Exercise 2: Clearly, there are many potential uses for genetically engineered plant and animal products.

Do some digging to identify an example of a genetically modified animal.

- Describe the traits selected for with this genetic engineering.
- What are the potential benefits of these traits to the environment? To society?
- What are some potential drawbacks of introducing this genetically altered organism through more widespread production?

Pros and Cons of Genetically Modified Organism Crops

1. **Human health concerns:**

   Pro: There is no fundamental difference in the makeup of GMO and non-GMO crops. Both contain DNA; in GMOs, a new gene is inserted to produce a desirable trait such as rapid growth (Fig. 4.7.6). If the GMO has received genes from an organism containing an allergen, the new product must be tested for consumer safety. There are no confirmed human deaths or illnesses resulting from GMO consumption.

FIGURE 4.7.6
Here it is easy to see how a transgenic salmon (next to an unmodified salmon of the same age) can grow far more quickly. *Source: USDA.*

Con: Well-designed, unbiased studies of the long-term health effects of GMOs are lacking. The possibility of health effects was perhaps first underscored by the accidental introduction of GMO StarLink corn, approved only for use in animal feeds, into food products for human consumption. Fearing allergic reactions to a protein in the corn, in 2000, the US FDA issued the first-ever recall of a GMO product. In addition, there have been a number of scientific experiments on laboratory animals that have suggested that consumption of GMOs could lead to long-term health problems in humans. The validity of these studies has been called into question.

2. **Environmental concerns:**

   Pro: Use of GMOs like Bt corn allows farmers to use fewer pesticides to control pests. GMO seeds that would require less water and fewer fertilizer applications are being developed. GMO proponents argue that environmental concerns concerning development of herbicide-resistant plants are overblown and can be easily dealt with.

   Con: Concerns include nontarget impacts, such as the possible effects of Bt corn on monarch butterfly larvae and the loss of their food, the milkweed, in areas where GMO corn and soybeans are being treated with herbicides. Resistance among weed species to herbicides like Roundup can require increased use of these compounds for control. Transfer of pesticide-resistant genes to nontarget species may lead to "superweeds," as we've seen with herbicide-resistant amaranth, which has become common in some areas. Finally, genetically modified crops could breed with wild species, threatening genetic integrity and biodiversity.

There are a number of other issues related to ethics, social justice, and economics associated with the use of GMO crops. Proponents and opponents attack each other's positions and cast doubt on research that doesn't support their viewpoint. Opposition to GMOs in Europe is widespread and strong, with considerable public distrust of both governments' and industries' stance on the issue. Restrictive consumer and regulatory policies regarding GMOs are the rule in Europe. In the US, public opposition is more muted, and GMO use gets comparatively little attention, although several states are attempting to require labeling of GMO-containing food products.

PUTTING SCIENCE TO WORK

Exercise 3: Issues surrounding the widespread use of GMO food products seem to be characterized by too little science, too much emotion, and widespread ignorance. But providing the science in a timely fashion is difficult when impacts may take a long time to manifest or are evidenced in complex ways.

Focus on a specific genetically engineered food crop that includes high rates of production. Pay particular attention to the traits being selected for in the modified organism.

- How might you design a study to assess the impacts of introducing a GMO version of this food crop on the surrounding natural environment?
- How might this study design be adjusted to test specifically for impacts on human health?

CRITICAL THINKING

Exercise 4: Assume you're the new Commissioner of Agriculture in a Sub-Saharan African nation with serious food supply and security issues. The newly elected Prime Minister, your boss, tells you that she is going to take a position on GMO crops, and she asks you to **develop two things: a one-page statement summarizing her position on growing and consuming GMO crops and an eight-page white paper justifying her choice**. The white paper will need to include specific scientific justifications for her position, including reputable research support. She gives you three options to investigate:

a. an outright ban on growing and consuming all GMO crops in your country;
b. a labeling requirement for all seeds and consumable food products containing GMOs; and
c. no restrictions on the use of GMO crops or sale of GMO-containing food products.

We've listed a number of links to articles and websites on your Resource Page to get you started. Review them, taking careful notes, and others you can find online. Once you've completed your review, write your white paper, justifying your choice, and then complete the one-page summary statement.

Here are some specific requirements for your white paper:

...It should be eight pages long.
...It should open with a 250-word abstract summarizing the main points.
...It should include one-half page of relevant background on your nation related to its food supply and the reasons for this development of GMO policy.
...It should include four pages of review of the relevant scientific research on the topic.
...It should include a two-page summary of the pros and cons of the three options.
...It should close with a one-page justification of your position.

Once all members of your class have completed their papers, you'll break up into three groups, one for each choice, and debate the topic.

SOLUTIONS

Exercise 5: There is nothing new about environmental controversies. We've long had debates about everything from the pros and cons of nuclear power to the most effective strategies for dealing with climate change. What is so striking about the GMO debate is the seemingly complete disconnect between the two sides.

- Given the knowledge you've gained while completing your assignment, discuss at least three reasons why this disconnect might exist in the case of GMOs.
- How might opposing groups find common ground?
- Are there any potential compromises that could satisfy both sides?

CONSIDER THIS

Exercise 6: While it may be easy to question the safety of GMOs in food production, the reality is that at the current rate of population growth, scientific advances, such as genetic engineering, will be required to avert widespread famine. There are many ethical issues that arise when considering future policies around GMOs.

- How can those who have sufficient food justify denying others access to higher yielding or more resistant versions of many staple crops?
- Many genetically modified seeds produce nonviable offspring, making farmers dependent on companies to provide new seed each year. Consider the economic, environmental, cultural and ethical aspects of introducing such nonviable seed sources into poorer nations.

REFERENCES

Frater, J., April 1, 2008. Top 10 Bizarre Genetically Modified Organisms. Listverse. http://listverse.com/2008/04/01/top-10-bizarre-genetically-modified-organisms/.

GMO Inside Blog, 2013. GMO Timeline: A History of Genetically Modified Foods. http://gmoinside.org/gmo-timeline-a-history-genetically-modified-foods/.

Moss, L., 2010. 12 Bizarre Examples of Genetic Engineering. Mother Nature Network. http://www.mnn.com/green-tech/research-innovations/photos/12-bizarre-examples-of-genetic-engineering/enviropig#top-desktop.

Pleasants, J.M., Oberhauser, K.S., 2012. Milkweed loss in agricultural fields because of herbicide use: effect on the monarch butterfly population. Insect Conserv. Divers. http://dx.doi.org/10.1111/j.1752-4598.2012.00196.x.

CHAPTER 5

Looking Ahead to a More Sustainable Future

Science and the Global Environment. http://dx.doi.org/10.1016/B978-0-12-801712-8.00005-6

Chapter 5.1

Introduction

FIGURE 5.1.1
Maintaining a sustainable society and healthy environment will require both innovation and attention. This is within our reach, but it will require each of us to contribute to the effort. *Photo by: mattwalker69 (CC-BY-SA 2.0) via Flickr.*

Our case studies have focused on some very challenging issues facing the environment from local to global scales. It's easy to get discouraged and feel overwhelmed by the magnitude of today's environmental problems. But it is also important to remember that there are many scientists, engineers, conservationists, community leaders, and others working hard to lessen the human impact on the environment, and we are making solid progress in a number of areas. In this section, we'll take a look ahead and consider some innovative approaches that are helping to meet the human need for resources, along with novel ways to reduce or mitigate our impacts on the environment (Fig. 5.1.1).

Much of the threat to the environment stems from more and more people expecting more and more goods and services from a finite resource, planet Earth. Today, global population growth is concentrated in developing nations. One challenge is to help these nations meet the needs of their populations in ways that avoid the damage to the environment that has occurred too often in developed nations. Can we provide an equitable and sustainable future for everyone that also protects and supports our natural world?

SUSTAINABILITY

Sustainability can mean different things to different people sustainability can be specific to a particular activity or product, such as ensuring that any resource extraction is done at a rate that allows those resources to be replaced at the same rate, or that an activity is not harmful to the environment and therefore helps guarantee long-term ecological balance.

Environmental scientists often define sustainability as being able to continue to meet basic human needs for resources without causing significant damage to the environment or depleting natural resources that future generations will need.

In the broadest sense, sustainability refers to the ability to maintain environmental quality, human rights, and equity among populations. Having a truly sustainable world would both insure that global natural resources are not degraded for future generations and guarantee that all humans would be treated in a just and equitable manner with equal access to basic necessities.

NEW WAYS TO MEET OUR NEEDS

In this section, we'll look at some new approaches that can help meet our basic needs in a more sustainable way. Many solutions exist to minimize the environmental impacts of our activities, from food consumption to energy usage. Let's break it down.

Chapter 5.2

Food Supply

The Issues: Supplying food for a growing world population has resulted in a host of environmental concerns, some of which we've discussed in our case studies. Direct effects include a variety of impacts from the use and misuse of pesticides and fertilizers to grow crops, the pollution threat posed by animal waste generated by intensive or "factory" farms, and the clearing of land, sometimes including critical habitats, to grow crops or support livestock.

Even providing fish and shellfish for human consumption may come at a price these days. Many wild fisheries are overtaxed, with more efficient harvesting techniques and higher consumer demands putting pressure on many fish stocks. Aquaculture, or fish farming (Fig. 5.2.1), is often practiced to raise species for which natural stocks are dwindling or disappearing. Water quality concerns, use of genetically altered strains, and other issues associated with some aquaculture facilities raise doubts about the sustainability of these practices and their impact on the surrounding environment.

FIGURE 5.2.1
A fish farm in Jian De, Hangzhou, Shanghai. *Photo by:* IvanWalsh.com *(CC-BY-SA 2.0) via Flickr.*

There are also indirect effects to consider. Food to supply the needs of major population centers often comes from great distances. One estimate is that the food consumed by the average US resident during one meal travels about 2415 km from farm to plate. Some items, like bananas grown in tropical climates, out of necessity must be transported great distances to markets in cold climates. Even items like apples may be imported from abroad more cheaply than grown locally.

Long-distance transport of food consumes fossil fuels and releases significant amounts of air pollutants, including CO_2, into the atmosphere. The quality of fruit and vegetables often suffers, as items may be picked before ripening due the long distance traveled to market. This contributes to food waste, with an estimated 25–40% of all food grown, processed, and transported in the US never even consumed. More food reaches landfills and incinerators than any other single material in municipal solid waste (feedingamerica.org).

A New Approach: Because of concerns about environmental impacts and food quality, the locavore movement, dedicated to buying and consuming food produced locally, usually within 160 km, has taken hold in many parts of the world. Consuming locally grown food can both reduce impacts on the environment resulting from the long-distance transport of food and yield a fresher product for the consumer. This approach also encourages small farmers, who generally practice more sustainable farming practices than their factory farm counterparts.

Obtaining food from local sources can work well in areas where there is sufficient land or the appropriate climate to support agriculture. It's not so easy in heavily developed urban areas. However, one approach is to practice "urban gardening." This can include larger scale commercial efforts using indoor, water-based systems, or aquaponics (Fig. 5.2.2). More and more city dwellers are also raising their own food in local community gardens or on rooftops and patios (Fig. 5.2.3).

An interesting example of urban gardening is FarmedHere in Chicago, IL. Built in an 8360 m^2 abandoned warehouse, the facility is one of the largest indoor aquaponics operations in the US. Because 8360 m^2 is still relatively small for a farm, FarmedHere uses vertical gardening techniques to maximize their usage of space for crop production. Located only 15 minutes from downtown Chicago, this facility is producing fresh produce, such as basil and leafy greens. The water used to grow these crops is enriched by the waste produced by the freshwater fish tilapia, which are raised in tanks in the same facility. The fish grow from fingerlings to adults in 18 months and are sold to local restaurants and markets. In this closed system, the waste from the fish becomes a source of nutrients to grow the produce. FarmedHere reuses 97% of its water and uses

FIGURE 5.2.2
An aquaponics greenhouse in Brooks, Alberta, Canada. *By Bryghtknyght (CC BY 3.0) via Wikimedia Commons.*

FIGURE 5.2.3
Cabbage growing in a New York rooftop garden. *By Howellboy (CC BY-SA 4.0) via Wikimedia Commons.*

less energy than traditional agriculture. No pesticides or commercial fertilizers are used in the process. For more information on this innovative company, visit the links on your Resource Page.

DIGGING DEEPER

Exercise 1: What's in your diet?

Make a list of the foods you eat today, and determine where they are produced or grown.

- Which item travels the greatest distance? The shortest?
- Estimate the "food miles" or total distance traveled by the items you consume.

In addition to the impacts resulting from transportation of your food, there are other environmental effects to consider.

- Choose one item from the list of foods you consume, and identify all the environmental impacts associated with its production.
- In addition to purchasing locally grown food, what are three additional steps you could take to lessen the impact of your diet on the environment?

Chapter 5.3

Energy for Electricity

The Issues: The past century has been the age of fossil fuel–generated electrical energy. Until the mid- to late 1800's, wood was heavily used, but since then, coal, oil and gas have been burned to meet most of our electrical needs. A full discussion of the impacts of these fossil fuels is beyond the scope of this text, but, as we saw in the tar sands case study, environmental impacts occur at all stages in the life cycle of fossil fuels, from exploration, extraction, and transport to the ultimate generation of electricity.

Alternatives to fossil fuels may come with their own drawbacks. In one of our water case studies, we looked at one of the world's great hydroelectric dams, China's Three Gorges, and considered some of the downsides of big hydro. Another common alternative to the burning of fossil fuels for electricity is nuclear power, and one only need to review meltdowns at Chernobyl and Fukushima to understand the concerns about nuclear power as a source of our electricity.

A New Approach: Many are working to develop and improve sources of renewable energy. Wind, solar, geothermal, biomass, and tidal power all are being viewed as possible contributors to a more sustainable energy future.

Photovoltaic (PV) solar panels are a technology that converts the sun's energy into an electrical current. Go to the site listed on your Resource Page to view the basics of how a PV cell works. PVs were initially used to power small electronics, but as the cost of solar electricity has fallen, the number of grid-connected solar PV systems has grown into the millions, with more utility-scale solar power stations coming online each year. The International Energy Agency projected that by 2050, solar PVs and concentrated solar power could supply more than 25% of the global demand, making solar the world's largest source of electricity.

While most of us are familiar with solar panels, research is now underway to develop PV surfaces that can take solar electricity from traditional panel systems to a wide variety of surfaces. The image below shows a solar parking lot with the hexagonal panels not only providing a parking surface, but also generating electricity from the sun's rays (Fig. 5.3.1). Recent tests on the solar roadway concept have focused on developing a glass surface that can withstand the weight and wear and tear caused by vehicles. To prevent snow from covering the surface and reducing sunlight, heating elements are built into the panels, as are LED lights. By using different panel shapes, even curves and hills can be paved.

FIGURE 5.3.1

A prototype of new solar surfaces that allow roads and parking lots to produce solar energy. *By Dan Walden (*http://www.solarroadways.com/hirespics.html*) (CC BY-SA 1.0) via Wikimedia Commons.*

FIGURE 5.3.2

Illuminated biking jacket powered by a PV panel in the back. *Source: Goose Design LTD* www.goose.london.

Another exciting new area of innovation is PV fabrics. Flexible PV panels are already being put into backpacks, hats, coats, and sunglasses. Goose Design and Product Design & Development (PDD) developed the Illum cycling jacket, which is an illuminated biking jacket powered by a PV panel (Fig. 5.3.2). Newer technologies are even making it possible to have PVs sprayed onto or woven into fabrics. The US Army is investigating whether PV fabrics could help power electronics carried by soldiers.

CRUNCHING THE NUMBERS

Exercise 2: As solar electricity becomes more economical, it has become easy to access for average homeowners. Assume you wanted to power your home entirely with PV technology. Visit the National Renewable Energy Laboratory's solar calculator linked on your Resource Page. This calculator estimates the power production and cost of grid-connected PV energy systems throughout the world. It allows homeowners, small business owners, installers, and manufacturers to easily develop estimates of the performance of potential PV installations.

- Work through this calculator to determine how much electricity a standard solar system (based on the default calculator settings) could generate annually for your home.
- Based on your current monthly electric bills, could this cover your full electrical needs?
- Play with the calculator settings to adjust the size of the solar system. What size system would you need to fully meet your electrical needs?
- What other steps could be taken to reduce your energy demands to make solar a viable option for your household? You can use the energy estimator from energy.gov linked on your Resource Page to identify items in your home with the highest energy demand. List some changes you could make to reduce your household energy consumption.

CRITICAL THINKING

Exercise 3: PV systems are becoming more common, but there are many other sources of renewable energy being developed. All have their pros and cons.

- Do some digging to brainstorm a list of pros and cons for the following alternative sources of electrical power generation:
  tidal, wind, hydroelectric, geothermal, and biomass
- Considering the pros and cons of each of these alternatives provides a general comparison, but some of these are more likely to be efficient (or even possible) in certain locations. For each of the following, which energy source would be the most sustainable and economical choice? Why?
  - Patagonia, Chile
  - Arizona, US
  - Oslo, Norway
  - Addis Ababa, Ethiopia
  - Rio de Janeiro, Brazil

Chapter 5.4

Transportation

The Issues: In terms of what each one of us produces individually, there probably is no single greater source of air pollution and greenhouse gas emissions than what we create through our use of internal combustion engines. Remember that for every gallon of gasoline burned, about 19.64 lbs (8.8 kg) of CO_2 are emitted. Considering that the average American drives 13,476 miles annually (US Department of Transportation's Federal Highway Administration) and the average car gets 25.4 miles per gallon, this adds up to a significant environmental impact.

A New Approach: There are now a number of alternatives to power our vehicles. Options like hybrids, electric, and fuel cell–powered cars are now commercially available. Clean diesel engines also provide significantly higher fuel efficiency than their gas model counterparts. One recent advance is the development of biofuels made from plant wastes and other cellulosic feedstocks like switchgrass. Such sources have the potential to yield a higher energy output per unit input than corn-based ethanol with substantially fewer environmental drawbacks.

FedEx has contracted to buy 3 million gallons of jet biofuel a year from a $200 million refinery designed to convert plant waste into fuels. At maximum production, the new refinery will turn about 140,000 tons of dry woody biomass into 15 million gallons of biofuel annually. In fiscal year 2015, FedEx used more than 1 billion gallons of jet fuel. With the addition of this refinery, they plan to meet 30% of their needs with renewable biofuel by 2030.

SOLUTIONS

Exercise 4: Even with standard gasoline engines, car manufacturers are regularly improving fuel efficiency in their fleet, usually driven by the need to meet increased government fuel efficiency standards. The technology to do this exists, but when fuel prices are low, the economics alone are not sufficient incentive for manufacturers to alter production lines to produce more fuel-efficient models or get consumers to purchase them.

- Working in groups, brainstorm all of the costs and benefits of increasing fuel efficiency in new cars. Consider both environmental and economic impacts.

SOLUTIONS—Cont'd

Despite the obvious benefits of increased fuel efficiency, the average fuel economy of new cars and light trucks in the US has decreased over the past 20 years due to increased sales of larger and more powerful vehicles.

- Why do you think this has this happened? What are some possible ways to encourage consumers to buy more fuel efficient cars?
- What are some possible ways the government could encourage car manufacturers to produce more fuel efficient vehicles?

Chapter 5.5

Green Buildings

The Issues: Many of our buildings were constructed at a time when energy from fossil fuels was thought to be limitless, and, as a result, relatively little effort was put into building energy efficient buildings. In addition, as large cities expanded, urban sprawl created its own problems, necessitating the construction of networks of highways and clearing vast tracts of land.

A New Approach: The green building movement, which views buildings as part of their ecosystem and seeks to minimize their impacts on the environment, has become increasingly popular throughout much of the developed world. Rating systems such as LEED (Leadership in Energy and Environmental Design) give green buildings different levels of achievement like gold and platinum, depending on how many points a building earns from an environmental checklist. Programs like LEED consider six credit categories: Sustainable Siting, Water Efficiency, Energy Usage and Atmospheric Impacts, Materials and Resources Consumed, Indoor Environmental Quality, and Innovation in Design. Up to 100 points can be earned based on the potential environmental improvements and human benefits identified in various categories of the US Environmental Protection Agency (EPA)'s Tools for the Reduction and Assessment of Chemical and Other Environmental Impacts.

In addition to green construction for individual buildings, there have also been efforts to build entire developments that focus on minimizing their environmental footprint. In the UK, the Beddington Zero Energy Development (Fig. 5.5.1), started in 2000, was designed to be the first housing development with zero carbon emissions. The buildings are constructed of materials that store heat during warm conditions and release heat at cooler times, and whenever possible, they have been built from natural, recycled, or reclaimed materials. While not all its green features have been successful, it has generally gotten good reviews from sustainability experts.

A newer, larger-scale effort is underway in North West Bicester (Fig. 5.5.2), one of the UK's eco-towns announced in 2007. In the first phase of development, 393 zero-carbon homes are being built. In addition, residents will be able to access a community hub via their mobile devices to check electric car availability, monitor home energy usage, and check on public transport information.

Primary schools will lie within 800 m of all homes, and jobs will be created within a short travel distance. Other features include bus service within 400 m of every home, charging stations for electric vehicles, and an electric car club. A key feature of sustainable towns like this is having stores, restaurants, and services close to residences.

SOLUTIONS

Exercise 5: Working in groups, focus on a local town of your choice. After reviewing some of the features of eco-towns like North West Bicester, consider your town of choice.

- Make a list of characteristics of your town that already contribute to its sustainability.
- What additional steps might your town take to move toward sustainability?
- Which step would be the easiest to accomplish? The hardest?
  What are some of the obstacles you'd face in achieving these steps?
- For one of your easiest steps, write up a one-page pamphlet to educate the public and encourage local government officials to take the steps to put this into action.
  Share these with other groups in your class.

FIGURE 5.5.1
The Beddington Zero Energy Development is the world's first carbon-neutral eco-community. *By Tom Chance from Peckham (CC BY-SA 2.0) via Wikimedia Commons.*

FIGURE 5.5.2
Greengrocer on Sheep Street, Bicester. *Source: Jonathan Billinger. (CC-BY-SA 2.0) via Wikimedia Commons.*

MINIMIZING OUR IMPACTS ON THE ENVIRONMENT

In addition to finding new ways to provide the resources humans need, we're working on innovative ways to reduce our environmental footprints. In our water case studies, we looked at restoring polluted rivers. Below we'll look at some additional areas where new technologies offer the promise of a more sustainable future.

Chapter 5.6

Wastewater Treatment

The Issues: Treating wastewater, whether industrial discharges or human sewage, is an important step in protecting the quality of our surface waters. If untreated, such waste discharges can damage aquatic ecosystems and threaten human health.

Traditional wastewater treatment, however, comes at a cost. It is energy-intensive, with such processes as aeration consuming large amounts of energy. It also often requires the addition of a number of chemicals, both to facilitate pollutant removal and, at many plants, to disinfect the wastewater prior to discharge to surface waters. In addition, the wastewater treatment process produces large volumes of solid residue called sludge or biosolids, which must be disposed of. While some facilities are able to apply biosolids to agricultural fields, often the material is simply disposed of in landfills.

A New Approach: By taking advantage of the ability of various organisms to break down wastes, ecologists and engineers are designing "living machines" that can purify wastewater and avoid many of the monetary and environmental costs incurred during traditional wastewater treatment. By establishing aquatic communities such as those shown below in an early living machine at Findhorn, Scotland (Fig. 5.6.1), living machines break down various types of waste, including sewage and some industrial effluents, using natural processes. The combination of microbes, snails, aquatic plants, and fish, when working together, can be very effective at processing waste materials.

An additional bonus is that the waste can be converted into useful products. Living machines can produce marketable flowers and fish like tilapia for consumption. Purified water from the system can be recirculated, reducing both water demands and water released to the environment. To learn more about living machines and their potential, watch the video linked from your Resource Page about the living machine installed at the Rubenstein School of Environment and Natural Resources at the University of Vermont.

DIGGING DEEPER

Exercise 6: While the technologies to naturally purify and reuse our wastewater exist, the challenge, of course, is one of cost and scale. Massive wastewater treatment plants have the advantage of scale; they are able to treat the millions of gallons of waste generated in large cities on a daily basis. Living machines built to date have a much more limited capacity. However, if the technology can be perfected to the point where it is easily replicated, perhaps living machines could be the answer for neighborhoods looking to lower their environmental footprint and move toward sustainability.

- Do some digging about your local wastewater treatment plant. Find out how much wastewater it treats, what costs are involved (e.g., energy used, chemicals added, disposal of biosolids), and how much effluent it discharges.
- Now visit several of the living machine websites listed on your Resource Page. What are the living machine options that might be most viable to replace your local wastewater treatment plant? If this isn't possible based on the size, how might this technology be incorporated on a smaller scale?
- Besides living machines, what are some other ways to reduce the amount of wastewater we produce in our communities?

FIGURE 5.6.1

The Living Machine wastewater treatment plant located at Findhorn Ecovillage, Scotland. *By L. Schnadt (Living Technologies Ltd) (CC-BY-SA-3.0) via Wikimedia Commons.*

Chapter 5.7

Curbing Greenhouse Gas Emissions

The Issues: A global challenge is how to reduce emissions of greenhouse gases into the atmosphere as quickly as possible. While the signing of the Paris Accord on climate change in late 2015 is a positive step on the policy front, there have also been some promising technological advances. Let's look at two of them:

A New Approach: **Carbon Capture and Underground Storage**: Several technologies to remove CO_2 from fossil fuel emissions have been developed. For example, a chemical solvent can be used to separate the CO_2 from the flue gas stream after fossil fuels have been burned. Once the CO_2 is captured by the solvent, it can be removed from the solvent and disposed underground or under the sea. Fig. 5.7.1 shows some of the equipment used to capture carbon at a coal-burning plant.

One of the world's largest carbon capture projects is underway in Texas. The Petra Nova Project, once completed, will capture 1.4 million metric tons of CO_2 annually from a coal-fired power plant. The captured CO_2 will be injected into the ground at an oil field to help extract additional crude from hard-to-access

FIGURE 5.7.1
Coal carbon capture technology. *By Peabody Energy, Inc. (CC BY 3.0) via Wikimedia Commons.*

FIGURE 5.7.2
Photobioreactors containing microalgae and other photosynthetic organisms. *Source: IVB Biotech. CC 3.0. By IGV Biotech. (CC BY-SA 3.0) via Wikimedia Commons.*

underground areas. After the CO_2 is separated from the oil, it will be reinjected underground for storage.

Photobioreactors: An alternate method for capturing CO_2 is to take advantage of the fact that primary producers take up CO_2 during photosynthesis. In this approach, algal populations maintained at power plants in tubes called photoreactors (Fig. 5.7.2) remove CO_2 from flue gas piped into the tubes. An added benefit is that other pollutants in the flue gas, such as NO_x, can also be extracted as they are used as nutrients by the algae.

The algal cells can be harvested and treated in an anaerobic digester that produces methane gas which can be used as a secondary energy source. The algal cells can also be dried and converted into animal or fish food. To view a US Department of Energy video on converting algae to fuel, use the link supplied on your Resource Page.

SYSTEMS CONNECTIONS

Exercise 7: Besides limiting our emissions of CO_2, another approach based on geoengineering is being considered to combat climate change. Called solar radiation management, these technologies are focused on reducing the amount of solar energy reaching the Earth's surface, thus slowing the rate of temperature increase.

- Do some research on these radiation management techniques, and identify and describe two that you find interesting.
- What are the pros and cons of each approach?
- What are some of the possible systems connections that might result if incoming solar radiation is reduced across the planet?
- Do you think that solar radiation management approaches are a sensible way to combat climate change? Why or why not?

Chapter 5.8

Solid Waste Disposal

The Issues: Given what you've learned in our E-waste case study, it's easy to understand why ours is often considered a "throwaway" society. According to the US EPA, each US resident generates about 2 kg of solid waste daily. We recycle or compost about 0.68 kg, just over one-third. What happens to the other two-thirds? Most of it ends up in one of approximately 3500 landfills in the US. In earlier times, dumping our waste into holes in the ground seemed like a good idea: "out of sight, out of mind." More recently, we've discovered that unlined landfills have contaminated groundwater in many areas, and these dump sites also produce substantial amounts of methane, a greenhouse gas. While we now are able to build "secure" landfills that reduce environmental damage, it is an expensive process.

There are also secondary effects of disposing of solid waste in the ground, including the environmental costs of trucking our trash to distant landfills. Alternatives like incineration at "trash-to-energy" plants are available, but they have their own drawbacks.

A New Approach: It makes sense for any number of reasons to reduce the amount of solid waste we throw away. Many school students learn the "3 Rs: reduce, reuse, and recycle" at an early age. We could consider buying less, reusing what we do buy, and recycling what we don't reuse. But that all refers to the product itself. What about the packaging required to get that product to us in the first place? In 2010, consumers in the US threw away more than 75 million tons of packaging waste. Plastic bottles, disposable sandwich boxes, and individually wrapped items like candy bars and gum are just a few examples of packaging that all too often ends up in our landfills.

What if we could make packaging that self-destructs or, better yet, could be consumed? Several firms are working to make biodegradable and edible packaging for food products more widely available. In 2012, D. Edwards, a bioengineer at Harvard University, started WikiCell Designs, which makes edible packaging. Binding together particles from foods such as chocolate and oranges with carbohydrates, Wikicells are a natural packaging that can keep food fresh for as long as conventional packaging, but then can also be consumed itself. WikiCells are ideal for transporting foods and drinks without plastic. Some innovative ideas from WikiCell include orange juice transported in an edible orange membrane and a chocolate membrane holding hot chocolate.

Other food scientists are working to develop edible food coatings derived from the dairy by-product whey. Such edible films could prevent spoilage and help retain vitamins and other nutrients. Scientists have even combined chitosan, found in crab and shrimp shells, with lysozyme, a protein in egg whites, to create an antimicrobial food wrap. Fascinating, but will consumers buy it?

Clearly, more work needs to be done before edible film replaces our current way of packaging and transporting food products. There is reason for optimism, however, that these new "bio" approaches to packaging will someday help ease the burden created by our current deluge of solid waste.

DIGGING DEEPER

Exercise 8: Go to your local grocery or "big box" store and identify one item that you think is overpackaged.

- What are the components of the packaging of this item?
- What is the environmental footprint of each packaging component?
- Suggest ways that the item could be packaged in more sustainable way.
- Why do you suppose manufacturers of such items resort to overpackaging?

CHALLENGE ASSIGNMENT

In this concluding section, we've looked at some more sustainable ways to provide resources for humans and reduce our environmental footprints. Now it's your turn to do some innovating.

Your Assignment: Assume you've been hired by a local developer who is planning to construct a new environmentally friendly neighborhood of 50 homes in your town. Your job is to develop a plan to maximize the sustainability of this development. The goal is to have all homes built with a net zero environmental footprint by 2025. Considering the items we've covered in this section and any others you think relevant, come up with a variety of steps to help the developer achieve her goal. You'll need to include several critical areas: **food, energy, water supply, wastewater treatment, infrastructure, and transportation**. To help you accomplish your goals, assume that the developer is willing to add $50 million to the standard development costs to get the job done.

- Your first task will be to determine the basics: Where is your community located? Where does the food in your area come from? How is the area's electricity generated? Where do local communities get their water and discharge their waste? Where do people work and how do they get there?

- Then, identify and list the various environmental impacts that will result from providing the critical resources to construct and sustain this community.

- Next, explore the sustainable options that may be available for the proposed community. Determine which make the most sense for the community based on its size, location, and local resources. For example, perhaps the community lies in an area served by a newly completed mass transit system, but the residents have to import their food from great distances. When you consider where the greatest improvements may be achieved, perhaps you'll conclude that going local with food production would do the most to get you close to a net zero ecological footprint.

- Budget how much of the $50 million will go toward each recommended option based on your estimate of costs and priorities to get the most "bang for the buck." This is a numbers exercise. You'll need to do some back-of-the-envelope calculations to support your proposed program.

- Summarize your final recommendations to the developer, and justify how each of your recommendations makes both economic and environmental sense.

- Create a pamphlet to summarize the environmental benefits of this planned community. This pamphlet will be used to market the properties to interested buyers.

CONSIDER THIS

Exercise 9: While new technologies are constantly coming online to minimize our impacts on the environment, restore degraded systems, and improve efficiencies in what we use, these technologies typically come at a high cost. Most companies won't implement such systems unless economic gain can be demonstrated.

- Consider technologies to clean emissions from power plants. Who should be asked to pay for installation of these technologies?
- What if impacts from our activities threaten populations or ecosystems far away? Consider the island nation of Tuvalu, which is slowly being inundated as sea levels rise as a result of global warming. Should their island be engineered to withstand sea level rise? Should the citizens be relocated? Who should pay for these activities?
- Many developing nations are struggling to bring basics like electricity and clean water supplies to their residents. Doing this is hard enough for economically struggling nations, but even more challenging to accomplish while incorporating the latest technologies to minimize environmental impacts.
    - Should developed nations assist poorer nations in such efforts?
    - Should these nations be held to the same emissions standards as developed nations that can afford to retrofit power plants or construct massive public transportation systems?
    - In what other ways can poorer nations be encouraged and supported in their efforts to reduce their environmental footprint without denying their populations an improved standard of living?

Index

'*Note:* Page numbers followed by "f" indicate figures, "t" indicate tables and "b" indicate boxes.'

F

L

M

N

O

R

S

W

Y

Z

Printed in the United States
By Bookmasters